Complexity and Industrial Clusters

Contributions to Economics

http://www.springer.de/cgi-bin/search_book.pl?series=1262

Michael Carlberg
Intertemporal Macroeconomics
1998. ISBN 3-7908-1096-7

Sabine Spangenberg
The Institutionalised Transformation of the East German Economy
1998. ISBN 3-7908-1103-3

Hagen Bobzin
Indivisibilities
1998. ISBN 3-7908-1123-8

Helmut Wagner (Ed.)
Current Issues in Monetary Economics
1998. ISBN 3-7908-1127-0

Peter Michaelis/Frank Stähler (Eds.)
Recent Policy Issues in Environmental and Resource Economics
1998. ISBN 3-7908-1137-8

Jessica de Wolff
The Political Economy of Fiscal Decisions
1998. ISBN 3-7908-1130-0

Georg Bol/Gholamreza Nakhaeizadeh/Karl-Heinz Vollmer (Eds.)
Risk Measurements, Econometrics and Neural Networks
1998. ISBN 3-7908-1152-1

Joachim Winter
Investment and Exit Decisions at the Plant Level
1998. ISBN 3-7908-1154-8

Bernd Meyer
Intertemporal Asset Pricing
1999. ISBN 3-7908-1159-9

Uwe Walz
Dynamics of Regional Integration
1999. ISBN 3-7908-1185-8

Michael Carlberg
European Monetary Union
1999. ISBN 3-7908-1191-2

Giovanni Galizzi/Luciano Venturini (Eds.)
Vertical Relationships and Coordination in the Food System
1999. ISBN 3-7908-1192-0

Gustav A. Horn/Wolfgang Scheremet/Rudolf Zwiener
Wages and the Euro
1999. ISBN 3-7908-1199-8

Dirk Willer
The Development of Equity Capital Markets in Transition Economies
1999. ISBN 3-7908-1198-X

Karl Matthias Weber
Innovation Diffusion and Political Control of Energy Technologies
1999. ISBN 3-7908-1205-6

Heike Link et al.
The Costs of Road Infrastructure and Congestion in Europe
1999. ISBN 3-7908-1201-3

Simon Duindam
Military Conscription
1999. ISBN 3-7908-1203-X

Bruno Jeitziner
Political Economy of the Swiss National Bank
1999. ISBN 3-7908-1209-9

Irene Ring et al. (Eds.)
Regional Sustainability
1999. ISBN 3-7908-1233-1

Katharina Müller/Andreas Ryll/Hans-Jürgen Wagener (Eds.)
Transformation of Social Security: Pensions in Central-Eastern Europe
1999. ISBN 3-7908-1210-2

Stefan Traub
Framing Effects in Taxation
1999. ISBN 3-7908-1240-4

Pablo Coto-Millán
Utility and Production
1999. ISBN 3-7908-1153-X

Frank Riedel
Imperfect Information and Investor Heterogeneity in the Bond Market
2000. ISBN 3-7908-1247-1

Kirsten Ralf
Business Cycles
2000. ISBN 3-7908-1245-5

continued on page 308

Alberto Quadrio Curzio · Marco Fortis
(Editors)

Complexity and Industrial Clusters

Dynamics and Models in Theory and Practice

With 23 Figures and 8 Tables

Physica-Verlag
A Springer-Verlag Company

Series Editors
Werner A. Müller
Martina Bihn

Editors
Professor Alberto Quadrio Curzio
Università Cattolica
CRANEC
Via Necchi, 5
I-20123 Milano
cranec@mi.unicatt.it

Professor Marco Fortis
Montedison
Piazzetta M. Bossi, 3
I-20121 Milano
m.fortis@montedison.it

ISBN 978-3-7908-1471-2 ISBN 978-3-642-50007-7 (eBook)
DOI 10.1007/978-3-642-50007-7

Cataloging-in-Publication Data applied for
Die Deutsche Bibliothek – CIP-Einheitsaufnahme
Complexity and industrial clusters: dynamics and models in theory and practice; 8 tables / ed.: Alberto Quadrio Curzio; Marco Fortis. – Heidelberg; New York: Physica-Verl., 2002
(Contributions to economics)

Physica-Verlag Heidelberg New York
a member of BertelsmannSpringer Science+Business Media GmbH

Softcover Design: Erich Kirchner, Heidelberg

SPIN 10865729 88/2202-5 4 3 2 1 0 – Printed on acid-free and non-aging paper

Preface

This volume contains the proceedings of the international conference "Complexity and Industrial Clusters: Dynamics and Models in Theory and Practice", organized by Fondazione Comunità e Innovazione and held in Milan on June 19 and 20, 2001 under the aegis of the Accademia Nazionale dei Lincei (founded in Rome in 1604), one of the oldest and most famous national academies of science in the world.

Fondazione Comunità e Innovazione encourages research and the dissemination of knowledge about social, economic, cultural and civil issues. It promotes research and innovation related to local production systems and industrial districts, with special reference to: the interactions between large companies and SMEs (small and medium-size enterprises), the effects of industrial districts on the development and welfare of their communities and of neighbouring areas, the effects of globalisation on these local systems of productions.

Fondazione Comunità e Innovazione was created in Milan in 1999. It supports studies, publications, and events, both on its own and in cooperation with corporations, research institutes, foundations, associations and universities. It also grants scientific sponsorship to research that is in line with its mission, as set forth in its by-laws. The founding member of the Fondazione is Edison (formerly Montedison). The other subscribing members, in historical order, are: Ausimont, Tecnimont, Eridania, Accenture, and PricewaterhouseCoopers.

The editors of this volume, personally and on behalf of Fondazione Comunità e Innovazione, wish to extend their heartfelt thanks to all those who have contributed to the success of the conference and particularly to Enrico Bondi, past Chairman of the Fondazione, Edoardo Vesentini, President of Lincei, which granted the scientific sponsorship, to all of the speakers, whose papers are presented following the order of the conference programme, and to David A. Lane and Franco Malerba, chairmen of the sessions who contributed greatly to the debate and the deeping of the issues.

Milan, December 2001

Alberto Quadrio Curzio
Marco Fortis

Contents

III. Success Cases Around the World

Introduction: From Specific Industrial Cases to a General Economic Model?

Alberto Quadrio Curzio and Marco Fortis[1]

1. Foreword

It seems fitting to begin this publication, which contains the proceedings of the international conference described in the preface, with a brief mention of two institutions: Fondazione Comunità e Innovazione, which organized the conference, and Accademia Nazionale dei Lincei, which gave the scientific sponsorship to it.

Both of these institutions are important for their different traditions in the fields of scientific and technological research.

Fondazione Comunità e Innovazione is closely associated with the historical roots of Edison, founded in 1884, of Montecatini, founded in 1888, and, later on, of Montedison. This is true primarily, but not exclusively, because these two companies were able to build on their "local" civil and economic base and then expand nationally and internationally by constantly focusing on technological innovation. It is a well-known fact that the discovery of polypropylene, for which Giulio Natta received the Nobel Prize in 1963, was made possible by the support of Montecatini in an outstanding example of collaboration between industry and research. Many other individuals throughout Montecatini's and Edison's histories deserve mention for their ability to combine entrepreneurial spirit and scientific and technological innovation, most notably Guido Donegani, Giacomo Fauser and Giuseppe Colombo.

Accademia Nazionale dei Lincei, which honoured our conference with its confidence and scientific sponsorship, was founded in 1603, making it the oldest Academy in the world. It is worth mentioning that its founding members included Galileo Galilei, who in 1613 published his *Historia e dimostrazioni intorno alle macchie solari* as part of the Academy's proceedings. This is not the place for a lengthy and detailed presentation of the extraordinary scientific merits of the Accademia dei Lincei and of its contributions to Italian and international scientific history. Interestingly, Giulio Natta, Nobel Prize, was a member of the Academy in

[1]Our heartfelt thanks to Mario A. Maggioni for his help in planning the conference and for his useful remarks about this essay. Many others who have contributed to the conference should be thanked, especially Biancamaria Frondoni for the organization of the conference and for the general editing of this volume. We also thank David A. Lane and Nicola W. Palmieri for the connections with the Santa Fe Institute, Monica Carminati, Cristiana Crenna and Franca Sapienza for their help in editing this book.

1947, as were Giuseppe Colombo, member since 1888 and Giacomo Fauser, member since 1948.

It seems to us, then, that the holding of this international conference and the publication of this book provide an ideal historical perspective linking these two institutions, which share a similar vision, even though their objectives may be different. And this vision is that science and technology must cooperate, that industry cannot prosper without research. It is for this reason that the next two paragraphs of this introduction present the points of view of two highly respected personalities in their own field: Enrico Bondi, at that time the Chairman of Fondazione Comunità e Innovazione and C.E.O. of Montedison, and Edoardo Vesentini, Chairman of the Accademia Nazionale dei Lincei. With each one of these personalities the editors of this book had numerous exchange of ideas while preparing this conference. And therefore the editors will refer their points of view, obviously taking some liberties in interpreting their thoughts, which are shared by the editors of this volume.

2. General Models, Economic Analysis and Applications

In the preface we have already said that the conference and this volume were made possible by the initiative of Fondazione Comunità e Innovazione, which for years has been promoting studies and research on the evolution of Italian industrial districts. Therefore, we will use this section of the introduction to present Enrico Bondi's point of view. The idea to connect complexity and industrial districts belongs to him. But here we will explain the facts about Italian districts that have much interested Enrico Bondi convincing him that this phenomenon was worthwhile of deep analysis also from the point of view of a big company.

While it may seem peculiar to begin the introduction to this book, which presents the contribution of eminent theoreticians (some of them Nobel laureates), with data on the Italian industrial system, we chose this approach to help the reader understand how a specific situation can be used to develop a general scientific approach.

In Italy, there are more than 200 main industrial districts, which account for over two million jobs, or about 40% of Italy's total manufacturing employment and 60% of all jobs in companies that make the most familiar made-in-Italy products (fashion, home furniture and furnishings, food and traditional mechanical engineering products). Made in Italy products are recognized worldwide for their top quality, and industrial districts are the leading manufacturers of these quality products.

As a result, these districts, which almost always enjoy global leadership in their respective industries (from textiles and high fashion to ceramics and machine tools) account for more than one-third of all Italian exports. The largest districts, such as Sassuolo for ceramics or Prato for textiles, generate revenues in excess of five billion euros. But there are dozens of districts with revenues of between 500 million euros and one billion euros and dozens more with revenues of 100 to 500 million euros.

There are also extremely interesting examples of cooperation between industrial districts, which contain small and medium-size enterprises, and large companies. The most striking example in this area is the cooperation between the "Fluorine Valley" district and Montedison. This district is located in the province of Bergamo along the border with the province of Brescia, with its local production system that manufactures specialized products using elastomers and fluorinated polymers. Another and more famous case of the collaborative relationship is that between the Maranello mechanical engineering district, which is host to such a high-technology marvel but small firm as Ferrari, and the big automobile firm that is Fiat.

Therefore, a study of local manufacturing systems and industrial districts, and of the interaction between small and medium-size enterprises, must also take into account the role of large companies and the relationship that these local systems have with universities and centres of research in order to achieve technological innovation, since these factors are essential in helping enterprises grow and compete internationally.

Enrico Bondi emphasized this theme in his interesting remarks at the beginning of the conference. Agreeing especially with the views of two Italian economists that he often met, Giorgio Fuà and Giacomo Becattini, and through the discussion with the editors of this volume, he emphasised that the Italian district model is a spontaneous one based mainly on the self-organizing ability of local communities, which in many cases is rooted in the artisan traditions of the free city-states of the Italian Renaissance. The Italian model is that of a district consisting of small and medium-size enterprises located away from large metropolitan areas, primarily in the heart of some of Italy's northern and central provinces. It is a development model that evolved without the support of public industrial policies.

In order to understand the question contained in the title of this section of the introduction, it is important to remember that new industrial districts, or clusters, have developed in other countries, and have become important engines of economic growth. Just think of Silicon Valley or Hsinchu-Taipei. These clusters are agglomerations of non-Fordist enterprises strongly rooted in local society and supported by a network of social and economic relations with strong links with a major university or research centre. Two are thus the relevant (and intertwined) questions. The first: can we consider these innovative clusters a sort of "second generation" industrial districts? The second: will the Italian industrial districts be able to make the transition from traditional products to high-technology ones?

In conclusion, the system of Italian industrial districts appears to have produced a "model of capitalism" that is different from the Anglo-Saxon and Rhine Valley ones. However, this model (which we can define as "horizontal or network capitalism") seems not to be limited to Italy, for new district systems, often involving highly sophisticated technologies, that are emerging throughout the world seem to point to the fact that this new type of capitalism does not possess limited national connotations.

3. Can General Theoretical Models Be Used to Understand Specific Economic Cases?

A reference to the other institution that lent its prestige to this conference is also important in placing this book within the context of the methodological input that Edoardo Vesentini, Chairman of the Accademia Nazionale dei Lincei, provided to Alberto Quadrio Curzio and which has been presented by the latter in the introduction to the conference.

These introductory remarks, after noting that the Accademia Nazionale dei Lincei was glad to support the conference because it deals with economic development and technological innovation and, therefore, has to do with aspects of scientific research in which Lincei are interested, will review the three main topics of the conference and of this book, paying special attention to certain methodological issues.

The first issue has to do with the following question: can general models be used to understand specific economic cases? This question was addressed during the first session of the conference and is dealt with in the first part of this book. The issue is whether or not general theories of complexity can be used to develop models that can, in turn, be used to analyse economic dynamics, particularly those that affect the various types of industrial districts. Vesentini, while not addressing the specific issue of the link between complexity and industrial districts, pointed out that, frequently, general scientific theories and mathematical models have found important applications in the economic field. Therefore, further studies in this area are always important for the scientific progress and valuable because economic theory has been a major beneficiary of mathematical methods. The most striking example in this area is that of John von Neumann. However, it is necessary to proceed with some caution in order to avoid misapplication of the theory in general and the economic application in particular.

The second question, which was addressed during the second session of the conference and is discussed in the second part of this book (which is more closely related to economics because it presents several models of district dynamics), is the following: are these alternative or complementary models? Vesentini pointed out that here the analysis focuses on the relationship between the theoretical approach and the empirical approach. This interaction has been extremely important for the success of economics, giving also rise to a separate branch, that of quantitative methods and econometrics, that was recognized with the first Nobel Prize ever awarded for economics. In this area, Vesentini offered a suggestion: do not try to force a complex economic reality into a single model, since complementary interpretations are often one of the remarkable advantages of economic research.

The third issue, which was discussed during the third session of the conference and is reviewed in the third part of this book (which analyses different types of industrial districts within different economic-institutional and industry contexts), has to do with the development of national paradigms of district dynamics. Vesentini pointed out that in this area the analysis has to deal with the relationship between economic research, research on institutions and social research, which is the third and equally important methodological component of economics. In other words, he emphasized that the conference, and this book, must use the three different methodological approaches that, in the field of economics, have produced indisputable gains.

Having provided these general introductory remarks, we shall now review the individual parts of this volume.

4. Complexity and Economic Dynamics

The first part of this book, which contains three papers by Murray Gell-Mann (*What Is Complexity?*), John H. Holland (*Complex Adaptive Systems and Spontaneous Emergence*) and Enzo Rullani (*The Industrial Cluster as a Complex Adaptive System*), deals with the relationship between complexity and economic dynamics and industrial districts in an effort to develop a general model.

The study of complexity is based on the notion that there is a general law that governs the formation of orderly configurations within systems in dynamic balance. Complex systems are formed by independent agents that interact in a non-linear fashion, adapting, evolving and developing a form of self-organization that enables them to acquire collective properties that are not attributable to specific agents.

These complex self-organizing systems are adaptive in the sense that they do not react passively to events, the way a stone would tumble in an earthquake. On the contrary, they make a deliberate effort to turn all circumstances to their advantage.

This theory has been applied in the most diverse areas, ranging from the natural sciences to such fields as meteorology and urban planning. In economics, the theory of complexity has been used to study stock market cycles and is now being considered with great interest by some of those who are studying industrial districts.

It is also important to remember that industrial districts of different types now exist in many part of the world. In most cases, they reflect self-organizing models.

Therefore, it was reasonable to believe that the theory of complexity could provide important general guidelines for understanding the birth and dynamics of these districts when they are viewed as a general manifestation of self-organizing systems.

With this in mind, top priority was given to investigating the relationship between complexity and economic dynamics. This task was entrusted to Murray Gell-Man and John Holland, key representatives of the Santa Fe Institute, where mathematicians, computer scientists, physicists, neurobiologists, evolutionary biologists, economists, environmentalists, political scientists, historians, archaeologists, linguists and many other scientists work. At the Santa Fe Institute, Gell-Mann and his colleagues, among them John Holland (the father of "genetic algorithms" and one of the world's leading computer scientists), take an interdisciplinary approach in studying the meaning of simplicity and complexity.

The purpose of the first part of the book is to stimulate a broad debate on the ways in which the analytical methods, developed by physics and biology, can be transferred to economics, where the phenomenon of Italian industrial districts appears to offer to complexity scholars an extremely fertile ground for analysis. The issue that, more or less implicitly, is being addressed here is to what extent the general theories of complexity and adaptive systems can provide a theoretical basis to explain the economic dynamics that characterize industrial districts.

The papers produced by these two researchers on complexity are particularly relevant with regard to this issue. It is not by accident that, even though one is a physicist and Nobel laureate and the other a computer scientist and the developer of genetic algorithms, their contribution has proven to be extremely valuable in analysing the economic phenomena that are the subject of this conference. It would seem appropriate at this point to specifically mention again John von Neumann, who, though not an economist, provided a fundamental contribution to the development of the economic sciences with his general theory of growth and his theory of games. This convergence underscores the unity of all sciences and the fact that often we perceive only with a considerable lag the extent of the cross-fertilization effort - as exemplified by this book - entailed by the courage to innovate.

The obvious relationship between complexity and districts, as seen by an economist, is the subject of the paper contributed by Rullani, an expert on Italy's industrial districts. As such, he was able to abstract universal principles from specific cases by projecting them onto the canvas of general typologies, without losing sight of their individual characteristics. Because of this, his paper can also be seen as an introduction both to the second and third parts of this book. Rullani's paper provides an excellent explanation of the effort to combine economic sciences and complex adaptive systems by drawing attention to the binary paradigm of exploiting what is known and exploring what is unknown. Exploitation has to do with refinement, choices and efficiency, while exploration means searching for something new that cannot be codified beforehand. Exploration entails complexity, which brings to the realm of economic research a whole series of variables that previously had been excluded precisely because they did not fit within the logical guidelines of economic computations based on just two main variables: price and quantity.

5. General Models of Industrial Cluster Dynamics

The second part of this book contains the following five papers: *Complexity and Local Interactions: Towards a Theory of Industrial Districts*, by David A. Lane; *From Marshall's to the Italian "Industrial Districts". A Brief Critical Reconstruction*, by Giacomo Becattini; *Diversity, Not Specialization: The Ties That Bind the (New) Industrial District*, by Charles F. Sabel; *Competitive and*

Synergic Behaviours in the Development of Industrial Clusters: Ecological Modelling and Empirical Evidence, by Marco Fortis and Mario A. Maggioni; *On the Ubiquitous Nature of Agglomeration Economies and Their Diverse Determinants: Some Notes*, by Giulio Bottazzi, Giovanni Dosi and Giorgio Fagiolo.

The second part deals with the general models of district dynamics. It also has to do, more or less explicitly, with the issue of whether district interpretation models are complementary or alternative. Lane's introductory paper, which develops a taxonomy of district operating logics in an attempt to provide all researchers of industrial districts (and, perhaps, also enterprises) with a common language and logic, is followed by four modelling studies that offer different analytical-interpretative systems.

We shall begin by mentioning the paper written by Becattini, who, using Marshall's approach as a starting point, develops a personal and original interpretation, viewing districts as the dual expression of social cohesion and technological innovation and as new systems of network companies that irreversibly overcome the Fordist paradigm by viewing the division of labour as a complementary relationship among district companies.

Sabel, who views districts through the lens of an approach where sociological and institutional approach also matters, provides a more pragmatic-descriptive approach to explaining how districts operate.

Fortis and Maggioni apply an original theoretical framework, derived from population ecology, to the analysis of the development path of Italian industrial districts. The paper underlines the interplay of agglomeration economies and diseconomies in the growth process of an industrial cluster, distinguishes three main phases of its development process and stresses the complex and different (i.e. synergic, competitive, etc.) interactions that exist between different industries within the same area, and between different areas within the same industry. An econometric exercise allows the authors to estimate the values of specific parameters and to forecast the long run evolution of some industrial districts in Italy.

Dosi et al. in their work attempt to study the multiple drivers of agglomeration phenomena in contemporary economies and propose a tentative taxonomy where the conditions of knowledge accumulation, often specific-to-specific locations and specific sectors, play a paramount role. They discuss the achievements and limitations of current theorizing on spatial location of economic activities, proposing a simple model, which is based on Italian data, that highlights the rich intersectional diversity of agglomeration forces, together with, in a few cases, the lack of them.

6. Success Cases Around the World

The third part of the book contains the following five papers: *The Entrepreneurial Spark: Individual Agents and the Formation of Innovative Clusters*, by Maryann

P. Feldman and Johanna Francis; *From Exogenous to Endogenous Local Development: The Cases of the Toulouse and Sophia Antipolis Technopoles*, by Christian Longhi; *Biotechnology Development in Germany: The Case of Nordrhein-Westfalen* by Francesco Salamini, Anke Sohn and Hartmut Thomas; *Support for Technology-Based Firms: The Role Played by Property-Based Science Parks*, by Don Siegel, Paul Westhead and Mike Wright; *High-Tech Industrial Clusters in the Asian Countries: The Case of the Electronics Industry in the Hsinchu-Taipei Region*, by Kung Wang.

The theme of the third part of the book is therefore the study of cases that have an emblematic value at the global level. This section includes five papers that analyse an equal number of systems that, while not all of the district type, can provide insights on technological innovation and entrepreneurship that are territorially based rather than concentrated in large companies. In this case as well, the implicit questions are: which elements are common to all of these success stories and which differentiate them, and can the clusters of high-tech companies that have sprung up all over the world as alternatives to large companies be classified as types of second-generation districts? This question therefore relates to the case of Silicon Valley in USA, the case of Toulouse in France, the BioRegio system in Germany, the British scientific parks and the Taiwan electronics story.

Feldman and Francis outline the development of a regional industrial cluster in the U.S. Capitol region, through the efforts of entrepreneurs who adapted to both constructive crises and new opportunities. The paper examines the initial spark of entrepreneurship and how it influences the formation of high-technology clusters. The perspective taken here is that entrepreneurs are a critical element in the formation of clusters, and their actions are important to the analysis of clusters as complex adaptive systems.

Longhi's paper analyses two French technopoles, Toulouse, in the southwest region of Midi-Pyrénées, and Sophia Antipolis, in the southeast French Riviera, both considered the result of the French national system of innovation and process of decentralisation initiated in the 1970s. The first part of the paper draws on recent developments in complexity, which emphasize openness, non-linearity and path dependence, providing an analytical framework that makes it possible to explain the development of these "technopolitan areas" (defined as areas developed around several knowledge elements, including but not limited to science parks). The second part of the paper analyses the cases of Toulouse and Sophia Antipolis along these lines. The recent development of these complexes allows us to consider their histories and the "small events" that have resulted in self-reinforcing expansion, although along highly distinctive paths.

Salamini et al. in their paper review the history and the players that have been influencing the development of biotechnologies in Nordrhein-Westfalen (NRW): the BioRegio project, the role of the Bio-Gen-Tec-NRW agency and the first steps toward biotech development in NRW.

Westhead et al. compare the performance of Science Park and off-Park firms in the U.K. with regard to several indicators, and explore the role played by the Science Park manager/director in the development of firms located in "managed" and "non-managed" Science Parks.

Finally, Kung Wang mainly discusses Taiwan innovation system (TIS), that is, the operation system by enterprises, government, universities and public research laboratories of science and technology (S&T) industry development, and also the most representative characteristics that can explain TIS, that is: the technology development program (TDP) of the Ministry of Economic Affairs, the Industry Technology Research Institute, the Hsinchu Science-based Industrial Park and bringing back overseas scholars.

7. Conclusions

The last paper, *Financial Markets, Industrial Clusters and Small and Medium-Size Enterprises*, by Angelo Tantazzi, is somewhat separate from the rest of the book, in that it opens a new area of study and provides guidelines for developing it on another occasion. In particular, he discusses the links that exist between the financial markets and small and medium-size enterprises within different economic and institutional contexts.

In conclusion, we would like to point out that the goals of firm's growth and profit are always paramount for firms irrespective of the organizational system adopted by an enterprise. Generally speaking an organizational system does not exist for its own purpose: it is an operational tool, a method of achieving economic objectives, two of which are, obviously, growth and profit.

However, districts pursue another objective, which, while not always explicit, is clearly evident in certain historical and social contexts. It is the goal of propagating across the generations the district community as a self-organizing system, even when this requires district participants to sacrifice a proportionate share of growth and profits. This justifies an interest in districts not only by economists and sociologists, but also by researchers of complexity, in light of districts' nature as complex adaptive systems in which enterprises and institutions interact, accumulating experiences that increase the efficiency of the districts. As we enter the 21st century, we can forecast that it will be characterized by two great new phenomena: globalisation and complexity. And perhaps in both cases the new type of industrial and economic organization that goes under the name of "indutrial clusters" or "industrial districts" will have an increasing role in its capacity to combine innovation and flexibility, to combine systems of small-medium size firms with giant corporations. The reality and the researches of the next years will tell us which is the answer.

I. Complexity and Economic Dynamics

What Is Complexity?

Murray Gell-Mann

Abstract. It would require many different concepts to capture all our notions of the meaning of complexity. The concept that comes closest to what we usually mean is effective complexity (EC). Roughly speaking, the EC of an entity is the length of a very concise description of its regularities. A novel is considered complex if it has a great many scenes, subplots, characters, and so forth. An elaborate hierarchy can contribute to complexity, as in the case of nested industrial clusters each composed of a great variety of firms and other institutions. In general, though, what are regularities? We encounter in many different situations the interplay between the regular and the random or incidental: music and static on the radio, specifications and tolerances in manufacturing, etc. But ultimately the distinction between the regular and the incidental depends on a judgment of what is important, although the judge need not be human or even alive. For instance, in the case of songs of a male bird in the nesting season, the identification of regularities is perhaps best left to the other birds of the same species - what features are essential in repelling other males from the territory or attracting a suitable female? A technical definition of EC involves the quantity called algorithmic information content (AIC). The description of an entity is converted to a bit string and a standard universal computer is programmed to print out that string and then halt. The length of the shortest such program (or, in a generalization, the shortest that executes within a given time) is the AIC. The AIC is expressed as the sum of two terms, one (the EC) referring to the regularities and the other to the random features. The regularities of a real entity are best expressed by embedding it conceptually in a set of comparable things, the rest of which are imagined. The EC can then be related to the AIC of the set, the choice of which is restricted by the conditions imposed by the judge. Theorists like to study highly simplified models of complex systems, often by computer modelling. What can be claimed for such models? Overall agreement with observation is hardly to be expected. However, in many cases simple regularities can be found in both the observational data and the model, which may then be helpful in understanding those regularities. Examples are given, involving scaling laws and also "implicational scales".

It would take a great many different concepts to cover all our intuitive notions of what is meant by complexity (and its opposite, simplicity), but the concept that agrees best with the meaning in ordinary conversation and in most scientific discourse is effective complexity (EC). Roughly, the EC of an entity is the length of a very concise description of its regularities. Thus we would call a novel complex if it had many different subplots, scenes, and characters. Likewise, an international conglomerate firm would be complex if it had many different branches in different countries with different products, management styles, and so forth. A hierarchical structure that takes a long time to describe would also contribute to complexity. It is easy to see that an industrial cluster comprising many different firms and other institutions in diverse locations within a region, perhaps grouped into sub clusters (which may, in turn, be composed of sub-sub clusters) is complex in this sense.

An amusing exercise consists of looking at the patterns of neckties to see which ones are simple (for example those with regimental stripes) and which ones are complex (for instance most hand-painted ones). Note that we are looking at how long it takes to describe the regularities, not the features treated as random or incidental. Of course the description length will typically depend on the "coarse graining" (level of detail), which depends in turn on how far away the tie is when viewed with the unaided eye.

We are constantly observing in the world around us the interplay between regularity and randomness. When we hear music and static on the radio, we identify the music as regular and the static as random noise, and we try to make the signal-to-noise ratio as high as we can. In cooking, we may use a recipe but there are always some variations in the quantities of ingredients, in cooking times and temperatures, etc. In manufacturing, we encounter specifications and tolerances.

But the distinction between the regular and the random is not absolute. Many decades ago scientists at the Bell Telephone Laboratories investigated the origins of static and found that some of it comes from sources in particular places among the constellations in the sky. That was the origin of radio astronomy, and hence a permanent association of some static with important regularities. When we looked at neckties, we were concerned only with the pattern, and we neglected possible wine or food stains, but to a dry-cleaner the stains might be the significant regularities, while the patterns are of little concern. We conclude, then, that what is a regularity depends on a judgment of what is important and what is not. Of course the judge need not be human, or even alive.

Suppose we are concerned with the regularities in the repeated, but somewhat variable territorial songs of a male bird of a particular species in the nesting season. The other birds of the same species would make a suitable judge. Their behaviour can reveal what features of the song serve to repel other males from the territory or perhaps to attract a suitable female.

The crude definition of effective complexity given above in terms of minimum description length needs to be refined and made more technical. First, one should recognize that the effective complexity is necessarily context-dependent in a number of ways. To begin with, the entity in question is described at some level of coarse graining and in some language, and so we have immediately two sources of context dependence. It is also true that some degree of knowledge and understanding of the world is assumed. To illustrate the importance of this point, imagine that you are entering the village of a group of hitherto uncontacted

Indians in the Brazilian jungle, and that you know their language from experience with another group speaking the same tongue. Now explain to your hosts the meaning of a tax-managed mutual fund. It should take quite a while.

Of course as we get more technical we must think in terms of a computer. The description of the entity is coded into a string of bits (zeroes and ones) according to some coding scheme, and then a particular standard universal computer is employed (we have here two more sources of context-dependence). The algorithmic information content (AIC) of the description is then defined as the length of the shortest program that will cause the computer to print out the bit string in question and then stop computing. In fact, we shall generalize that definition by considering a variable time T and asking for the length of the shortest program that will perform this task within time T. The AIC is the limit of this quantity as T goes to infinity.

In discussing K, the AIC of the entity, or the corresponding quantity $K(T)$ for finite values of T, we are not yet dealing with the effective complexity. The EC has to do with the length of a very brief description of the *regularities* of the entity, not the features treated as random or incidental. Thus we need to exhibit the AIC (or its generalization to time T) as the sum of two terms, one applying to the regularities and the other to the features treated as random.

The best way to represent particular regularities of an entity is to embed that entity conceptually in a set of comparable things, all the rest of which are imagined, and to assign probabilities or weights to the members of the set. Such a set, with probabilities for the members, is called an ensemble, and it embodies the regularities in question.

The probabilities allow us to define a measure I of "Shannon information" or ignorance for the ensemble (These two apparent opposites are really two facets of the same thing: the amount of ignorance about the contents of a letter before one receives it is the same as the amount of information learned when it is received and read).

Since the ensemble serves to describe the regularities of our entity, the AIC of the ensemble (call it Y) is a candidate for the effective complexity. We just have to find the right ensemble. It turns out that the prescription for that is a straightforward one:

- First, minimize the "total information" $Y+I$. The minimum value can easily be shown to be about equal to K, the AIC of the entity. Thus K is exhibited as the sum of two terms referring to the regular and random features respectively. That is just what we needed.
- Second, following a famous principle, maximize the ignorance I or - what is the same thing - minimize Y, subject to the conditions imposed by the judge. The imposition of those conditions, based on a judgment of what is important, is an essential part of the prescription. One can show that otherwise Y would always be very small and the ignorance I would correspondingly use up almost all of K. Every entity would be simple.

In the arts, we are accustomed to the use of the imagined alongside the real, as in fiction and drama, and we are not surprised that light is often thrown in that way on the regularities of the real.

The break-up of K - the AIC of the entity - into the effective complexity Y and the ignorance I can be conceived in terms of a basic program and a lot of data to be fed into that program. The effective complexity is related to the length of the basic program, while the ignorance measures the information content of the additional data needed to specify the entity with which we are dealing.

The famous computer scientist, psychologist, and economist Herbert Simon used to call attention to the path of an ant, which has a high AIC and appears complex at first sight. But when we realize that the ant is following a rather simple program, into which are fed the incidental features of the landscape and the pheromone trails laid down by the other ants for the transport of food, we understand that the path is fundamentally not very complex. Herb said, "I got a lot of mileage out of that ant". And now it is helping me to emphasize the difference between total AIC and effective complexity.

A significant advantage of generalizing the measures K and Y to $K(T)$ and $Y(T)$ is that we can then discuss situations such as the following:

When I was a graduate student, more than fifty years ago, we wondered what kind of dynamics underlay the structure of atomic nuclei. It appeared that the laws that govern the energy levels of nuclei might be extremely complex. Today, however, we have the theory of quantum chromodynamics, the field theory of quarks and gluons, supplemented by quantum electrodynamics, which governs the electromagnetic interaction. We now believe those simple theories are correct (and many of their predictions have been verified by experiment). They must yield the energy levels of nuclei to an excellent approximation, but the calculations are so elaborate that even today's machines and techniques can't handle them properly. The standard universal computer would take a very long time to execute the short program corresponding to these simple theories and thus calculate the energy levels. If we insist on execution in a much shorter time T, then the program would have to be exceedingly long. So, the energy levels of nuclei have very little effective complexity Y in the limit of T going to infinity, but huge values of Y if T is taken to be small.

There can be no finite procedure for finding all the possible regularities of an arbitrary entity. We may ask, then, what kinds of things engage in identifying and using particular sets of regularities. The answer is complex adaptive systems, including all living organisms on Earth.

A complex adaptive system receives a stream of data about itself and its surroundings. In that stream, it identifies certain regularities and compresses them into a concise "schema", one of many possible schemata related by mutation or substitution. In the presence of further data from the stream, the schema can supply descriptions of certain aspects of the real world, predictions of events that are to happen in the real world, and prescriptions for behaviour of the complex adaptive system in the real world. In all these cases, there are real world consequences: the descriptions can turn out to be more accurate or less accurate, the predictions can turn out to be more reliable or less reliable, and the prescriptions for behaviour can turn out to lead to favourable or unfavourable outcomes. All these consequences then feed back to exert "selection pressures" on the competition among various schemata, so that there is a strong tendency for more successful schemata to survive and for less successful ones to disappear or at least to be demoted in some sense.

Take the human scientific enterprise as an example. The schemata are theories. A theory in science compresses into a brief law (say a set of equations) the regularities in a vast, even indefinitely large body of data. Maxwell's equations, for instance, yield the electric and magnetic fields in any region of the universe if the special circumstances there - electric charges and currents and boundary conditions - are specified (we see how the schema plus additional information from the data stream leads to a description or prediction).

In biological evolution, the schemata are genotypes. The genotype, together with all the additional information supplied by the process of development - for higher animals, from the sperm and egg to the adult organism - determines the character, the "phenotype" of the individual adult. Survival to adulthood of that individual, sexual selection, and success or failure in producing surviving progeny all exert selection pressures on the competition of genotypes, since they affect the transmission to future generations of genotypes resembling that of the individual in question.

In the case of societal evolution, the schemata consist of laws, customs, myths, traditions, and so forth. The pieces of such a schema are often called "memes", a term introduced by Richard Dawkins by analogy with genes in the case of biological evolution.

For a business firm, strategies and practices form the schemata. In the presence of day-to-day events, a schema affects the success of the firm, as measured by return to the stockholders in the form of dividends and share prices. The results feed back to affect whether the schema is retained or a different one substituted (often under a new CEO).

A complex adaptive system (CAS) may be an integral part of another CAS and it may, in turn, contain smaller complex adaptive systems. In fact, a CAS has a tendency to give rise to others.

On Earth, all complex adaptive systems seem to have some connection with life. To begin with, there was the set of prebiotic chemical reactions that gave rise to the earliest life. Then the process of biological evolution, as we have indicated, is an example of a CAS. Likewise each living organism, insofar as it can learn, is a CAS. In a mammal, such as a human being, the immune system is a complex adaptive system too. Its operation is something like that of biological evolution, but on a much faster time scale (if it took hundreds of thousands of years for us to develop antibodies to invading microbes, we would be in serious trouble). The process of learning and thinking in a human individual is also a complex adaptive system. In fact, the term "schema" is taken from psychology, where it refers to a pattern used by the mind to grasp an aspect of reality. Aggregations of human beings can also be complex adaptive systems, as we have seen: societies, business firms, the scientific enterprise, and so forth.

Nowadays, we have computer-based complex adaptive systems, such as "neural nets" and "genetic algorithms". While they may sometimes involve new, dedicated hardware, they are usually implemented on conventional hardware with special software. Their only connection with life is that they were developed by human beings, such as that outstanding human John Holland. Once they are put into operation, they can, for example, invent new strategies for winning at games that no person has ever discovered.

It is probably helpful to mention at this point that John Holland and I use somewhat different terminology. In fact, we both heard at a meeting some years ago that a scientist would rather use someone else's toothbrush than another scientist's nomenclature. What I call a schema is called an internal model by John Holland, and what I call a complex adaptive system is close to what he calls an adaptive agent. He uses the expression "complex adaptive system" to mean a loose aggregation of complex adaptive systems adapting to one another's behaviour, for example investors in a market or organisms in an ecological system.

According to my system of nomenclature, an industrial district would have to have a schema of its own to qualify as a complex adaptive system. From what little I know, I would say that may not generally be the case. Using John Holland's meaning of complex adaptive system, we would ask whether the component institutions (such as firms or universities) have schemata, and in general they do. In my notation, then, an industrial district may or may not be a complex adaptive system itself, but it certainly is an aggregation of complex adaptive systems, adapting to one another and to external influences.

Science fiction writers and others may speculate that in the distant future a new kind of complex adaptive system might be created, a truly composite human being, by wiring together the brains of a number of people. They would communicate not through language, which - Voltaire is supposed to have said - is used by men to conceal their thoughts, but instead through sharing all their mental processes. My friend Shirley Hufstedler says she would not recommend this procedure to couples about to be married.

The behaviour of a complex adaptive system, with its variable schemata undergoing evolution through selection pressures from the real world, may be contrasted with "simple" or "direct" adaptation, which does not involve a variable schema, but utilizes instead a fixed pattern of response to external changes. A good example of direct adaptation is the operation of a thermostat, which simply turns on the heat when the temperature falls below a fixed value and turns it off when the temperature rises above the same value. It just keeps mumbling "It's too hot, it's too hot, it's just right, it's too cold", and so forth.

In the study of a human organization, such as a tribal society or a business firm, one may encounter at least three different levels of adaptation, on three different time scales.

1) On a short time scale, we may see a prevailing schema prescribing that the organization react to particular external changes in specified ways. As long as that schema is fixed, we are dealing with direct adaptation.
2) On a longer time scale, the real world consequences of a prevailing schema exert (in the presence of events that occur) selection pressures on the competition of schemata and may result in the replacement of one schema by another.

3) On a still longer time scale, we may witness the disappearance of some organizations and the survival of others, in a Darwinian process. The evolution of schemata was inadequate in the former cases, but adequate in the latter cases, to cope with the changes in circumstances.

It is worth making the elementary point about the existence of these levels of adaptation because they are often confused with one another. As an example of the three levels, we might consider a prehistoric society in the U.S. Southwest that had the custom (1) of moving to higher elevations in times of unusual heat and drought. In the event of failure of this pattern, the society might try alternative schemata (2) such as planting different crops or constructing an irrigation system using water from far away. In the event of failure of all the schemata that are tried, the society may disappear (3), say with some members dying and the rest dispersed among other societies that survive. We see that in many cases failure to cope can be viewed in terms of the evolutionary process not being able to keep pace with change.

Individual human beings in a large organization or society must be treated by the historical sciences as playing a dual role. To some extent they can be regarded statistically, as units in a system. But in many cases a particular person must be treated as an individual, with a personal influence on history. Those historians who tolerate discussion of contingent history (meaning counterfactual histories in addition to the history we experience) have long argued about the extent to which broad historical forces eventually "heal" many of the changes caused by individual achievements - including negative ones, such as assassinations.

A history of the U.S. Constitutional Convention of 1787 may make much of the conflicting interests of small states and large states, slave states and free states, debtors and creditors, agricultural and urban populations, and so forth. But the compromises invented by particular individuals and the role that such individuals played in the eventual ratification of the Constitution would also be stressed. The outcome could have been different if certain particular people had died in an epidemic just before the Convention, even though the big issues would have been the same.

How do we think about alternative histories in science? Is the notion of alternative histories a fundamental concept?

The fundamental laws of nature are:

(1) the dynamical law of the elementary particles - the building blocks of all matter - along with their interactions and

(2) the initial condition of the universe near the beginning of its expansion some ten billion years ago.

Theoretical physicists seem to be approaching a real understanding of the first of these laws, as well as gaining some inklings about the second one. It may well be that both are rather simple and knowable, but even if we learn what they are, that would not permit us, even in principle, to calculate the history of the universe. The reason is that fundamental theory is probabilistic in character (contrary to what one might have thought a century ago). The theory, even if perfectly known, predicts not one history of the universe but probabilities for a huge array of alternative histories, which we may conceive as forming a branching tree, with probabilities at all the branchings. In a short story by the great Argentine writer Jorge Luis Borges, a character creates a model of these branching histories in the form of a garden of forking paths.

The particular history we experience is co-determined, then, by the fundamental laws and by an inconceivably long sequence of chance events, each of which could turn out in various ways. This fundamental indeterminacy is exacerbated for any observer - or set of observers, such as the human race - by ignorance of the outcomes of most of the chance events that have already occurred, since only a very limited set of observations is available. Any observer sees only an extremely coarse-grained history.

The phenomenon of *chaos* in certain non-linear systems is a very sensitive dependence of the outcome of a process on tiny details of what happened earlier. When chaos is present, it still further amplifies the indeterminacy we have been discussing.

Some years ago, at the science museum in Barcelona, I saw an exhibit that beautifully illustrated chaos. A non-linear version of a pendulum was set up in such a way that the visitor could hold the bob and start it out in a chosen position and with a chosen velocity. One could then watch the subsequent motion, which was also recorded with a pen on a sheet of paper. The visitor was then invited to seize the bob again and try to imitate exactly the previous initial position and velocity. No matter how carefully that was done, the subsequent motion was quite different from what it was the first time. Comparing the records on paper confirmed the difference in a striking way.

I asked the museum director what the two men were doing who were standing in a corner watching us. He replied, "Oh, those are two Dutchmen waiting to take away the chaos". Apparently, the exhibit was about to be dismantled and taken to Amsterdam. But I have wondered ever since whether the services of those two Dutchmen would not be in great demand across the globe, by organizations that wanted their chaos taken away.

Once we view alternative histories as forming a branching tree, with the history we experience co-determined by the fundamental laws and a huge number of accidents, we can ponder the accidents that gave rise to the people assembled at this conference. A fluctuation many billions of years ago produced our galaxy, and it was followed by the accidents that contributed to the formation of the solar system, including the planet Earth. Then there were the accidents that led to the appearance of the first life on this planet, and the very many additional accidents that, along with natural selection, have shaped the course of biological evolution, including the characteristics of our own subspecies, which we call, somewhat optimistically, *Homo sapiens sapiens*. Finally we may consider the accidents of genetics and sexual selection that helped to produce the genotypes of all the individuals here, and the accidents in the womb, in childhood, and since that have helped to make us what we are today.

Now most accidents in the history of the universe don't make much difference to the coarse-grained histories with which we are concerned. If two oxygen molecules in the atmosphere collide and then go off in one pair of opposite directions or another, it usually makes little difference. But the fluctuation that produced our galaxy, while it too may have been insignificant on a cosmic scale, was of enormous importance to anything *in* our galaxy. Some of us call such a chance event a "frozen accident". It produces substantial effects, if only in a limited region of space and time. Once it happens and its outcome is determined, an accident no longer belongs to the realm of the random. Frozen accidents, affecting significantly things judged to be important, give rise to regularities. Of course the fundamental laws also contribute to regularities, but those laws are thought to be simple, and so effective complexity comes mainly from frozen accidents.

We can take an example from modern human history to illustrate the idea of a frozen accident. It was mentioned earlier that a few historians are willing to discuss "contingent history", in which one asks "What if .....?". One of their favourite incidents occurred when Buffalo Bill's Wild West Show was touring Europe in 1889. Of course a star attraction of the show was the female sharpshooter Annie Oakley, who would ask for a male volunteer so that she could knock the ash off his cigar with a bullet. Normally there were no volunteers, and her husband, himself a famous marksman, would step forward and stand still while Annie shot, hitting the ash but not her husband. On this occasion, however, there was a volunteer, the Kaiser - Wilhelm der Zweite - with an expensive Havana cigar. Annie was a little worried, having drunk heavily the night before, but she fired and we know the result. The Kaiser survived, fired Bismarck, cancelled the reinsurance treaty with Russia, engaged in competition in naval construction with Great Britain, and in other ways laid the groundwork for the First World War.

The dominant role played by frozen accidents in generating effective complexity can help us to understand the tendency for more and more complex entities to appear as time goes on. Of course there is no rule that everything must increase in complexity. Any individual entity may increase or decrease in effective complexity or stay the same. When an organism dies or a civilization dies out, it suffers a dramatic decrease in complexity. But the envelope of effective complexity keeps getting pushed out, as more and more complex things arise.

The reason is that as time goes on the results of frozen accidents can keep accumulating. If that process outstrips the erasure of the consequences of frozen accidents, then more and more effective complexity can arise. That is so even for non-adaptive evolution, as in galaxies, stars, planets, rocks, and so forth. It is well known to be true of biological evolution, where in some cases higher effective complexity probably confers an advantage. And we see all around us the appearance of more and more complex regulations, instruments, computer software packages, and so forth, even though in many cases certain things are simplified.

The tendency of more and more complex forms to appear in no way contradicts the famous second law of thermodynamics, which states that for a closed (isolated) system, the average disorder ("entropy") keeps increasing. There is nothing in the second law to prevent local order from increasing, through various mechanisms of self-organization, at the expense of greater disorder elsewhere. (one simple and widespread mechanism of self-organization on a cosmic scale is provided by gravitation, which has caused material to condense into the familiar structures with which astronomy is concerned, including our own planet).

Here on Earth, once it was formed, systems of increasing complexity have arisen as a consequence of the physical evolution of the planet over some four and half billion years, biological evolution over nearly four billion years, and - over a very short period on a geological time scale - human cultural evolution.

The process has gone so far that we human beings are now confronted with immensely complex problems and we are in urgent need of better ways of dealing with them (the role of industrial clusters and their potential constitute, of course, an example of a complex issue). When we attempt to tackle such difficult problems, we naturally tend to break them up into more manageable pieces. That is a useful practice, but it has serious limitations.

When dealing with any non-linear system, especially a complex one, it is not sufficient to think of the system in terms of parts or aspects identified in advance, then to analyse those parts or aspects separately, and finally to combine those analyses in an attempt to describe the entire system. Such an approach is not, by itself, a successful way to understand the behaviour of the system. In this sense there is truth in the old adage that the whole is more than the sum of its parts.

Unfortunately, in a great many places in our society, including academia and most bureaucracies, prestige accrues principally to those who study carefully some aspect of a problem, while discussion of the big picture is relegated to cocktail parties. It is of crucial importance that we learn to supplement those specialized studies with what I call a crude look at the whole.

Now the chief of an organization, say a head of government or a CEO, has to behave as if he or she is taking into account all the aspects of a situation, including the interactions among them, which are often strong. It is not so easy, however, for the chief to take a crude look at the whole if everyone else in the organization is concerned only with a partial view.

Even if some people are assigned to look at the big picture, it doesn't always work out. Some time ago, the CEO of a gigantic corporation told me that he had a strategic planning staff to help him think about the future of the business, but that the members of that staff suffered from three defects:

1) They seemed largely disconnected from the rest of the company.
2) No one could understand what they said.
3) Everyone else in the company seemed to hate them.

Despite such experiences, it is vitally important that we supplement our specialized studies with serious attempts to take a crude look at the whole.

I am sure that observation applies to industrial clusters.

At many research institutions, theorists are busily engaged in constructing simple models of quite complex real systems. Today, those are usually computer models. How, though, does one rate the success of such a model in helping us learn about the real world? Normally, a theory is compared to observation and the value of the theory is judged by the degree of agreement with observed facts, along with criteria such as elegance, consistency, and compatibility with well-established principles. However, when we are dealing with a highly simplified model of a very complex system, detailed agreement with observation could well be embarrassing rather than encouraging.

At the Santa Fe Institute, we have had a number of discussions of this question: what can be claimed on the basis of a simplified model? One outcome of those discussions is the conclusion that one should look for rather simple regularities in the data and see if the model predicts some or all of them. This semi-empirical "middle level theory" can be enormously valuable. Often, the regularities in question will hold along the whole spectrum from the complex real world down through successive simplifications of that world to the model being examined. If that is the case, then the reason for the regularity in the simple model may be easy to understand and it can throw light on how the regularity arises in the real world.

Examples of such simple regularities certainly include scaling laws, in which one quantity of interest appears as a power of another such quantity. These laws crop up in all fields of science: physical, biological, and behavioural. A striking example is the rule, verified with increasing confidence over most of a century, but unexplained until recently, that the metabolic rate of a mammal species goes like the three-quarters power of the mass. The mammal in question can range from a tiny shrew to an elephant or a whale. If one mammal weighs ten thousand times as much as another, it will process energy at a rate one thousand times higher. Similar laws (with the same power but different coefficients) apply to birds and to higher plants.

This law was explained by Geoffrey West, a theoretical physicist at the Los Alamos National Laboratory, and James Brown, an ecology professor at the University of New Mexico in Albuquerque, along with his student Brian Enquist. These researchers met and collaborated at the Santa Fe Institute, in a perfect example of how that institute is supposed to function.

They modelled (analytically rather than on a computer) the cardiovascular and respiratory systems of the animals and the vascular system of the plants. They based their model on three-dimensional fractals, and they used two facts:

1) in the branchings of the tubes, areas are preserved and

2) the thinnest tubes all have about the same diameter, independent of the mass.

The scaling law then followed, with the fraction 3/4 arising from the number of spatial dimensions divided by that number plus one.

Sometimes scaling laws are laws of distribution, for example the famous Zipf's law - found in fact by Auerbach in 1913 - for the populations of "cities" (actually metropolitan areas). It states that if we rank the cities in order of decreasing population, then the population is roughly inversely proportional to the numerical rank. The data are a bit ragged for the very largest cities, but they agree quite well with the law for the next several hundred cities. Thus the 300th city has about one-third the population of the 100th city.

Recently Xavier Gabaix published an explanation of Zipf's law starting from Gibrat's law, which is known to be true and states that the mean growth rate and mean square deviation of growth rates for metropolitan areas are independent of population to a good approximation. Gabaix showed that the steady state distribution of populations then obeys Zipf's law.

Besides scaling laws, we can look at other examples of simple middle level semi-empirical laws, for instance implicational ladders. Around 1940, when the USA was still not a belligerent in the Second World War, some sociologists were asked to measure the attitudes of Americans toward Great Britain. They formulated a long list of questions with yes or no answers, and those questions were posed to a representative sample of American citizens. The results showed a fascinating regularity. If the questions were arranged in a certain order, then most of the returned questionnaires showed responses that were affirmative down to a certain point and then negative after that. Thus most of the sets of answers to the questions could be characterized by a single parameter, the point in the list where the answers switched from positive to negative. As far as the questionnaire went, attitude toward the United Kingdom could be summarized in a single number. Such a result is often called an implicational scale (not to be mixed up with scaling laws, as described earlier). To avoid confusion, I like to call it an implicational ladder.

Another example can be cited from linguistics. There are some countries, such as Haiti and Jamaica, where a Creole language is spoken that gets its words mainly from the official language of the country. In Haiti, nearly everyone speaks a French-based Creole language at home, while the official language is, or was until recently, French. Likewise, in Jamaica, the official language is English but many people speak an English-based Creole at home (this is in contrast to the situation on the island of St. Lucia, for example, where the official language is English but the Creole is French-based).

In 1961, the linguist DeCamp investigated the situation in Jamaica, where, under the influence of official English in the schools, there is a tendency for certain people to substitute usages from regular English for those in the broadest or deepest Creole. He chose six cases of such usages, four involving vocabulary and two involving pronunciation. He discovered that those usages could be arranged in a particular order such that some speakers used the standard English version in all six cases, some in the first five, some in the first four, some in the first three, and so forth. Other patterns were negligibly rare. Thus, out of all the 64 possible patterns for choices in the six cases between Creole usage and Standard English usage, only 7 patterns existed with appreciable frequency among the population, and those seven formed an implicational ladder.

If I had to model industrial clusters, I would search carefully for "middle-level" semi-empirical rules that might persist all the way from the real world down to the highly simplified model, and test the model by seeing if it clarifies how the rules arise.

Complex Adaptive Systems and Spontaneous Emergence

John H. Holland

Abstract. This paper introduces concepts associated with complex adaptive systems (*cas*), linking those concepts at some points to economic planning. The paper begins (section 1) with an informal description of the notion of a *cas* and then (section 2) discusses the critical role of "building blocks" in understanding *cas*. Using these ideas, the paper goes on (section 3) to discuss the phenomenon of "emergence", wherein the whole of the system's behaviour goes beyond the simple sum of the behaviours of its parts. The body of the paper (sections 4 through 6) looks at the role of modelling in predicting the behaviour of *cas*, examining the kinds of model that will serve this purpose. The paper concludes (section 7) with a brief discussion of the relevance of these ideas to economic planning.

1. Complex Adaptive Systems

A complex adaptive system (*cas*) exhibits three distinguishing characteristics:

(i) A *cas* consists of a large number of interacting components, usually called *agents*. The agents may range from firms in an economy or participants in a market to antibodies in the immune system or signalling proteins in a biological cell.

(ii) The agents in a *cas* interact in non-additive (non-linear) ways. The interactions can be specified by associating a set of condition/action rules with each agent, where each agent's rules describe its strategy for interacting with other agents (the counterpart of a strategy for playing a game like chess). Such rules range from a simple stimulus-response form, IF stimulus x THEN make response y, to message processing rules, IF message x THEN send message y. Because message processing rules can implement any program that can be written for a general-purpose computer, this way of specifying interactions is fully general. Often, the rules are organized to form an internal model of the agent's external world, allowing the agent to anticipate the future.

(iii) The agents in a *cas* adapt or learn. That is, they modify their rules as experience accumulates, searching for improvements. Learning to play a complicated game, such as chess, provides an example: as one learns to play, some rules begin to implement sub goals ("develop your pieces during the opening stages") and look ahead ("if I move that bishop, I will lose a rook to a knight's fork three moves from now"). As we will see (section 2) discovering new, plausible rules depends upon finding appropriate "building blocks" for describing parts of the *cas* agents.

2. Building Blocks

Building blocks are the pervasive, critical foundation of an ability to act with insight in a complex world. Human perception, for example, consists primarily in combining well-known, simple components to describe familiar phenomena. Different trees are described by different arrangements of familiar parts: leaves, stems, branches, and trunks. Similarly, human faces are composed of variants of standard parts - hair, forehead, eyebrows, eyes, nose, and so on. A little thought shows that we approach all objects, familiar and unfamiliar, via combinations of familiar building blocks.

We can also extract building blocks as the generators of the dynamic situations that characterize the interaction of agents. Consider the rules of chess: the rules are building blocks that generate a miniature artificial world in which two agents interact. Though chess is simply defined - less than a dozen rules suffice - it is an artificial world of perpetual novelty. We never see the same game played twice unless some previous game is deliberately recorded and repeated. The game is so complex that, after centuries of study, we have no idea of what a "best strategy" would be for playing the game. Nevertheless, in this perpetually novel world, planning and anticipation are possible, indeed essential.

Planning in chess depends upon extracting and exploiting certain patterns (configurations of pieces) that occur repeatedly in the play of the game. These patterns become higher-level building blocks from which to construct strategies (plans) for playing the game. Over the centuries, chess players have repeatedly discovered new repeating-pattern building blocks that make possible ever more sophisticated strategies. For example, early in the 20th century, chess players discovered that certain interlocking pawn structures could exhibit strong influence on the play of the game. As a result the pawns in today's game do a lot more than they did in the 19th century. By adding this pawn-structure building block to the repertoire, a present-day master chess player can easily outplay a 19th century player of equal rank.

Lest this discussion of games seem frivolous, note that the rules of a game are not different in kind from the axioms that generate Euclidean geometry, or the set of computer instructions that define a flight simulator for a new aircraft. Indeed if we look to that most complex of organizations, the biological cell, we see layer after layer of building blocks. At a low level we have the 4 nucleotides that form the building blocks from which all chromosomes are constructed (Note that a physicist would add several levels of building blocks *below* this level, ranging from atoms down to quarks). One level up we have the 20 amino acids, coded by triplets of nucleotides, that form the building blocks for proteins. The proteins, in turn, have substructures such as alpha-helices and beta-sheets that determine their three dimensional structure. From there we go to small interacting groups of proteins, such as the Krebs cycle, on to building blocks for membranes and tubules, to organelles, and on up through the hierarchy of living organisms.

If we turn from organisms to artefacts, we see that most major innovations in science and technology amount to new combinations of old building blocks. For example, the internal combustion engine, that great source of mobile power that transformed the 20th century, was simply a new combination of building blocks that had been around for a long time: the carburettor was a modification of Venturi's perfume sprayer, the spark plug was presaged by Volta's sparking device, gear wheels and pistons had been known from antiquity, and so on. The innovation was the combination of these known building blocks in a way that provided new interactions between them.

While innovations typically come from new combinations of familiar building blocks, once in a while there is a genuinely new building block. Though such events are rare, they can have dramatic impact: the transistor revolutionized the design and production of electronic circuits. However, even the transistor is a new combination of well-known building blocks one level down; there was a theory of semi-conduction many decades before that idea was combined with other ideas to yield this new switching device. New building blocks often arise through "cross-breeding" successful lower-level building blocks, blocks drawn from previously unassociated areas or disciplines.

In sum: whether it be natural or artificial systems, building blocks are almost always the basis of understanding and innovation. Moreover, the resulting structures almost always have a hierarchical format, with selected combinations of building blocks at one level serving as building blocks one level up.

3. Emergence

It is a commonplace that a small number of well-chosen building blocks can be combined to generate a vast array of interesting, non-random structures. The 26 letters of the English alphabet suffice to generate the vast literature in English, past and future. The 20 amino acids generate the unending array of proteins that form the basis of life. Chess, defined by fewer than a dozen rules, offers new patterns of play after centuries of study, and the geometry defined by Euclid's five axioms surprises us with new theorems after two millennia of study. This perpetual novelty generated by finite means puts us at the threshold of the phenomenon called *emergence*.

There is no accepted definition of emergence, even among scientists, but few who have seriously studied such phenomena believe it to be an "eye-of-the-beholder" effect. Indeed, it is possible to list criteria that go far toward distinguishing some observation as emergent, regardless of the observer and the time of discovery.

First among these criteria is the requirement that the phenomenon be a repeating pattern in a system that exhibits perpetual novelty. This repeating pattern is a feature or property common to some subset of the unending building-block combinations. In mathematical terms, the repeating pattern is an *equivalence class* within the subset of combinations. Let us call such a repeating pattern a *regularity*.

In the dynamics of a *cas*, regularities typically distinguish commonalities among some of the agents. When these regularities are reinforced by interactions among the agents, they become persistent, providing possibilities for "speciation" and selection. For example, the Krebs (citric acid) cycle is a regularity common to all living aerobic organisms. This dynamic cycle, mediated by eight enzymes (rule-like building blocks), underpins the production of all the basic components of the cell: amino acids, lipids, and the like. The persistent regularity we call the Krebs cycle becomes an emergent building block making possible higher levels of organization, ultimately yielding the organelle-cell-organ hierarchy common to all metazoans.

This example suggests the second criterion: emergent phenomena exhibit a hierarchical organization wherein selected combinations of building blocks at one level become building blocks at a higher level of organization. Herb Simon's famous "watchmaker" parable (Simon 1996) makes clear the great advantages hierarchical organization confers on dynamic systems undergoing selection. The "adaptive" part of a cas assures the prevalence of this kind of organization under emergence.

Once we allow levels of organization, it becomes natural to ask if, or how, the different levels interact. Clearly living organisms and complex artefacts, such as computers or airplanes, exhibit such organization, with interactive signalling in both bottom-up and top-down directions. For example, a single gross movement of the control column in an airplane sends signals that affect dozens of devices throughout the airplane (a top-down effect), while the malfunction of a single critical part, say a warning light, can ground the plane (a bottom-up effect). Similarly, though an overall market indicator (say the Dow-Jones average) is an average of the actions of individual investors (a bottom-up effect), a change in that indicator can exert a top-down influence on the actions of individual investors (agents). In general, agents at higher levels in a *cas* hierarchy are composed of agents at lower levels. Here then is a third criterion: the overall form and persistence of an emergent regularity depends upon both bottom-up and top-down effects.

The fourth criterion sounds mysterious, but is quite simple when understood: the whole emergent regularity is more than the *sum* of its parts. This criterion contrasts with the simple reductionist strategy for parsing phenomena: study the building blocks, and then *sum* their behaviours to get the behaviour of the whole. Throughout the history of science this simple reductionist strategy has worked over and over again, from the description of gases as colliding particles to the superposition of waves of different frequencies to describe complex sounds. Trends, polls, and averages, as predictors, all exemplify this additive approach. However, the non-linear interactions of the agents in a *cas* short-circuit simple reductionism. We cannot *add* up the behaviours of the agents to get the behaviour of the whole because the interactions are *not* additive. Reduction still works, but it is not simple because the interactions are not simple. The behaviour of the whole cannot be obtained by simply adding together the behaviours of the parts.

These four criteria go a long way toward separating the phenomena I would like to call emergent from most other phenomena. The criteria are subject to explicit testing procedures, so emergent phenomena satisfying these criteria are not an eye-of-the-beholder effect.

4. Models and Prediction

Though modelling is the subject of an extensive literature with rigorous mathematical underpinnings (homomorphism), I will only comment informally on a few aspects relevant to the modelling of *cas* and agents.

A first principal: models, like theories, suggest where to look for answers to some previously formulated question or conjecture. Accordingly, anything considered irrelevant to the question should be considered a detail that can be eliminated from the model. A model much like a political cartoon, makes its point by exaggerating certain features while eliminating incidentals. And a model, like a map, is only serviceable when it contains a limited amount of detail.

When we model *cas*, the models are perforce-dynamic models. The object is to find a few mechanisms and laws (similar to the pieces and rules of a game) that generate a description of the *cas*'s changing states (configurations). A typical guiding question will concern the occurrence of some set of regularities (see section 3) in these unfolding configurations. For example, how does a vaccine confer lifetime immunity on the continually changing immune system? Or, more generally, why do almost all *cas* exhibit *lever point* phenomena, where small "inexpensive" inputs cause major directed effects in the *cas* dynamics?

It is an art to select a level of detail that allows simple laws. Such laws, when selected with insight, generate state-sequences (configuration-sequences) that faithfully describe, or at least approximate, the behaviour of the system at that level of detail. In weather forecasting, discovering the laws of movement for regularities called *fronts* made possible dynamic models yielding predictions an order of magnitude more accurate than previous statistical models.

The quintessential exhibits of this art form are the physicist's *gedanken* experiments. Einstein's formulation of thought experiments illustrating the difficult questions in quantum mechanics are classic (Jammer 1974). In such models the details are pared to the point that the model can be examined by thought alone.

Cas are generally too complex to allow such elegant formulations, but we can build computer-based models that serve a similar exploratory purpose. Computers, of course, allow the construction and execution of models of remarkable detail. However, this facility also poses a great danger. Because the computer can handle the detail, there is a strong impulse to add detail upon detail. It is possible to wind up in the position of Borges' king (Borges 1964) with a map the size of the country. The essence of elegant science is providing just enough detail to explore the question posed.

Once a question has been posed, it is typical to start with a *very* simple model, quite likely an over-simple model. You then see how far the model will take you toward answering the question, only adding a new mechanism or law (see pieces and rules in a game) when there is a clear need. This process of going as far a you can with a simple model, in order to see what's missing, can be quite productive. And it does produce models that are just detailed enough to answer the question.

5. Modelling Agents

In a *cas* each agent is surrounded by, and interacts with, a multitude of other agents. It is convenient to think of all of an agent's environment as represented by agents, some of which are passive and do not adapt (such as a rock or a chair). There is no real loss of generality in such a conceit.

Each adaptive agent receives information about its environment via a set of detectors (receptors, gauges, or the like). We can think of the output of a set of detectors as a standardized packet of information, called a *message*. An agent's internal information processing can also be treated in terms of messages (much like the memos circulated in an office). Some messages may be directed to other agents or to effectors (machines, muscles, or the like), providing the interactions that change the agent's environment. In short, we can look upon an agent as a message processing device. In consequence the overall *cas* becomes a complex signalling network.

One formalization of this outlook is provided by the adaptive rule-based systems called *classifier systems* (Holland 1995, 2000). The building blocks for a classifier system are IF/THEN rules called *classifiers*. The condition (IF) part of the rule "looks for" certain kinds of messages; when the rule's conditions are satisfied, the action (THEN) part specifies a message to be sent. For example, if the agent has a visual system, one of the rules might be IF there is (a message from the detectors indicating) an object left of centre in the field of vision, THEN (by issuing a message to the effectors) cause the eyes to look left.

In a classifier system many rules may be active simultaneously because additional active rules simply means more messages being issued at a given instant. From a computational point of view, it is convenient to think of the messages present at a given instant as collected in a list called a *message list*. As a rough analogy, think of the rules in a classifier system as standard operating procedures. Then think of an agent as an office containing many desks, where each desk (rule) is responsible for executing a particular operating procedure. In this analogy, the message list is analogous to a bulletin board that contains all the memos (messages) directing the activities of the office that day. The action of each desk, when there is a memo for it on the bulletin board, is to produce a memo for the next day. In classifier system terms, *each* rule is looking to the message list, to see if there are messages satisfying its conditions (the IF part). If the rule's conditions are satisfied by messages on the list, it generates a new message (the THEN part) to be added to the list.

It is often desirable to direct a message to a given rule or set of rules, much like the header on an email message selects a set of destinations for the email message. To select destinations in a classifier system we use *tags*. Tags for this purpose are quite common in the natural world. For example, in molecular biology, we have amino acid sequences that bind to a specific locus in DNA. The DNA loci are the counterparts of rules that are repressed (cease to act) by proteins (messages) having particular tags. Tags play similar roles in identifying antigens for the immune system, they also serve to define active sites in an enzyme, and tags serve as "headers" for the signalling proteins that mediate reaction cascades and complicated feedback loops in biological cells. Rules using tags may represent DNA loci that are repressed by signalling proteins (messages) having particular amino acid sequences (tags) that provide binding action, or rules may serve as steps in a cascade wherein a protein initiates the formation of a new protein in the cascade (sending a signal), or the rules may interact in much more complicated feedback loops.

In a classifier system using tags rules typically have the form IF (signal with apropos tag present) THEN (send signal with new tags).

The condition part of a classifier rule will often specify several conditions, so that the rule requires the simultaneous presence of different tagged messages before it acts. If an agent has a modular, hierarchical structure, the rules in each module can be directly identified by a common tag. That is, to send a message to a module you attach to it the identifying tag of the module. An appropriately tag-linked collection of classifier rules can implement any subroutine or program that can be written for a computer. As a result any hierarchical *cas* can be implemented as a classifier system.

6. *Cas* Models Based on Classifier Systems

Classifier systems address three basic problems in the study of *cas*:

(1) Parallelism and coordination.
Should an agent use a single, monolithic rule for each situation it could encounter, the demands on its capacity to store rules would be overwhelming. No system can afford to retain enough rules to handle unique situations such as "a red Saab by the side of the road with a flat tire". On the other hand, such a situation is easily handled by simultaneously activating rules for the building blocks of the situation: "car", "roadside", "flat tire", and the like. In short, a rule-based system can handle a broad range of novel situations if it can act on them with combinations of "building block" rules. Combinatorics then work for the

system instead of against it. Moreover, by definition, useful building blocks are frequently used, so they are frequently tested and confirmed. The object, then, is to provide for the simultaneous interaction and coordination of well-confirmed rules that, together, describe the new situation.

Classifier systems address this coordination problem by restricting rule action to the emission of tagged messages. Complicated logical questions about "rule consistency" and "consistency maintenance" are avoided because messages only serve to activate rules. Messages have no other intrinsic meaning. More messages simply mean more active rules, and vice-versa.

(2) Credit assignment.

There is no secure way to determine a priori which rules will serve as good building blocks for generating system behaviour. This means there must be some means of rating the usefulness of rules as the system accumulates experience. This is a difficult problem when many rules are active simultaneously, some helping and some obstructing. The problem is exacerbated when the environment requires stage-setting actions in order to achieve some desired outcome (as when one sacrifices a piece in chess to set the stage for the capture of a major piece later). Solution of the credit assignment problem is a sine qua non for adaptation.

In realistic situations, exhaustive exploration of all possible action paths is not feasible, so the credit assignment scheme must be "local". Samuel's early work (Samuel 1959), using prediction of the value of future options, points the way, but few have exploited his insights. In classifier systems, credit assignment is handled by setting up a market situation.

Each rule in the classifier system is assigned a strength that is a measure of its usefulness. In the credit assignment procedure, this strength serves as a kind of cash-in-hand. When the condition part of a rule is satisfied it makes a bid to become active, a bid proportional to its strength. If the rule is a winner in the bidding process, its strength is reduced by the amount of the bid, and the strengths of rules sending activating messages to it, its "suppliers", are increased by that amount. That is, the newly active rule makes a payment that increases the strength of its "suppliers". Then the newly active rule stands to profit from bids for its newly posted message, the bids of its "consumers". A rule increases its strength if it makes a profit in this market (its payments to its suppliers are less than the payments it receives from its consumers).

In effect, a rule acts as a go-between (broker, middleman) in rule-chains leading from the current situation to (possibly favourable) future outcomes. At intervals, some rules have their strength increased when the system acquires direct payments from the environment (reward, payoff, reinforcement - e.g., "food", or some payment from an "ultimate consumer"). The newly strengthened direct-paid rules will make larger bids the next time this particular chain is executed. These larger bids will strengthen supplier rules in the chain that set the stage for the direct-paid rules. And so on back through the chain on successive executions. As rules become stronger they are more likely to win the bidding process, so their messages are more likely to influence system behaviour. In short, rules which make a profit in rule-chains leading to direct environmental payments will come to control the system.

(3) Adaptation via rule discovery.
Rule discovery (or its equivalent) is the most recondite problem the adaptation of classifier systems. It was not even wellhandled in Samuel's remarkable work. We know that rules receiving little credit (low strength) should be replaced, but random generation of new rules can only work for the simplest problems. The key to effectiveness is use of past experience to generate plausible new rules for situations as yet poorly understood.

Classifier systems exploit a *genetic algorithm's* (Mitchell 1996) ability to discover and recombine building blocks. Classifier systems have "building blocks" at two levels: (i) the parts (called *schemata*) from which the condition and action parts of individual rules are constructed, and (ii) the rules themselves, as components of the overall system. A genetic algorithm is designed to "crossbreed" well-rated (strong, fit) elements in a population, so it is wellsuited to working on this task at both levels. The strength of a rule determines its likelihood of being a parent in the crossbreeding process.

Each of the mechanisms used by the classifier system has been designed to enable the system to continue to adapt to its environment, while using its extant capabilities to respond instant-by-instant to that environment. In so doing the system is constantly balancing exploration (acquisition of new information and capabilities) with exploitation (the efficient use of information and capabilities already available).

7. Relevance to Organizations (such as Industrial Districts)

If we think of the agents in a cas as building blocks, then the discussion of hierarchy in section 3 suggests that we should see hierarchical organization in most, if not all, *cas*. Certain *combinations* of agents, because of efficient exchanges of resources and signals, acquire a greater persistence. In effect they out compete other combinations, absorbing resources that might otherwise be available to those other combinations. Persistent combinations of agents (say firms) become candidates for agents at a higher level of organization (say industrial districts).

Agents in a *cas* typically are *not* indefinitely persistent, a fact that strongly affects the organizing process. For example, consider the mammalian body as a *cas*: every atom in a mammal is turned over in less than two years, so that its body is a persistent pattern imposed upon a flow of atoms. This persistent pattern is a complex version of the standing wave imposed on a flow of water molecules past a rock in a white-water river. The same is true of larger human organizations, be they firms or cities. Firms typically persist over time spans much longer than the time of employment of an average employee, and cities persist over spans of time much longer than the lifespan of their inhabitants. A *cas*, rather than being a "bricks and mortar" structure, is primarily a pattern imposed on flows of resources and signals.

Under this view, adaptation in a *cas* is facilitated by a redirection of flows to provide enhanced persistence. Resource exchanges with surrounding agents (the equivalent of mutualism and symbiosis in ecosystems) provide for exploitation of strategic discoveries already made, while the modification of extant agents, and the creation of new agents, provides for further exploration and discoveries. Because the agents in a *cas*, at whatever level, are continually adapting, innovation is a regular feature. This continual innovation makes the balance between exploitation and exploration a critical factor in the persistence of agents.

One of the least understood, but pervasive, *cas* properties is the "lever point". A "lever point" is a kind of flow redirection that causes long-term directed changes at a small expenditure of resources. For example, an inexpensive vaccine can cause a lifelong change in the mammalian immune system. All *cas* that have been examined closely to date exhibit lever points. However, we have no principled way of searching for lever points. It is here that good models and, ultimately, theory should tell us "where to look". As with vaccines, lever points offer opportunities for making substantial changes in "locked-in" situations. Without such knowledge, we often wind up "throwing money" at a *cas*, with little or no effect on the outcome.

In economic planning, knowledge of the relevant lever points can make the difference between resounding success and stagnation or failure.

References

Borges, J. L. (1964), "On Rigor in Science", *Dream Tigers*, Austin, U. Texas Press.
Holland, J. H. (1995), *Hidden Order*, Reading, MA, Addison-Wesley.
Holland, J. H. (2000), "What is a Learning Classifier System?", in: *Learning Classifier Systems*, Lanzi, P. L., Stolzmann, W. and Wilson, S. W. (eds), Berlin, Springer, pp. 3-6.
Jammer, M. (1974), *The Philosophy of Quantum Mechanics*, New York, Wiley.
Mitchell, M. (1996), *An Introduction to Genetic Algorithms*, Cambridge, MA, MIT Press.
Samuel, A. L. (1959), "Some Studies in Machine Learning Using the Game of Checkers", in: *Computers and Thought*, Feigenbaum, E. A. and Feldman, J. (eds), New York, McGraw-Hill.
Simon, H. A. (1996), *The Sciences of the Artificial*, Cambridge, MA, MIT Press, 3rd ed..

The Industrial Cluster as a Complex Adaptive System

Enzo Rullani

Abstract. In the economic debate of recent years, the rediscovery of *industrial clusters* is concurrent with the rediscovery of *complexity*.
And with good reason.
These two issues regained popularity at the same time, after years of neglect, because they both underscore, in different ways, the failure of the idea of modernity derived from Fordism and the need for a new vision in which reason is unshackled from the *deterministic prejudice* that results from the extraordinary success enjoyed until now by a system based on the control of technology and financial success. *Neo-modern* reasoning, or post-modern reasoning, as some prefer to call it, should salvage the worthwhile aspects of modern tradition - rational criticism, dialog communications and universalism of values and rights - integrating these values with an attitude that is more open to the *exploration, interpretation and invention* of possibilities that, while outside our control, can be important and yield the benefit of a rational exploitation of what we already know and control (Rullani 1996).
The notion of modernity that shaped the imaginary edifice of Fordism was the expression of a *deterministic reasoning* that sought to *exploit* to the fullest what was already known. The step that is being proposed today, which brings complexity back into play, is to develop alongside this line of reasoning another one, more dangerous but more promising, based on the *exploration* of the world of possibilities (Axelrod and Cohen 1999, p. 43).
According to J. March (1991, p. 71):

> "exploration includes things captured by terms such as search, variation, risk taking, experimentation, play, flexibility, discovery, innovation. Exploitation includes such things as refinement, choice, production, efficiency, selection, implementation, execution".

In order to govern complexity, we must explore and capture what is new, but also exploit the results of what has been learned each time. If we were to just explore, we would have to bear all the costs of the activities needed to research, interpret and test that which is new, but we would reap only a fraction of the benefits that could be obtained from the resulting innovations. The choice of focusing only on exploitation, with a minimal interest in exploration, would be equally unsatisfactory, because if we were to limit ourselves to exploiting rationally what we already know, we would find ourselves trapped in some sub optimum equilibrium zone from which we could not escape even if we were confronted with exceptional opportunities.

Between exploitation (of what is known) and exploration (of what is new) there is a border that neo-modern reasoning must cross over repeatedly as it seeks useful opportunities, without ever stopping at that methodological borderline beyond which is a land that until a short while ago was marked "Danger! Unknown".
The relationship with complexity - and, as it concerns us here, with industrial clusters - brings into relief a basic alternative between rationalizing that which is old and exploring that which is new one that all systems of learning must face sooner or later and that, not accidentally, represents a "place" (*topos*) that is typical of the more general studies of complexity (Holland 1988, 1995).
During the past century, when deterministic reasoning was prevalent, the focus was primarily on computing and optimising what was already known, reducing the exploration of that which was new to a routine that administered and controlled all innovative acts, neutralizing their subversive energy. In this sense, the century of Fordism was characterized by *hostility to complexity*, which was seen as a source of uncertainty and risk and, therefore, an obstacle to rationalization. Increasingly codified and rigid forms of *forecasting, planning and control* were developed in order to minimize the exposure to complexity, which artificially reduced opportunities and interest in exploring that which is new. In both theory and practice, the emphasis was on rationalizing that which existed by relying on bureaucracies and automatisms. Complexity was tackled by breaking it up into fragments and forcing these fragments into predetermined computation (technology, markets), control (plans) and negotiation (management) procedures.
To do this, economic behaviour, which could be put through computation and optimisation procedures more easily than other activities, was separated from everything else. Economics became the science of efficiency, specialized in exploitation and removed from exploration. Its complexity was drastically *reduced* by assigning all its anthropological, social and institutional variables to the realm of *exogenous* variables, which left economics with just one task: maximize the efficient use of known means to achieve equally known ends.
In this context, the *territory* was inevitably pushed to the margins of economics studies. Since it is an elective place where economics, society and history intersect, the territory contains information and complex relationships that exceed the sphere of calculable actions. One could even say that thinking in the past century, because it was consumed with reducing complexity, was ultimately also an *enemy of the territory*, which was perceived as a system that creates, preserves and reproduces the complexity of history and experience.
The reason why we are now re-examining with such great interest industrial clusters and other local phenomena that are rooted in geographical locations and are rediscovering old studies of territorial economies that were gathering dust in local archives is because we need to understand how to tackle high levels of complexity without first breaking it down with computations, controls and negotiations, since time has shown all these methods to be blunt and increasingly ineffective instruments.
The *rediscovery of complexity entails the rediscovery of the territory* and vice-versa. Choosing to take a *pragmatic*, experimental approach that allows rationality to explore the open world of possibilities also means finding a foothold in the

stratified complexity of an archipelago of many and varied territories (Sabel 1989, 1999). The reference to the plurality of experiences contained in the different types of territories helps reject a reductive idea of rationality that is limited to the ability to forecast, compute and control. It also opens the way for the spirit of *exploratory rationality*, thereby contradicting the legacy of deterministic reasoning from the Fordist century.

In hindsight, we can say that the ambition to reduce the economic world to a computable tool was an *illusion* from the start. Complexity, which was being pushed out the door, re-entered shortly afterward through the window. Out of the limelight, complexity continued to exist unhindered, *despite Fordism*, in the economic reality of daily life, even though, unquestionably, it was mostly ignored by economic theory, which relegated it to the margins of attention, labelling it deviation, exception, error, anomaly, uncertainty, etc. Complexity, although viewed as a *negative* factor, continued to play a role in the practical history of businesses and individuals, actually becoming a problems' problem, the beginning and end of the concerns of a techno structure that denies its existence and fears its impact at the same time.

The true story of modernity has two faces: an *official* one, which shows the work of players who understand their goals and are rational in the computation of their means, and an *unofficial* one, in which complexity resurfaces, pushing through the meshes of computation and escaping the control of planning. When we look at the history of great organizations, of business systems, of the webs of alliances and of territorial formations - *even during the long season of Fordism* - the dynamics of these relationships clearly show the complex nature of an *experimental type of reasoning* that does not go from goals to means or from computation to result, choosing instead to intertwine goals and means in a maze of relationships and circularities in which intentional action follows the path of evolutionary learning, which proceeds in a zigzag pattern instead of acting in accordance with predetermined objectives and computations (Di Bernardo and Rullani 1991).

Today, both complexity and clusters are the subject of attention and discussion because one season of modernity has ended and another one is beginning. The *new modernity* entails a reconciliation with complexity and with distributed-intelligence forms of organization, such as clusters, that allow the exploration and governance of complexity at a reasonable cost.

1. Reducing Complexity: The Goal of Deterministic Reasoning

The main goal of deterministic reasoning, as it developed during the long reign of Fordism, was the efficient exploitation of what was known. From this standpoint, the first order of business was to neutralize the world's complexity, since it hindered computation and rationalization efforts. In other words, it was necessary to narrow the spaces occupied by *variance* and *indeterminateness*, since they were the reasons for the unpredictability of natural events, individual choices, meanings perceived by different people and organizations and behaviours occurring in the course of economic and social life.

It was clear from the outset - in the years straddling the eighteenth and nineteenth centuries, which saw the birth of modernity - that these spaces, which in the premodern world were relatively wide and unregulated, were incompatible with the yearning of modern reason for full control of destiny.

Consequently, they were regulated and narrowed with the use of the tools created by the industrial revolution.

With the use of the *science* and *technology* incorporated in the machines and technical systems of factories, which were characterized by the repetitiveness and standardization of procedures and products.

With reliance on the *automatism of the marketplace*, which made *price* the universal measure of everything and, therefore, forced productive and social life through a powerful process of quantification and abstraction.

During the 20th century, Fordism accelerated the drive toward technical rationalization, adopting work methods that were consistent with the scientific organization developed by F. Taylor and generalizing the use of machines in the production process in all areas of business, even the most complex.

Thanks to this approach, even inherently complex activities could be modernized because complexity could be efficiently minimized with the technique of *segmentation*: a complex task could be broken down into a sequence of basic steps, each sufficiently simple to be delegated to a computational algorithm or a machine. An assembly line or production program could then *reassemble the fragments*, increasing complexity only to the level strictly necessary, which, however, was significantly lower than it had been before (Di Bernardo and Rullani 1991).

In a large Fordist enterprise, planning, by reassembling a puzzle, generated an artificial environment of reduced complexity. This environment had the fundamental advantage of making the behaviour of all players *controllable*, since the consequences of their actions were computable and the relationships between the different actions programmable ahead of time.

In this way, *computation* became wedded to *power*: what can be computed can be easily controlled. The reverse is also true: what can be controlled constitutes an excellent input - free of ambiguities and surprises - for computational procedures. Modern reasoning thus became *technocratic* reasoning, which combines the powers of computation and control. Thanks to these twin powers, technocratic modernity was able to bring a growing range of events and behaviours into compliance with predetermined standards and computable algorithms. In those areas where variance or indeterminateness remained too high, *forecasting and control tools* were supplemented with *negotiation procedures* between organized interests. To achieve this goal, working against the backdrop of managerial capitalism triumphant, as it existed until the 1970s, great public and private institutions ensured the stability and predictability of behaviours using the power of scientific and technological knowledge, the power of ownership control and the more flexible and adaptive power of negotiations with stakeholders to reduce complexity (Galbraith 1967).

2. What Does Territory Mean?

The notion of modernity in which the ability to produce is based on reducing the "natural" complexity of human and social behaviour has very little in common with the notion of *territory*, which represents a synthesis, sedimented at one location, of history, culture and interpersonal relationships. This synthesis was too rich in variances and had too many levels of indeterminateness to fit into the framework of modernity and, therefore, was ignored or completely removed from the horizon.

For all intents and purposes, the idea of territory as a complex synthesis of life stratified in one place disappeared from the theory and practice of modern economics. In its place, modernity offered an *artificial space* devoid of complexity where players can easily perform opportunity computations. Obviously, this metamorphosis caused the notion of territory to lose all the relevance it had gained in premodern times. This was also because, in the Fordist vision of modernity, the complexity of local societies could not be used in any way to create economic value. On the contrary, it was an obstacle and an impediment. The modernity we have been familiar with until now attempted to render *abstract* the available physical space, thereby suppressing the uniqueness, identity and history of different locations.

When viewed through this filter, the territory becomes an easily *fungible* space, indifferent to the quality of the places it contains. In other words, it becomes something that can be measured in acres of space and miles of distance.

The distribution of activities over the territory - and the very idea of industrial cluster - thus becomes the product of a *localizational algorithm* that produces geographical agglomerations and dispersions that mirror the distribution of costs and revenues over the different locations, in accordance with a routine computation of opportunities that is indifferent to the subject matter being optimised. There is no indication of the existence of places with economic identities and histories that are different from those of all other comparable places.

The disappearance of complexity from the ideal and technical horizon of modernity also marked the disappearance of the concept of industrial cluster and the notion of territory that it entailed. This was unavoidable, but we must be aware of the theoretical and practical consequences of such a disappearance.

A territory without complexity is a place *without qualities*. It is just another place, just another geographical agglomeration or dispersion produced by a computational algorithm. Consequently, it is easy to see why the scientific and practical program of modernity would virtually ignore the existence of *industrial clusters* despite the prodding of an influential author such as A. Marshall and a wealth of relatively successful territorial clusters or similar areas that developed on repeated occasions during the past century.

Industrial clusters were viewed as little more than a geographic oddity or a sitting anomaly: "typical" areas, or system areas, that for some reason had spawned a mono-industry culture instead of a well-balanced and diversified multi-industry economic system. The hidden bond that, in actuality, linked the quality of the cluster to the quality of the site remained hidden and was basically uninteresting. At best, it aroused the curiosity of a few, anomalous, theoretical economists (in Italy, mainly Giacomo Becattini and Sebastiano Brusco).

Today, we know that this small group of scholars was not studying an irrelevant or backward-looking phenomenon, but that it had uncovered one of the anomalies that made it possible to observe complexity escaping from the net of deterministic reasoning. Clusters did not have an authority or official belief that limited the opportunity space available to their members. On the contrary, everyone was allowed, at his own risk, to explore a segment of the opportunity space. Each member, operating within an environment characterized by a great deal of variance and indetermination, learned to manage complexity whenever it arose, without attempting to minimize it beforehand. Industrial clusters, with their minute specializations and strong interactions, can metabolise complexity, transforming it into learning, rules and collective identities.

3. Rediscovering Industrial Clusters

An industrial cluster develops when a significant number of businesses in the same area of manufacturing, or in related areas, cluster at the same location, using their *territorial contiguity* as a means of interaction and exchange. The territory, viewed as the place where the *shared culture, history and institutions* of the cluster operators are being sedimented, functions as a *relational and communicational frame* that can integrate thousands of decentralized and interdependent intelligences that interact with each other in an aggregate pattern of behaviour that is *organized and efficient* - efficient enough to make the clusters competitive in numerous industries in the modern economy.

The most successful industrial clusters occurred in what used to be called *traditional* industries: textiles and garments, footwear, furniture, tiles, light mechanical engineering (production of machines, equipment and components, metalworking systems, surface treatment systems, micro-mechanical engineering). In reality, these industries are far from "traditional" because they do not embody methods and meanings handed down through tradition and preserved largely unchanged until the present. Today, considering the technologies and organizational models employed, we could easily classify these areas of industry as *innovative* businesses and as *complex* businesses, i.e., businesses open to unpredictability and ready to experiment with the communications and symbols they offer their customers. Think, for example, of the density and volubility of the messages that fashion trends associate with objects of daily use that are an integral part of people's lives.

It is important to clear up a misunderstanding caused by the habitual taxonomies of industrial economics. Even though industrial clusters helped revitalize artisan types of professional endeavours and businesses that are far removed from the Fordist stereotype, they represent a kind of industrial modernity that is rooted in the contemporary world, even when clusters manufacture products that hark back to consumer needs that already existed in the pre-industrial world or use manufacturing techniques based on an intensive use of human beings (entrepreneur or worker) and their skills.

It is totally misleading to associate industrial clusters with *traditional* forms of production. Clusters are not competitive in those areas in which tradition is repeated with unchanging and aproblematic forms, because in the modern world, production of this type is handled more cost effectively by large companies that can lower costs by standardizing their products and optimising their manufacturing cycles. Clusters are effective in those areas in which it is difficult to standardize and plan production far in advance. They excel in those areas where operators must adapt quickly and creatively to changes that they cannot control, exploring new meanings and building around them a structured network of specializations, transactions and exchanges. Clusters are efficient at this task not because there is a *mastermind* that decides what to do and orders the necessary behaviours, but because they are able to develop *decentralized forms of exploratory and constructive learning* as a result of attempts, imitations, mutual adaptations and communications among the different independent intelligences that comprise the cluster. At each production phase or function, competitors work independently but are capable of learning (copying, imitating, duplicating) from each other. Suppliers organize experimental and flexible relationships with their customers along the value chain, using the relational channels sedimented locally over time. Even though it entails a certain amount of "evolutionary waste" (unsuccessful attempts, redundancies, lock-in circuits), the result is that the web of relationships is quickly reshaped and remodelled each time, as it adjusts to environmental changes and to endogenous learning dynamics.

Cluster companies are "extended companies" in the sense that, because they are part of a broad learning circuit involving the supply chain and local society, they optimise the use of internal intellectual and financial resources, specializing in the performance of narrowly defined functions that cannot be performed without a cluster-type learning (Varaldo and Ferrucci 1997).

The self-organizing that occurs through cluster dynamics does not confer significant competitive advantages when the task at hand is the repetition of traditional models and forms. It is useful when it becomes necessary to respond to unexpected situations, explore unknown paths and follow untested intuitions. To sum it up: industrial clusters, even when they are rooted in traditional consumer needs and craftsmanship, constitute a modern way of producing and competing that is particularly suited to conditions of high complexity.

After all, cluster formations (territorial aggregation and integration of distributed intelligence) have also occurred in such high-tech industries as information technology, telecommunications and biotechnology, with the flowering of new enterprises around large centres to produce the intellectual capital (higher research and learning) needed to fuel growth and innovation in these industries.

In Italy, almost one third of manufacturing activity is carried out in ways that, to varying degrees, reflect the cluster model. More importantly, the prevailing trends of international trade are leading to a further specialization of Italian industry in those fields where industrial clusters are more present.

In other countries, industrial clusters are less common. But cluster-related issues are equally important, because communities everywhere are interested in deriving the maximum benefits from the connection between production and territory, between proprietary business resources and resources that are not proprietary but can become accessible through territorial localization. This is true for the availability of infrastructures, highly qualified or particularly inexpensive human resources, or regulatory or tax incentives or advantages. These considerations are especially true for knowledge, a resource that is of key importance for modern competition. An important part of the knowledge used in manufacturing has a *localized* nature, i.e., it cannot be transferred from one place to another. It is embedded in a specific place (or, more often, in a specific society), so that it can be accessed only by establishing a direct relationship, of variable intensity, with that place (Corò 1998).

4. Out of the Limelight, an Economic Theory of Industrial Clusters Begins to Develop

The development of industrial clusters, as a physical event, was not recognized for quite some time. There was an even longer delay in translating the "teachings of facts" into theoretical awareness.

There is a basic reason for the failure to develop a theoretical explanation for the economic growth visible in real life in developing clusters. The growth of industrial clusters is triggered by mobilizing resources that are doubly anomalous for standard theory: first of all, because they are local resources, unique and very difficult to duplicate and, secondly, because they are complex resources that represent the summation of history, culture, relationships and interactions with unpredictable results and, therefore, create patterns of behaviour that are far from simple and virtually impossible to calculate.

To acknowledge that industrial clusters are part of the modern economy and not vestiges of the pre-industrial economy is not an easy thing to do, because such an acknowledgment entails a conclusion difficult to accept: that modernity can take complex forms that escape instrumental reasoning and deterministic calculations.

At first, the handful of pioneers in this field received limited attention from mainstream economists. For some time, clusters were seen as empirical and theoretical variants that existed only in Italy. Later, it would become apparent that these variants carried the seeds of a conceptual revolution that was much broader in scope and extended beyond Italy.

The first clear theoretical demarcation, separating the *economy of clusters*, which is location-based, from the classic *economy of industries* appeared in 1961 in a work by Giacomo Becattini on the relationship between economic value and industries. Becattini criticized the theory of value based on the notion of industry because it lacked all those concrete references - relationships with society, history and culture - that exist in localized production processes. Industries are abstractions that enshrine the primacy of technology over all other determinations: all factors of production become useful inputs to the extent that they can be utilized by the technology employed to generate the industry's output. All else is irrelevant, a mere accident or obstacle to manufacturing efficiency. But real history suggests a totally different interpretation: the quality of the people, the places where they live and the societies that shape their actions are of fundamental importance in suggesting which product should be manufactured and which process should be employed. Looking beyond the notion of industry, Becattini saw a capitalism that is different from the one described by conventional theory, a capitalism that produces value by mobilizing the intelligence of individuals and their ability to interact within appropriate contexts.

Since then, Becattini has never relinquished this critical approach, challenging the vast majority of the academic and political establishment. At the centre of his theory is the industrial cluster, seen as a vital form that facilitates a synthesis between economy and society. The empirical proof of these theories is provided by a rise in competitiveness, which, beginning in the 1970s, produced an increase in the importance and wealth of almost all Italian industrial clusters.

But the hidden subject of this dispute was, in reality, another one. The question was whether or not its was appropriate to include within the boundaries of economic theory phenomena such as the industrial cluster that, instead of being just a system for the efficient allocation of resources, represents an organized form of learning in action that explores complexity and evolves as a result of this exploration. Clusters are fuelled by complexity, which they regenerate and increase, adding new varieties and possibilities.

So the question is whether this is still an economic process or one concerned with history, sociology, geography and that oblique form of knowledge that has to do with "totality".

Clearly, regional economies regained importance because the territory, with its economic, social and cultural complexity, once again became a resource for the creation of value and competitive advantages (Sabel 1989).

It was precisely the *circular* nature of the process that creates the cluster as a total system that encouraged the study of the different industrial clusters with their array of relationships and histories. At the same time, this circularity was also the reason for the resistance offered by orthodox economists. The circular process of self-production (we could, perhaps, call it *autopoiesis*) by the cluster cannot be limited to the economic sphere, since the process does not apply just to machines and professional skills, but also to a social context, institutional rules and the values and attitudes that are inside people's heads. Self-production is an issue that is viewed with suspicion not only because, by virtue of its circular nature, it escapes static optimisation, but also because society and the people who populate it can no longer be measured with given *exogenous* variables, becoming instead both actors in and products of the economic process.

The main appeal of the industrial cluster is not its differential efficiency nor its competitive success, but the fact that the industrial cluster brings to the fore the nexus through which "that which is economic-productive and that which is sociocultural fuel and condition each other's development" (Becattini 2000c, p. 16), placing at the centre of economic analysis the *people* that create enterprises and those who work at those enterprises. These people, even when they seem to be following a merely efficientistic and instrumentalistic logic, have, in fact, intertwined their life goals inextricably with the work they perform at their enterprises, thereby creating a type of economy that may seem perfectly normal at a given location, but could not exist anywhere else (pp. 19-21). An industrial cluster is a *local system* with characteristics that are different from those of the global economy: even though it creates a steady exchange of goods and means of production with the rest of the world, a cluster reflects its borders and local identity when it has to regenerate the conditions that underpin its existence as a complex system, correcting and reinventing the balance of relationships it inherited from the past (Becattini 2000b, p. 107). The problem with reproducing a given cluster at another location stems from the fact that what makes up a cluster is not just institutions, contractual rules, factories or material infrastructures, but also people of flesh and bones, with their anthropology, way of life and interactions. The Prato experience cannot be exported because the people from Prato do not exist anywhere else, and it would be hard to create a cluster without its main raw material (Becattini 2000a).

Sebastiano Brusco (1982, 1989) agrees that an industrial cluster is a place where the causes of manufacturing efficiency, social balance and knowledge creation coexist in a dynamic equilibrium. A cluster must reproduce not just the efficiency model that impacts the costs and revenues of its businesses, but also the "environmental" conditions that allow the model to function at a given location but not elsewhere. The environment includes institutions, social culture, fiduciary relationships between people, networks of family members and friends, and local identities and ethics - in other words, all that the cluster has generated during its history to keep in balance the delicate relationships that exist between economic and social forces, individual and collective needs, the past and the present.

Within industrial clusters, history proceeds with the rhythms and ways of evolutionary learning, including breaks that occur when one of the original points of equilibrium vanishes or there is a sudden change in the competitive environment to which a cluster is linked. Local systems, when they are able to find sufficient energy to take up the challenge, answer these developments by cumulating a thousand small steps, a thousand minuscule explorations that, when combined, end up producing a visible change. It is the so-called "development without fractures" (Fuà and Zacchia 1983), which does not mean without breaks, dramas and big or small revolutions, magnified by the narrow confines of a local environment.

The works published by Becattini, Brusco and Fuà paved the way for innumerable theoretical and empirical essays that viewed industrial clusters not just as examples of local economies, but as manifestations of a complex and self-organizing system that provides a "key to understanding capitalism" and opens up for exploration certain aspects of reality that had been closed to conventional theory (Becattini 2000c, p. 16).

We thus discover that there is a "Third Italy" (Bagnasco 1977) studded with industrial clusters; that, in recent years, clusters are starting to appear in Southern Italy as well (Viesti 2000a, 2000b); that a significant portion of the "Made in Italy" products that give strength to the Italian balance of trade are manufactured in industrial clusters (Fortis 1996, 1998); and that all industrial clusters are different and must be studied individually. This realization gave rise to a new branch of economic literature consisting of an increasingly large number of empirical studies (I will not even attempt to produce a list of Italian cluster "cases" because it would be too long and would inevitably be incomplete). By becoming an "ideal type", in Weber's sense, the cluster economy is transmogrified into a flexible specialization - an alternative to mass production (Piore and Sabel 1984) - thereby achieving a level of international recognition that was underscored by the success of Silicon Valley in the United States (Saxenian 1994, Zagnoli 1991). This evolution has progressed to such an extent that recent broad-based studies have focused on comparing the performance of economies organized into industrial clusters with other economic forms (Signorini 2000).

In recent years, the issues surrounding industrial clusters have received international attention not just because clusters represent a type of organization that is attractive in all those cases where the goal is to trigger the proliferation of neo-entrepreneurship, but also because they can help explain the localized formations created by economic development. With the theory of clustering (Porter 1989) and the new "economic geography" of Krugman and Venables (1990, 1995), economic theory took a closer look at the territory, with the goal of assimilating the complex geometry of its forms and histories. Even though these formal approaches contained an echo of the old economy of localization that paid scant attention to the anthropological and social quality of locations[1], a connection between the Italian theory of clusters and the new economy of the territory that has gained acceptance on an international scale was firmly established.

5. Meta-theory of Clusters: Self-organized Complexity

The distinguishing feature of industrial clusters is their extraordinary ability to allow an efficient organization of production, with the full range of specializations and exchanges that are typical of a modern economy, to coexist with high levels of variety (in space), variability (in time) and indeterminateness (in the realm of possibilities). In other words, clusters represent the practical refutation of the "central dogma" of modernity, which postulates the *incompatibility of organization and complexity* and opts for organizations that are doomed to maintain at an artificially low level the complexity that can exist within them and

---

[1]On this issue see the discussion that appeared in *Economia e Politica Industriale*, (2000), Nos. 106 and 107, following the publication of an article by Viesti (2000a).

in their transactions with the outside world, the only alternative being a situation devoid of organization in which complexity increases freely and chaotically.

Industrial clusters offer a practical example of how to achieve a reasonable synthesis of organization and complexity. Their market success is proof positive that a good level of organizational efficiency can be achieved even in the absence of artificial systems that use techniques, command and negotiations to reduce complexity. In other words, clusters deal with complexity *in another way* (without reducing it beforehand) and with *other systems* (different from techniques, command and negotiations).

What are these systems?

Industrial clusters metabolise complexity through a dynamic process of *distributed learning* that mobilizes the intelligence of many separate players who communicate and interact with each other (Micelli 2000). None of these players has sufficient technical, regulatory or negotiating power to force the others to behave in accordance with a common action framework. Instead, each player learns in the course of action how to explore his/her segment of complexity, accumulating personal experiences and at the same time benefiting from the experience of others. The result is a circuit of *self-organization* (Silverberger, Dosi and Orsenigo 1988) that slowly, through repeated attempts, "learns" how to develop a system with an efficient internal organization that can respond flexibly to the demands of the outside world.

The organizational tools that can be used to manage complexity are those that allow:

- the *autonomous and intelligent* exploration of a fraction of complexity (small business);
- the taking of personal *risk* with respect to the results of the exploration (entrepreneurial spirit);
- the ability to *learn from the experience of others* through the circuits of communication and local emulation (imitation of competitors, integration of customers and suppliers, switching of work from one company to another, spin-offs of new businesses, etc.).

The behaviour of a cluster as a whole is the product of the combination of a series of partial organizational choices that let each enterprise find its position vis-à-vis the other enterprises in accordance with classic swarm dynamics (Kelly 1994, 1998). The sum of these semi-independent behaviours of limited rationality produces an orderly result that, in many cases, is more efficient than the result that could have been obtained through centralized planning.

From this standpoint, industrial clusters have all of the characteristics typical of a complex adaptive system, or CAS (Axelrod and Cohen 1999), because:

i. they are ruled by *self-organizing* processes that occur as a result of changes, selections and retentions of the variables that are best suited to each contingent situation and may or may not be the product of an intentional design and the work of conscious players;
ii. in the development of alternatives and the selection and retention of these alternatives, clusters use an intellectual and relational capital sedimented

through experience and assimilated in the specific and original structures of each cluster (*languages, logistics circuits, fiduciary and guarantee systems*);

iii. they are characterized by *non linear* processes of growth and dynamics, due to the multipliers that disseminate throughout the clusters the most effective solutions and block those that are less successful;

iv. they harbour a wealth of environmental variances and niches that is significantly greater than is required to meet current needs, thereby creating a *reservoir of diversity* that permits flexible and creative responses to all contingencies.

6. Rediscovering Complexity

As I have already stated, after years of neglect, the physical and social sciences rediscovered complexity. In this field of learning, as in many others, there is a tendency to generalize the work of a few pioneers who laid the foundations for future advances.

The theory of complexity has put forth certain very general *formal* principles and noted the existence of *complex adaptive systems* in many different fields, from physics to biology, from social evolution to economics (Axelrod and Cohen 1999, p. 16).

The key element that justifies accepting the very notion of complexity in the physical, biological and human-social world is the idea that the formal structure of complexity has certain common characteristics that make it possible to frame it within a general, transdisciplinary context.

There are many definitions of complexity, and this is not the place to review them. For our purposes, we can say that a field of variance that exists within a space of possibilities can be deemed to be complex if:

- the space of possibilities in question is *broad* and *undetermined*, and
- the space of possibilities is not flat or just plain chaotic; it is *curved*, due to the impact of attractors, breaks, slopes and the intertwining of different phenomena that facilitate or hinder learning.

In a complex situation, the space of possibilities can be described as an extremely *rugged* and *non-linear* landscape, where small causes can have big effects or, quite to the contrary, major forces can implode without any apparent consequences, and as an environment in which there are levers that can trigger chain reactions with unpredictable effects at remote spots on the playing field.

Consequently, any attempt to explore complexity will have to deal with the effects of magnification (positive feedback circuits) or implosion (negative feedback circuits), with rising and descending paths and with catastrophic events that render the progress made irreversible.

In order to identify useful or efficient solutions, any exploratory work must take advantage of favourable curvatures while resisting unfavourable forces. This approach entails complex forms of learning that can help overcome the most obvious attractors (local optimum points that block the path to remote optimum points) and the ability to guess the positions of the most favourable possibilities.

The general idea of complexity is shaped by the varied and undetermined nature of the space of possibilities that is being explored and by the "curvature" of that space. In a flat or perfectly casualistic space, the exploration can proceed stochastically on the basis of the degree of probability assigned to the different possible events.

The general model for the exploration of the curved space of possibilities is derived from the evolutionary paradigm and "tested" through computer simulations (Dosi and Marengo 1993, Axelrod and Cohen 1999). It includes a sequence of *variation, selection and differential reproduction* for the variables that are found to be suitable (having a minimum degree of adaptability) and allows a large number of different modalities in the execution of these activities. The use of this form of learning - which is not the only one, but is the one that is more general in scope because it does not presuppose any previous knowledge of the space of possibilities that is being explored - is important because adaptive learning alters the frequency and structure of the chosen systems. In other words, the complexity of the environment becomes "imprinted" in the internal structure of the organism or organization, thereby ceasing to be an exogenous variable that is fundamentally unpredictable and uncontrollable. Evolutionary learning creates a link, a structural coupling of the (adapted) *internal structure* and the possibilities that exist in the external environment. The complexity of the environment is thus metabolised by the system as it learns and evolves into an organization, with an implicit match between what is internal (complex system) and what is external (environmental complexity).

We can then speak of complexity with respect to the *environment* (latent form of the space of possibilities) or the *system* that explores it (its internal structure that metabolises the complexity of the external environment and makes it manageable). These two definitions are different but convergent: we expect the exploration of a complex environment to produce the construction of a structure of equal complexity within the system, and the internal complexity of a system to be useful only if it refers to an environment that is in some way related to the structure of the system.

The learning process removes *free complexity* from the environment and metabolises it into the structure of the system as *governed complexity*, i.e., as variance, indeterminateness and curvature of the space of possibilities that the system can control with variances or adjustments that are quick and inexpensive but not deterministic (Di Bernardo and Rullani 1990).

The internal structure is organized *by levels*, based on the *emerging properties* that, for example, differentiate a macro-system from its elemental components.

The discovery of *emerging properties* that, for example, are typical of cells and not of molecules, of social groups and not of the individuals within the groups, permits the allocations of the governance of complexity among different *levels* of the organization, from micro to macro. The governance of complexity can be achieved by regulating the behaviour of the elemental components. However, the control of all those aspects that concern properties that emerge at the macro level can become very difficult and expensive unless governance mechanisms based on the principle of *functional equivalence* are established at the macro level. Elemental components that perform the same function, because they do not alter properties that emerge at the macro level, are deemed to be equivalent and, therefore, interchangeable. Control and feedback mechanisms that correct any variances, regulations and standards that guarantee a high level of process and product consistency, and shared meanings that permit the joint testing of virtual possibilities ensure that all of the parties involved are functionally equivalent, thereby preserving the emerging properties of the macro-system. As a result, the macro level becomes largely independent of the fate of its elemental components, which can be easily exchanged or replaced with others, when needed.

Complex systems cannot be reduced to the sums of their component parts because the ability to maintain the emerging properties depends more on the services that guarantee functional equivalences than on the behaviour of individual elemental components.

7. Different Forms of Complexity

Clearly, beyond this general outline, there are profound differences between the different levels of organization and the different fields in which complexity occurs. *A different system of variation/selection/differential reproduction* is adopted for every level or field.

Because of this, the complexity of economics is different from the complexity of, for example, physical or biological sciences. This occurs because:

- the mechanisms of variation that explore what is new are specific (deliberateness, innovation);
- the mechanisms of selection are tied to the internal choices of businesses and the competitiveness of the marketplace;
- the mechanisms for the expanded reproduction of the "winning" or "correct" variances are tied not only to the competitive success of the winning companies, but also to processes involving contractual transfers (licensing, franchising), reputation (brand, image), copying, imitation, communication and learning.

Therefore, even in the area of complexity, there is an economic specificity that cannot be overlooked. On the economic plane, the cluster has its own specificity as a special complex adaptive system.

In the field of economic studies, the theory of complexity has been applied to the solution of specific problems, almost all of them involving the "invisible hand" of the market and the changes that occur in response to attempts at finding equilibrium between prices and quantities.

The market can be considered an example of CAS and was treated as such by a tradition of research that viewed it as a cognitive mechanism (Hayek 1937; Rizzello 1997).

The marketplace represents the summation of the behaviours of many independent operators who try to maximize their profits. As they move and line up on the sides of demand or supply they help identify the unknown variable: the price that can level demand and supply simultaneously in all markets.

A special field for implementing the ideas derived from the theory of complexity is that of the financial markets, which often display dynamics of research that are the summations of independent valuations by different operators who supply and demand funds. According to the theory of repeated games, the rules of the game that develop among the players incorporate the links of reciprocity and the expectations that are necessary for the proper operation of the markets by metabolising the complexity that develops on each separate occasion.

Another field suitable for the use of complex adaptive systems is that of the transformation of the rigid forecasting and planning systems used by large Fordist companies, which have become inefficient due to the inescapable complexity of the environments in which they have to operate. For example, in the organization of distributed logistics, when setting up appropriate systems to respond to market changes, manage the production cycle and handle all problems that exist in a nonreducible space of variance and indetermination, experimental learning paths that adjust choices to the results achieved can prove more effective than conventional systems, which are based on a predetermined representation of reality and unchangeable selection procedures.

The superiority of adaptive solutions, which do not require drastic measures to reduce complexity, can be tested with computer simulations that show the greater efficiency of on-the-job adaptation compared with planning systems based on rigid forecasts or commands that cannot learn from experience.

In all of these cases, a circuit of learning that proceeds by trial-and-error uses repeated experiences to generate a structure that can respond to complexity with research procedures that are adaptive without being deterministic. In such an environment, market operators or nodes in a logistics network can transform the limited intelligence of each operator into a form of collective intelligence that represents the summation of a plurality of experiences and develops solutions that can ensure a reasonably high level of efficiency within a given range of variance.

8. Peculiarities of Clusters as Complex Adaptive Systems

Compared with other examples, industrial clusters present certain peculiarities that make them especially interesting as a means of organizing and managing complexity. In particular:

1) Industrial clusters are *natural self-organizing systems*. They are not technical or computational devices designed by someone to use self-organization as a problem-solving tool, but *are themselves products of self-organization*. As opposed to what occurs in the market or in large organizations during periods of change, the rules of the game by which clusters govern

complexity are not dictated "from above". On the contrary, they *form spontaneously* over time without any intervention. In addition, these rules are updated through a dynamic process of collective learning that follows the swarm model rather than being preordained or mandated by a higher order-making power.

2) Clusters are *localized ecosystems* of relationships created through history and organized in accordance with an internal logic that is largely inaccessible to observation. Much like *neural networks*, clusters are *black boxes* that metabolise experience and develop specific integrations of resources and relationships that are unique and nonreproducible because they escape logical representation, even though they exist in the different visions of the individual players. In the same way that a neural network modifies the weight and form of the relationships between the different nodes, cluster ecology continuously modifies the structural coupling established between individuals and the existing meanings, physical assets and products. In their activity, clusters do not just "process" information and technical solutions; they develop social structures, collective identities and value scales, and, ultimately, shape *anthropologies* and *mindsets* that exist only at one location and that are virtually impossible to re-create elsewhere, even when very similar learning paths are followed.

3) Industrial clusters are *living systems*, not just more or less efficient problem-solving devices. Or, rather, clusters acquire a problem-solving ability only when they can reproduce themselves as living systems consisting of thousands of people who use these cluster environments to develop not just efficient and profitable solutions, but also answers to their need to interact, communicate and feel, which goes beyond a narrow economic vision. All of these people *live* permanently within each cluster and lack an alternative world where they can satisfy their noneconomic needs. Therefore, cluster micro societies cannot be efficient only from an economic standpoint; they must also meet meta-economic and extra-economic needs that are equally important to ensure the system's ability to reproduce over time.

Because they are natural self-organizing systems, localized ecosystems, living systems that are required to reproduce society in its entirety and not just an instrumental portion of it (the economy), clusters do not fit easily into the classic categories of economic science. For example, their anthropology is broader, by definition, than the one applicable to *homo oeconomicus* and their rationality is always more than just instrumental. The ties that link clusters to the history, culture and mindset of the people who populate them are never contingent or superficial, binding instead deep structures and distinctive differences. A cluster is not just a logical category; it is an *organism* made of past experiences, collective identities and relationships based on mutual recognition. When studying cluster dynamics, there is a limit to which extent the evidence grid can be dissected analytically, because if the dissection is not done with sufficient care, one runs the risk of separating that which is inseparable and loosing sight of the fact that clusters function like systems of integrated parts that interact with each other and not like complexes of isolated elements that are joined together strictly for contingent reasons.

Clusters and the economy belong to two different regions of knowledge: the former belong by full right to the region of complex phenomena and systems; the latter belongs methodologically to the region of computing and determinism. Consequently, clusters can be rightfully included in the realm of economics only if the field of interest of economics is enlarged, leaving determinism behind and entering the open sea of complexity.

It is only by taking this step, which many economists still refuse to do, that clusters become meaningful examples, serving as *prototypes* and *living laboratories* for systems in which society metabolises complexity, using it as a productive resource capable of creating economic value and competitive advantages. In a certain sense, clusters are *in vitro* displays of the learning system of an economy that has discovered the most effective way of moulding people, economic developments and social organizations for the purpose of harnessing complexity (Axelrod and Cohen 1999).

9. Clusters as Cognitive Systems

The localized nature of learning, and of the resulting knowledge, is not the product of a historic accident or a geographic anomaly. On the contrary, it derives from the very way in which knowledge is produced, each piece of knowledge being generated in a *context* that includes a web of personal, business, local and industry relationships, each of which can be essential in order to obtain the desired effect. Consequently, each piece of knowledge retains its validity only as long as its context remains the same, losing validity (and value) as the conditions of the original context begin to change. Obviously, there are certain activities - abstraction, codification, standardization, measuring - that serve the specific purpose of extending the validity of knowledge beyond the original context to a much broader field. However, it is also true that these activities involve only a portion of the knowledge acquired through experience. The greatest portion of knowledge remains embedded and sedimented in personal abilities, business routines and territorial contexts (Corò 1998).

People, businesses and territories *know* more than they can make explicit or transfer to others. More importantly, *they know how to do* more than they can explain because they use the embedded knowledge that is sedimented in the contexts. In complex situations, where there are many and undetermined variances, the processes of abstraction, codification, standardization and measurement are too slow and costly to produce significant quantities of this implied knowledge and make it transferable from one context to another. When complexity is high, conditions change so rapidly that they require a continuous process of interpreting and modifying the existing knowledge. As a result, there is not enough time or potential benefit to justify extracting from the existing contexts a transferable knowledge that can become obsolete in a short time. It becomes more effective to mobilize people, businesses and territories in a process of invention and productive organization, because they are bearers of a practical knowledge sedimented into experience (Becattini and Rullani 1993).

In this way, complexity develops an organic relationship with industrial clusters: thanks to the *sharing of the context of experience*, which involves many people, businesses, social participants and institutions gravitating around a single productive hub (furniture, footwear, tiles, household appliances, etc.), clusters use a type of organization that is particularly suited to mobilizing the embedded intelligence of many independent actors for the purpose of accomplishing a difficult but rewarding task: *making governable the complexity* of modern industrial production, thereby transforming a problem - the problem of unpredictability, uncertainty and risk - into a valuable competitive resource.

This is true in part for any form of territorial organization, but in industrial clusters the sharing of context between people, businesses, social participants and institutions is particularly meaningful because at the cluster level, everyone, or nearly everyone, can draw from experience gained from life, work and relationships *within the same industrial sector*. There is a centre around which the different learning curves coalesce and become integrated. The sharing of the same territorial context is also the sharing of ideas, experiments, expertise and innovations, and this sharing makes it easy to integrate individual fractions of experience quickly, because they derive from a common matrix.

10. Will Clusters Survive the Increase in Complexity that Is Bound to Occur in the Near Future?

For the foreseeable future, the economy will continue to display a growing level of complexity, if for no other reason than that market globalisation and communications networking will inevitably multiply connections and interactions, thereby increasing significantly the variety of nodes and the indeterminateness of their behaviour. Any research into the type of organization that will be best suited to handle this growing complexity must take into account the cluster - i.e., the localized specialization of a nucleus of autonomous businesses that interact with each other - as *one of the systems* capable of providing an effective answer (Grandinetti and Rullani 1996, Vaccà 1997).

In this endeavour, the cluster will have to compete with other solutions, including three alternative systems that must be reckoned with in assessing how competitive the cluster system will be (Rullani 2000):

The *market*, i.e., a *decontextualized* relationship among businesses that does not presuppose the existence of a context of shared experience and that, therefore, can apply to exchanges and dissemination of the highest possible multipliers.

Hierarchy, i.e., the development of forms of *shared ownership* of knowledge and skills, organized around the common membership in a single corporate organization.

Networks, which are based on interaction and the sharing of skills by different businesses that develop away from each other, at different locations, with the help of modern communication technologies.

There is one specific feature that distinguishes the cluster from these three systems: in a cluster, the communication and interaction medium linking all parties involved (customers, the full range of suppliers, entrepreneurs, employees, banks, local institutions, etc.) is in fact the *territory* and whatever became sedimented in the territory through history.

In the market, it is *price* that provides an interface between the parties. In a hierarchy, the links between the various companies are provided by communications, logistics and assurance systems that must be built and maintained through a collaborative effort that entails sharing investments, risks and, if nothing else, beneficial behaviours.

Obviously, these are very different relationships and exchange media.

In a competitive comparison with other systems, the cluster is aided by the advantages provided by territorial contiguity. These advantages are not insignificant, considering that most of our complexity-governing skills are the product of localized learning dynamics that are strongly influenced by the context in which the learning takes place.

In successful clusters that have steadily increased sales to their terminal markets, territorial continuity has produced a marked *division of labour* among different businesses characterized by:

Mutual specialization, which allows companies to focus their investments and risks on a narrowly defined core business, relying on other local specialists (subcontractors, components manufacturers, providers of specialized services and production, logistics suppliers, distributors, consultants, etc.) for everything else.

Weaker entry barriers, because new companies can be established easily and enter the market by specializing in very narrow fields of expertise, thereby minimizing capital requirements and risk, and relying on the skills, manufacturing capacity and services of outsiders that are available locally.

The continuous birth of new companies, the creation of a pool of *specific professional skills* (related to a specific industry) that are both highly developed and plentiful, the organization of flows of exchange among specialists and the resulting interdependencies.

The growth of a cluster is fuelled by the innovations made by its enterprises, provided that these innovations are at the same time *inclusive* internally (open to other workers and entrepreneurs within the cluster) and *exclusive* externally.

Proprietary innovations developed by large companies that invest in assets that cannot be "socialized" within a cluster (for example, proprietary brands, commercial chains, etc.) fuel the growth of individual companies and not of local systems.

Codified innovations, which can be incorporated into easily transferable machines or technologies, can provide only a minimal contribution to the growth of a cluster. Innovations of this type can be quickly disseminated outside a cluster because they are not exclusive externally: after a short while, they become accessible to competitors located outside the cluster.

11. Inclusion and Exclusion: A Battle on Two Fronts

The innovations that are best suited to producing an increase in the knowledge that becomes sedimented within a cluster and fuels the growth of the businesses that use this knowledge are those that have to do with *products* and *production processes*. A cluster works best when the task at hand is identifying niche needs, meeting latent consumer requirements and bringing innovations to the production process incrementally through on-the-job field tests. In all of these cases, innovations are based on an unspoken knowledge that is not easily transferred outside the cluster. As a result, these innovations have the characteristics typical of *localized innovations*: they are *inclusive internally* (they can be easily imitated) and *exclusive externally* (they cut out competitors located outside the cluster).

The growth of a cluster occurs along the often-narrow divider that separates localized innovations from proprietary innovations (*non inclusive*) and codified innovations (*non exclusive*). In order for a cluster to replicate its competitive position and protect the quantity and quality of its product supply from competitors, the flows of innovations that occur within the cluster at a given moment in time must be sufficient to maintain the balance of this *twin frontier* - both internal and external. When innovations serve to disseminate good ideas at the local level while at the same time excluding potential external competitors, a cluster expands. If, on the other hand, there are obstacles that hinder internal propagation and external defence, a cluster contracts, and if it shrinks below a certain critical threshold, it can even implode to the point of disappearing as an organized system of learning and localized production.

Therefore, the future of industrial clusters will be predicated on a critical factor: the manner in which they handle the growing complexity they will be faced with, i.e., the innovations that will emerge in the economy in the near future (Arora and Gambardella 1993; Arora, Gambardella and Rullani 1997). In the case of codified innovations, the division of labour could use the market to bypass the cluster. In proprietary innovations, the division of labour within a major enterprise could be more effective than the one occurring externally, within the cluster. Under these new conditions, the survival of clusters is endangered simultaneously on two fronts: clusters risk losing business activity to the market, on the one hand, and to delocalised or multilocalized large companies, on the other.

Nevertheless, in the specific field of innovation, clusters continue to enjoy important competitive advantages in the division of innovative labour in noncommoditized fields that resist codification, but can, at the same time, be imitated, transferred and reproduced at the cluster level. For example, the work of *exploring what is new*, which will ultimately result in *new products* and *new processes*, is often carried out in ways that permit good permeability at the local level, but provides a relatively solid defensive barrier against external imitators. Clearly, this type of work can be carried out by *cohesive local networks* that include many independent operators capable of working together. However, it is also important to keep in mind that companies are required to make certain types of investments that, for a number of reasons, pull in opposite directions. For example:

a) *Investments in technology, mechanization and process standardization* tend to expand the market for commoditized products that will be sold in the *market* at low cost and in large quantities. Here, too, it will be unlikely that clusters will play a major role, considering that the resulting innovations do not exist exclusively in the local domain and can easily be transmitted to external competitors, who will tend to gravitate toward locations with low labour costs.

b) *Investments made in marketing* to expand downstream along the value chain tend to fuel the growth of *proprietary transnational systems* that are controlled directly or gravitate strongly around a leader company. The produced innovations increase the division of labour within a single enterprise (large company) and can easily be duplicated or imitated within the cluster.

In the first case (commoditizing), there will be an expansion of the global *markets* that will result in increasingly *impersonal* transactions between producers and users, who may be separated by great distances. The share of the market segments held today by industrial clusters will tend to contract, paving the way for distribution systems that operate from multiple locations within the global economy that are occasionally linked for trading purposes. From this perspective, clusters will survive if new complexity can develop on the cusp between manufacturing innovation and consumption, as those areas that are better known and controllable become standardized and commoditized. When complexity decreases in certain areas of production and consumption, it can be recreated in other areas if the quest for innovation drives producers and consumers toward exploring new possibilities. There are good reasons to believe that the resources saved through commoditization will be used by producers and consumers to explore new forms of production and new forms of consumption. This will allow clusters to survive the pressure of commoditization, provided they position themselves decisively on the cutting edge of innovation, where complexity is re-created every time a portion of it is neutralized and absorbed by the market.

In the second case mentioned above (proprietary investments), large multinational companies that set up manufacturing systems in different countries generally prevail over other types of economic organizations. This occurs because they are better prepared to assume investment risks, plan interdependencies and link production cycles that are carried out at different, distant locations. Large post-Fordist companies have gained experience in managing complexity through their ability to forecast, invest and take risks. Clusters will be able to continue to perform these functions competitively only if local societies are able to:

a) Invest in reproducing and expanding the social (intellectual and relational) capital required to tackle increasing levels of complexity. Today, this is one of the main shortcomings of the cluster system, which until now has relied on spontaneous learning processes to make up for the lack of investments and limited propensity to assume collective responsibility. To compete with large multinational companies in complex innovations, clusters must learn to rely less on the spontaneity and individualism of the past, choosing instead to leverage their ability to share projects, risks and identities.

b) Maintain a greater intellectual and relational vitality than the one produced by the direct investments of large enterprises. In this area, a critical contribution will come from those knowledge, trust and communication resources that in rich and vital local societies exceed basic needs and help create a wealth of variety and social intelligence that can be used to perform new tasks and explore unusual possibilities. Clusters can compete with large multinational companies if, for example, local societies become more open to the multicultural environments that have become prevalent in the global economy, more effective in interpreting what is different and surprising, and more tolerant of transgressions, anomalies and errors.

In the economies of the near future, the ability to manage complexity will be the decisive factor in the survival of industrial clusters. Clusters will not survive on the strength of "external economies" or on the magical effect of economies produced by "externalities" and "rising returns". They will survive if they are able to field their key resource: an intelligent local society, open to innovative experimentation and willing to take risks and assume collective responsibilities.

12. Clusters and Networks: Competition and Complementarity

As I have explained before, clusters manage the complexity inherent in the organization of innovative labour using as an "aggregant" the sharing of the same local context by companies that work alongside each other and belong to the same experience system. In this sense, clusters are nothing more than localized networks that use the territory as a relationship medium to facilitate the division of labour and the dissemination of successful ideas.

On the other hand, translocal networks, which cannot use the territory as a relationship medium, must build their own relationship media, investing resources in communication systems, logistics channels and assurance systems to link a certain number of companies that, in this way, are able to work together and overcome the distance handicap. The new information and communication technologies available today can play a decisive role in making remote relationships quite rich and helping translocal networks become competitive with those that develop at the local level and use the territory as an aggregant.

Building a communications, logistics and assurance network that links distant companies requires a significant amount of resources and time because it entails designing and operating an *artificial system*. This is not the case for industrial clusters that use a "natural platform" and have developed spontaneously, without special investments and deliberate efforts.

Nevertheless, a network has many advantages over spontaneous (local) systems for the division of cognitive work:

A network can use a pre-existing system of skills and specializations, combining the types of knowledge and relationships that are required for each situation without the handicap of physical contiguity that can significantly restrict the range of available possibilities.

A network can control what companies are added to or removed from an organized context of relationships and exchanges, setting forth explicit and enforceable conditions for the sharing of knowledge and the use of common resources.

Networks can proliferate in localized form whenever physical contiguity proves to be useful or can expand into translocal systems (the limit of which is the global system) whenever it becomes necessary to capture useful varieties and differences that are distributed on a vast scale.

Clusters and networks are two ways of managing complexity through the interaction of different companies. To succeed they must overcome the competition of multinational and/or market hierarchies. They can do this when the task at hand is organizing complex solutions that, with the help of contextual knowledge (of the location or of the network), can be *disseminated within* the system and, at the same time, *protected from external competitors*.

Both clusters and networks must defend their competitive positions on two fronts: from commoditization (market) and from proprietary control (hierarchy). In view of the fact that the innovations that are prevalent today are already moving in these two directions, the result of which is the codification of the production knowledge (which favours the markets) and the creation of a need for proprietary investments (which favours hierarchies), it is easy to see how, over the coming years, clusters and networks will have to reposition themselves away from today's production and systems organization.

Competition between clusters and networks will also increase. The competitive space gained at the expense of the market and the hierarchy will provide a venue for comparing the effectiveness of a managed sharing of the territory (physical contiguity) with a managed sharing of remote interaction (virtual contiguity). In the first instance, the act of sharing is a by-product of contiguity and can even be involuntary (when an entrepreneur copies other entrepreneurs and is copied by them). In the second instance, sharing occurs only when it is planned and supported by adequate investments. Up to a certain point, a cluster can survive even without an organized system. A network cannot.

However, as I have said before, a translocal network does have the advantage of being able to multiply the frequency and opportunities of experience by linking different locations and cultures.

The decisive element in comparing these two systems is the possibility of *gaining experience in the sharing of innovations*. The shared context of a cluster (*local circuit*) and the shared context of the network (*remote cooperation*) can be *complementary* resources - for example, when a cluster establishes a series of relationships with the delocalised nodes of a broader experience sharing network - or *competing resources*, when one can replace the other.

The most probable evolution is a mutual repositioning based on the assumption that existing clusters will use a whole host of functions available through the translocal networks to which they are linked. At the same time, networks could become denser at certain locations, creating quasi-clusters that exchange goods and knowledge over a finely meshed network that comprises light and highly mobile nodes with minimal territorial roots.

13. Conclusions

The most realistic possibility is that all four of the systems I have discussed (global market, hierarchy of the multinational corporation, cluster-type local network, translocal remote interaction network) will occupy different regions of complexity, in accordance with their respective characteristics.

In order to survive, clusters must avoid becoming prisoners of the businesses in which they have engaged thus far, embracing without reservation or delay the changes that are starting to reshape post-Fordist complexity. For example, they must quickly surrender to other systems those areas that are indefensible, concentrating instead on that for which they are best suited: managing complex solutions that can be developed with the resources of intelligence and self-organization possessed by local societies.

We cannot tell whether clusters will account for a large or small percentage of total production and employment, since we cannot know if they will increase in number and importance or shrink quantitatively compared with competing transterritorial systems (market, multinationals, remote interaction networks).

What we do know is that in order to foster the competitive vitality of the existing clusters, we must invest in *localized innovations* (inclusive internally and exclusive externally) that take advantage of the skills of interpretation, invention and sharing that are typical of local societies.

The future of industrial clusters is tied first of all to their ability to exploit promptly - without being excessively defensive - the strengths of the markets, hierarchies and networks, borrowing the functions that give these systems irresistible competitive advantages. It will also be predicated on the ability of local companies to manage complexity better than the multipliers of global markets, the investments of large multinational corporations and the selective communications of transterritorial networks.

The destiny of industrial clusters and of complexity is intertwined, as is their past. It will be up to our intelligence to discover how to continue forward on a path already taken, giving new meaning to roots and identities sedimented in the places we inhabit.

References

Arora, A. and Gambardella, A. (1993), *Division of Labour and Inventive Activity*, H. John Heinz III School of Management, Carnegie Mellon, 93-3, January.

Arora, A., Gambardella, A. and Rullani, E. (1997), "Division of Labour and the Locus of Inventive Activity", *Journal of Management and Governance*, No. 1, Autumn.
Arthur, W.B. (1988), "Self-reinforcing Mechanisms in Economics", in: *The Economy as an Evolving Complex System*, Anderson, P.W., Arrow, K.J. and Pines, D. (eds), Santa Fe Institute, Studies in the Sciences of Complexity, Redwood City, CA, Addison-Wesley.
Axelrod, R., and Cohen, M.D. (1999), *Harnessing Complexity: Organizational Implications of a Scientific Frontier*, New York, The Free Press.
Bagnasco, A. (1977), *Tre Italie. Problematiche dello sviluppo italiano*, Bologna, Il Mulino.
Becattini, G. (1961), *Il concetto di industria e la teoria del valore*, Torino, Boringhieri.
Becattini, G. (2000a), *Il bruco e la farfalla. Prato: una storia esemplare dell'Italia dei distretti*, Firenze, Le Monnier.
Becattini, G. (2000b), *Il distretto industriale. Un nuovo modo di interpretare il cambiamento economico*, Torino, Rosenberg e Sellier.
Becattini, G. (2000c), *Dal distretto industriale allo sviluppo locale. Svolgimento e difesa di un'idea*, Torino, Bollati Boringhieri.
Becattini, G. and Rullani, E. (1993), "Sistema locale e mercato globale", *Economia e Politica Industriale*, 80, December, pp. 25-48.
Brusco, S. (1982), "The Emilian Model: Productive Disintegration and Social Integration", *Cambridge Journal of Economics*, 6, 2, pp. 167-184.
Brusco, S. (1989), *Piccola impresa e distretti industriali*, Torino, Rosenberg e Sellier.
Brusco, S. and Paba, S. (1997), "Per una storia dei distretti industriali italiani dal secondo dopoguerra agli anni novanta", in: *Storia del capitalismo italiano dal dopoguerra a oggi*, Barca, F. (ed.), Roma, Donzelli.
Corò, G. (1998), "Distretti e sistemi di piccola impresa nella transizione", in: *Il postfordismo. Idee per il capitalismo prossimo venturo*, Rullani, E. and Romano, L. (eds), Milano, Etaslibri.
Di Bernardo, B., and Rullani, E. (1990), *Il management e le macchine. Teoria evolutiva dell'impresa*, Bologna, Il Mulino.
Dosi, G., and Marengo, L. (1993), "Some Elements of an Evolutionary Theory of Organizational Competences", in: *Evolutionary Concepts in Contemporary Economics*, England, R.W. (ed.), Ann Arbor, University of Michigan Press.
Fortis, M. (1996), *Crescita economica e specializzazioni produttive*, Milano, Vita e Pensiero.
Fortis, M. (1998), *Il "made in Italy" – Quando stile e creatività non sono solo moda*, Bologna, Il Mulino.
Fuà, G. and Zacchia, C. (eds) (1983), *Industrializzazione senza fratture*, Bologna, Il Mulino.
Galbraith, J. K. (1967), *The New Industrial State*, Boston, MA, Houghton Mifflin. Italian translation: *Il nuovo stato industriale*, Torino, Einaudi, 1968.
Grandinetti, R., and Rullani, E. (1996), *Impresa transnazionale ed economia globale*, Roma, La Nuova Italia Scientifica.
Hayek, F.A. von (1937), "Economics and Knowledge", *Economica*, n.s. IV, No. 13, pp. 96-105.
Holland, J. (1995), *Hidden Order: How Adaptation Builds Complexity*, Reading, MA, Addison-Welsey.
Holland, J. H. (1988), "The Global Economy as an Adaptive Process", in: *The Economy as an Evolving Complex System*, Anderson, P.W., Arrow, K.J. and Pines, D. (eds), Santa Fe Institute, Studies in the Science of Complexity, Redwood City, CA, Addison-Wesley, Proceedings.
Kelly, K. (1994), *Out of Control*, Addison-Wesley, Italian translation: *La nuova biologia delle macchine, dei sistemi sociali e del mondo dell'economia*, Milano Apogeo, 1996.

Kelly, K. (1998), *New Rules for the New Economy. Ten Ways the Network Economy is Changing Everything*, Londra, Fourth Estate. Italian translation: *Nuove regole per un nuovo mondo. Un decalogo per chi vuole cavalcare le nuove regole e non esserne scavalcato*, Firenze, Ponte delle Grazie, 1999.

Krugman, P. (1995), *Development, Geography and Economic Theory*, Cambridge, MA, MIT Press.

Krugman, P., and Venables, A. J. (1990), "Integration and the Competitiveness of Peripheral Industry", in: *Unity with Diversity in the European Economy*, Bliss, C. and Braga de Macedo, J. (eds), Cambridge MA, Cambridge University Press.

Krugman, P., and Venables, A. J. (1995), "Integration, Specialization and Adjustment", *European Economic Review*, 40.

March, J. G. (1991), "Exploration and Exploitation in Organizational Learning", *Organization Science*, vol. 2, No. 1, pp. 71-87.

Micelli, S. (2000), *Imprese, reti, comunità*, Milano, Etas.

Piore, M. J. and Sabel, C. F. (1984), *The Second Industrial Divide*, New York, Basic Books, Italian translation: *Le due vie dello sviluppo industriale*, Torino, Isedi, 1987.

Porter, M. E. (1989), *The Competitive Advantage of Nations*. Italian translation: *Il vantaggio competitivo delle nazioni*, Milano, Mondadori, 1991.

Rizzello, S. (1997), *L'economia della mente*, Roma, Laterza. English translation: *The Economics of the Mind*, Elgar, Aldershot, 1999.

Rullani, E. (1996), "Complessità e informazione nella scienza economica", *Pluriverso* 2, RCS Libri.

Rullani, E. (2000), "Sistemi locali e produzione di conoscenza", presentation at the OECD Conference on *Enhancing the Competitiveness of SMEs in the Global Economy: Strategies and Policies*, Bologna.

Sabel, C. F. (1989), "Flexible Specialization and the Reemergence of Regional Economies", in: *Reversing Industrial Decline*, Hirst, P. and Zeitlin, J., Oxford, Berg.

Sabel, C. F. (1999), "Regionalismo sperimentale e i dilemmi della politica economica regionale in Europa", in: *Neoregionalismo. L'economia-arcipelago*, Perulli, P., Torino, Bollati Boringhieri.

Saxenian, A. (1994), *Regional Advantage: Culture and Competition in Silicon Valley and Route 128*, Cambridge, MA, Harvard University Press.

Signorini, F. L. (2000), *Lo sviluppo locale. Un'indagine della Banca d'Italia sui distretti industriali*, Cosenza, Meridiana Libri.

Silverberg, G., Dosi, G., and Orsenigo, L. (1988), "Innovation, Diversity and Diffusion: A Self-Organizing Model", *Economic Journal*, 98, pp. 1032-1054.

Vaccà, S. (1997), "Le differenze socio-culturali e istituzionali nello sviluppo delle imprese: il ruolo del capitale immateriale e del capitale umano", *Economia e Politica Industriale*, No. 94.

Varaldo, R., and Ferrucci, L. (eds) (1997), *Il distretto industriale tra logiche di impresa e logiche di sistema*, Milano, Angeli.

Viesti, G. (2000a), "Le strade dello sviluppo: come sono nati i distretti industriali del Made in Italy nel Mezzogiorno", *Economia e Politica Industriale*, No. 106, June.

Viesti, G. (2000b), *Come nascono i distretti industriali*, Bari, Laterza.

Zagnoli, P. (1991), *I rapporti tra imprese nei settori ad alta tecnologia. Il caso della Silicon Valley*, Torino, Giappichelli.

II. General Models of Industrial Cluster Dynamics

Complexity and Local Interactions: Towards a Theory of Industrial Districts

David A. Lane[1]

Abstract. In this talk, I will sketch an outline of a theory of industrial districts from a complexity perspective. The theory addresses three principle problems. The first is ontological: what is an industrial district? I argue that a district may be defined in terms of two kinds of structures: networks and scaffolds. At the nodes of the networks are individual entrepreneurs/artisans and firms. The links between the nodes consist of processes of recurring interactions. The networks carry the competences through which the district gathers and interprets information about products, production technologies and markets; produces and sells artefacts; and develops concepts for new artefact functionality, new markets, and new artefact types. The network structure undergoes constant transformation as the district generates new products and explores new markets. The transformation processes rely on various scaffolding structures, which may be regarded as the institutions that provide both a meta-stable identity *and* the possibilities for renewal and change to the district. I will discuss and illustrate two particularly important kinds of scaffolding structures: interaction loci and emergent rules and roles. Then, I will address two puzzles about districts from the viewpoint of the network-scaffold ontology: 1) How do districts achieve coordination in their distributed organization of production (or, as rephrased by Sebastiano Brusco, what is the difference between Carpi and Benetton)? 2) What exactly is the role of geographical space in the structure and functioning of an industrial district? That is, just what kind of links in a production system's network structure need to be constructed from physical rather than virtual interaction?
The second problem I discuss is: how do districts innovate? I begin by outlining two perspectives on innovation that have been developed in the complexity literature: recombination and generative processes. I then locate recombination and generative processes in the network-scaffold framework. For recombination, the role of spatially concentrated local interactions is fundamental, while I argue for the importance of scaffolding structures in increasing the generative potential of a district by organizing the processes whereby generative relations within and beyond district boundaries may form. I then use these ideas to explain changing innovation patterns in machines for tile manufacture.

---

[1]In preparing this paper, I profited from discussions with Robert Maxfield, Margherita Russo and David Stark, who kindly shared with me some of their knowledge about Silicon Valley, Sassuolo and Silicon Alley respectively. In addition, I learned much from my discussions over the years with my colleagues Andrea Ginzburg and Sebastiano Brusco about the ideas presented here and their relation with district phenomenology.

Finally, I turn to the problem of theory development through modelling and empirical research, and discuss two new emergent phenomena in academic district research: the use of agent-based models to provide insight into various aspects of district phenomenology, and the development of district-oriented ethnographic methods.
The talk concludes with a comparison of the complexity-based approaches I have outlined with other perspectives in the district literature.

1. Introduction

In this paper, I sketch an approach to a theory of industrial districts from a complexity perspective. The theory would address three principal problems:

- How can we describe the organization of an industrial district?
- How can we characterize the kinds of production activities that we would expect to benefit from a district organization?
- What kinds of processes and structures might account for the persistence or failure of a district organization?

Proposed answers to all three of these questions are presented in the penultimate section of the paper. The concepts in which these answers are framed are developed in the first six sections, while the last section suggests what kinds of empirical and modelling methods might be applied to help inform, develop and evaluate the theory I am proposing here. In this introductory section, I provide a road map for the rest of the paper.

What I mean by a complexity perspective is explained in section 2. I then turn to the task of defining a district and its organization. Since I want to define a district as a special kind of subsystem of what I call a market system, I have first to introduce this concept. That is the task of section 3, where I distinguish market systems from the more familiar notion of markets.

In section 4, I introduce two kinds of structures that together provide the organization of a market: networks and scaffolds. At the nodes of a network are agents - firms, project groups, individual human beings. The links between agents consist of processes of recurring interactions. Networks carry competences, through which the artefacts around which a market system is organized are produced and sold, new attributions for artefact functionality and new designs for artefact types and production processes are generated, and new markets developed. As new products come into being and new markets explored, the system's network structure is transformed. The transformation processes are supported by various scaffolding structures, institutions that provide both a meta-stable identity for system agents *and* the possibility for renewal and change for the system itself. In my discussion of scaffolds, I will focus on two particularly important kinds: *interaction loci* and *emergent rules* and *agent roles*.

Now I can define an industrial district, as a geographically concentrated production subsystem of a market system, composed primarily of small and medium firms, with a decentralized, non-hierarchic organization. As a market system subsystem, a district's organization can be described in terms of the competence networks of which it is comprised and the scaffolds through which these networks are formed and reconfigured, and the district identity is maintained and modified.

Two interesting questions need to be addressed about the particularity of districts in the class of all subsystems of market systems. The first has to do with the first defining characteristic of the district: its geographic concentration. Geographic concentration enables[2] local interactions among agents. I claim that local interactions are essential for agents to respond when the structure of their market system is undergoing a cascade of rapid changes, which affect both the artefact family around which the system is organized and the network of agents and agent relations that carries the system's economic functionality. In section 5, I argue that effective action in situations of rapid transformation of the structure of agent and artefact space depends upon the *generative potential* of an agent's relationships with the other agents with whom it interacts. As I argue in section 6, the generative potential of a relationship is determined by a set of characteristics that may be strongly affected by the locality of the participating agents' interactions. Section 7 argues that districts are one solution to the problem of organizing effective local interactions in systems that "inherently" undergo cascades of rapid change.

The second question concerns an important feature of a number of geographical regions in which districts are located: the same geographical *region* contains more than one *district*, in the sense that there are clusters of firms based in the area that participate in different market systems. For example, both Silicon Valley and the Washington Beltway are the homes of both an information and communication technology cluster, as well as a biotechnology cluster, while the province of Modena has clusters in precision machinery, textiles and tiles, among other artefact families. How might we explain the existence of such "mega-districts", or districts of districts?[3] As I show in section 7, the concept of scaffolding structures helps to answer this question: certain kinds of scaffolding structures provide support services to agents that participate in a district, *independently* of the market system in which the agent operates. Thus, these scaffolds can serve to promote the creation of new firm clusters that share the services they provide with existing clusters - and may even promote the integration of these clusters into pre-existing market systems, as new industrial districts.

---

[2]But of course does not necessarily product the right kinds of interaction structures. Whether these form or not depends on many factors - in particular, through the kinds of scaffolding structures that emerge to channel and generate local interactions.

[3]Before taking the question too seriously, it would be a good idea to establish the empirical validity of the proposition that districts tend to be located near other districts, rather than having a random geographical distribution, in the appropriate sense of "random" for this issue. I believe that districts do cluster, for the reason I provide in the text. Of course, if they don't, it is still the case that districts that are clustered may still share certain scaffolding structures!

Characterizing district organization in terms of competence networks and scaffolding structures may provide theoretical advantages, but it certainly introduces empirical difficulties. Consider for example one important implication of this idea: the firm is no longer a primary unit of analysis, but merely one type of node in the competence networks that carry out the district's economic functionality. In the Carpi knitwear district, a man might be a worker one year, an entrepreneur the next, and a worker the following year, without changing in any substantial way his productive activities. A Silicon Valley entrepreneur might leave one company he founded, take off a year or two, then found another in a different market system - and immediately hire the same engineering team responsible for the success of his first company. Examples like these help us to understand that an agent-network perspective is potentially much more powerful in describing how districts do what they do than a firm-based perspective. The difficulty is that most available data refer only to firms, and not to individuals, project groups, research collaborations, strategic alliances or other organizational forms that may be nodes in district competence networks. Moreover, the very definition I propose for network links has a temporal dimension, as it refers to "recurring patterns of interaction", not static relations or one-shot events. What kinds of empirical research can shed light on competence networks and the processes whereby they form, carry functionality, and change structure? Similarly, how can scaffolding structures be identified and their functions observed? A related issue concerns the role of modeling in developing a complexity-based theory of districts and, in particular, examining district innovation processes. The concluding section of this paper briefly considers two emerging answers to these methodological problems: ethnographic methods and agent-based modeling.

2. Perspectives on Complexity

The papers in this volume by Gell-Mann and Holland start from two different perspectives on complexity. Gell-Mann takes a theoretical perspective. The complexity of a system, he tells us, can be measured by the length of the minimal description of the system's regularities. Thus, if we want to address the question of whether an industrial district is a complex system, we are forced to confront some hard ontological issues. How do we characterize a district in system terms - that is, what are its component entities, and through which kinds of processes do they interact with one another? Even harder: taking on the role of Gell-Mann's judge, we need to decide what constitute a district's regularities - and to frame a vocabulary in which we can describe them.

Holland's approach to complexity begins phenomenologically rather than theoretically. For him, a complex system is one that behaves in a particular way. Roughly, the system is composed of a number of entities that have properties that in general differ from entity to entity; the entities interact with one another and the environment they inhabit; as a result of these interactions, their properties, the environment's, and even their interaction modes may change. In the examples

Holland presents, the entities' *structures* - that is, the set of their properties - determine how they *function* - that is, the transformations they effect through their interactions with other entities.

A complex system, then, is a particular kind of dynamical process, determined by a succession of interactions among entities. The phenomena of interest in the dynamics of complex systems are often characterized as *emergent*: they refer to features that can be concisely described only by reference to a higher level of aggregation than the individual entities themselves and that persist for periods of time considerably longer than those in which individual entities' interactions are denominated. Often, emergent structures *self-organize*, and this emergent organization of the system may constrain and channel the entities' interaction patterns. To apply the phenomenological perspective on complexity to industrial districts, one must describe the district in terms of the agents that compose them and their modes of interaction, then search for "global" emergent structures and self-organization at the level of the district itself. We might even ask to what extent the district itself might be regarded as an emergent entity.

Both Holland and Gell-Mann emphasize an aspect of complex systems that could be described as *cognitive*. Holland refers to "adaptive agents" that adjust their actions on the basis of what they have learned from their previous history of interactions with their world; Gell-Mann calls the representation of the world on the basis of which such adjustments are made a "schema", and prefers to reserve the term "complex system" itself for an entity with a schema, which can be thought of as its minimal description of the regularities in its world. This circle of ideas prompts a question rather novel to the district literature: are we justified in viewing the *district itself* as what Holland calls an adaptive agent and Gell-Mann a complex system? Or to put it another way: as we have already seen, a complexity perspective requires that we treat a district as a time-varying structure and that we try to understand how its structure determines its function; should we also view the district as a cognitive entity, and understand how its cognitive processes give rise to transformations in its structure?

Summarizing, the hallmarks of a complexity perspective include commitments to

- process and change, not stasis and equilibrium;
- a multilevel organization of entities;
- entity function determined by entity structure;
- distributed control and information-processing; and
- emergence and self-organization.

Whether it is useful to apply a complexity perspective to industrial districts depends on how rich an account of district phenomenology a theory founded on these commitments can produce.

3. Setting the Context: Markets and Market Systems

"The market" is an abstract entity defined formally by economists and employed informally by journalists, politicians and just about everyone else. It is a locus of impersonal exchange activities, where agents buy and sell products with defined characteristics, at prices that - according to standard economic theory - reflect supply-and-demand induced equilibria. Economic theory accords these prices the role of principal communication media between agents, who use the information prices convey to drive the actions they take in the economy. Relationships between agents do not count for much in "the market". What matters is how each agent separately values each artefact in "the market", values that "the market" then aggregates into prices for these artefacts. Frequently, in popular narratives about developments taking place in business and the economy, "the market" is assigned the central causal role.

By a market system, I mean a set of agents that engage with one another in recurring patterns of interaction, organized around an evolving family of artefacts. Through their interactions, the agents produce, buy and sell, deliver, install, commission, use and maintain artefacts in the family; generate new attributions about functionality for these artefacts; develop new artefacts to deliver the attributed functionality; and construct, augment and maintain new agents and patterns of agent interaction, to ensure that all these processes continue to be carried out, over time, even as the circumstances in which they take place are changing in response to perturbations from inside and outside the market system itself.

In a market system, the meaning of artefacts are up for negotiation. As a result of these negotiations, the artefacts take on value - not just in individual heads, but through a social process that takes place in concrete social settings. Agents learn more from each other than they do from prices, and they do not merely exchange information, they jointly develop interpretations. These interpretations drive action in new and hitherto unexplored directions.

Standard economic theory starts with the concept of "the market". Of course, something like "the market" will play a role, often an essential one, in many of the interactions in which market system participants engage. However, in the most interesting district phenomenology, agent relationships often hold centre stage, and many modalities of communication, from discourse based upon shared understandings through joint action coordinated by tacit knowledge, shape these relationships. Moreover, district functioning relies on non-market structures - like entrepreneur associations, user groups, trade fairs and standards organization - as well as shared attributions about agent roles and artefact functionality and rules that determine how agents may interact. These elements do not arise from market transactions, but from a complex network of agent interactions, whose description and analysis require a different set of ideas than standard economic theory has to offer.

4. Market System Organization: Networks and Scaffolds

A market system can be viewed as a collection of transformation processes: for example, producing, selling, installing, maintaining, designing artefacts in the family around which the system is organized. In the course of carrying out these processes, other processes are enacted by agents in the system - gathering and interpreting information, setting standards, establishing new entities like strategic alliances or trade associations. All these processes are achieved through interactions among agents - individuals, working groups, divisions, firms, alliances. Since these interactions taken together deliver the functionality that permit the system to endure over time, they tend to be organized into recurring patterns, with each pattern identifiable by its set of participants, interaction modes, and frequency and duration of interactions. Each recurring pattern of interaction defines a network; each network may be said to carry a system competence; as they are enacted, these competences generate the transformation processes that deliver the system's functionalities. Over time, of course, the transformation processes, the competences that enact them, and the networks that carry these competences change. But in the relatively short term, we may describe the system's organization in agent space as the crosscutting network of these competence networks.

If we view a market system from a somewhat longer time perspective, we notice that it is always undergoing perturbations, which may be generated from processes taking place within the system itself or may come from outside the system, as for example from large-scale macroeconomic shifts. In response, new networks are constructed and others change their structure, either by replacing or adding nodes or by altering the modes, duration or frequency of the linking interactions. Some of these changes may seem to happen "spontaneously", but for most of them we can identify structures that have provided the opportunity and the means for them to happen. Thus, the fluid network organization of a market system is constructed, renewed, and extended by a longer-lasting set of structures that serves to keep the system functioning. We call these structure *scaffolds*, since they shape and support the competence network structure of the system as it undergoes construction and reconstruction.

Scaffolds come in two basic flavours - physical and cognitive. Examples of physical scaffolds include user groups and trade fairs, trade and professional organizations, and standards bodies, as well as communication media like trade and professional journals, company and organizational newsletters, and websites. Some of these scaffolds are generated inside the system itself, others come from elsewhere, for example from other market systems, from professions, or from government agencies. *Interaction loci* are a particularly important kind of physical scaffold. Their function is to provide a space within which particular types of interactions may take place[4]. All the examples listed above are or provide

[4]Of course, the space need not physically contain all the participants in the interaction. Instead, it may be mediated by artefacts, as happens with electronic bulletin boards or

interaction loci. It is important to understand that all interactions are spatially located - and the way the space in which they happen is structured can have significant effects on the form the interactions take. Thus, interaction loci are crucial for constructing and maintaining agent relationships within a market system, and the kinds and organization of its interaction loci go a long way in determining the structure of the system's agent-artefact space.

Cognitive scaffolds are a bit more difficult to define and identify than their physical counterparts. Interacting clearly has important cognitive components: agents *intend* to derive certain consequences from the actions in which they engage; they have attributions about the *identity* of the agents with whom they interact that in part determine the way in which they act and the interpretations they make of the actions of the others, as do their attributions of the *functionality* and *value* of the artefacts around which their actions are oriented. For a system to function, there must be an alignment of some of these attributions and some degree of shared expectations about the relation between intention and consequence, so that recurring patterns of interaction can form and be maintained. In particular, *role structures* emerge within a market system, by means of which agents in the system classify one another and on the basis of which they generate expectations for the kinds of interactions in which others may engage and for the ways in which they may interact. Shared attributions of role structures then channel agents into particular interaction streams, which provide some continuity and stability to system processes. The more stable are the collective attributions of agent roles and artefact functionality and value that emerge in a market system, the more "rule-like" and stable are the patterns of interaction that deliver system functionality - that is, the network structure of the system itself. Naturally, there is a connection between cognitive and physical scaffolds: many of the processes through which cognitive scaffolds emerge take place within system interaction loci.

5. Generative Relationships, Generative Potential and Complex Foresight Horizons

Attributions play a fundamental role in market system processes. Agents represent the regions of agent-artefact space in which they operate by means of attributions. Agents' attributions of their own and each others' identities - what the agent does and how it does it - mediate how agents interact with one another, and their attributions of an artefact's functionality mediate the ways in which they make and use the artefact. We have seen how the emergence of shared attributions underlies the formation of stable interaction patterns without which a market system cannot endure for long. On the other hand, we have also seen that market systems change,

video-conferencing. As these two examples indicate, however, even such "virtual spaces" can be identified with physical spaces that determine the form of the interactions they support.

as new agent structures form and new artefact types are developed. In general, new entities are preceded by new attributions. For example, a new artefact type is developed in order to achieve a new kind of imagined functionality. So to understand how a market system changes, it is important to consider how new attributions come into being.

Robert Maxfield and I (Maxfield and Lane 1997) have argued elsewhere that new attributions generally arise in the context of *generative relationships* among agents. The key idea is that individual agents' attributions tend to be organized into closed systems, within which it may be possible to recombine existing elements to develop "better, faster, cheaper" versions of previously represented types, but which resist extension into new classificatory dimensions. To open an attributional system to new conceptualisations usually requires the realization that its current representations are incomplete, a realization that can generally emerge only in the context of discursive relationships with others, whose attributional systems highlight different aspects of the identity or functionality of some particular entity under discussion. Often, the realization begins with the uncovering of an ambiguity in the sense in which different participants in the discussion use the same terms. This ambiguity in categories that each participant regarded as both well defined and shared opens a space for innovation in the attributional systems of each of them. To fill that space generally requires further joint exploration, in which each may adjust boundaries and introduce new concepts and categories in one's own set of attributions as they all jointly construct what they come to regard as a shared attribution for the entities that have entered into their conversations.

Clearly, not every relationship between agents is generative in the sense described above. Maxfield and I identified five characteristics that together determine the *generative potential* of an agent relationship:

- aligned directedness - The participants in the relationship need to orient their activities towards a common zone of agent-artefact space. For example, the same kind of artefact might be the focus of each of their activities, although the participants need not have the same relationship to the focal artefact.
- heterogeneity - Generativeness requires that the participating agents differ from one another in key respects. They may have different attributional systems, competences or access to particular agents or artefacts.
- mutual directedness - Agents need more than common interests and different perspectives to form a generative relationship. They also must seek each other out and develop a recurring pattern of interactions out of which a relationship can emerge. Their willingness to do this depends on the attributions each has of the other's identity. It helps, but is not necessary, for the participants to start by trusting one another. Frequently, rather than a precondition, trust is an emergent property of generative relationships: it grows as participants come to realize the unforeseen benefits that the relationship is generating.
- permissions - Discursive relationships are based on permissions for the participants to talk to one another about particular themes in particular illocutionary modes (requests, orders, declarations, etc.). These permissions

are granted explicitly or implicitly by superordinate agents and social institutions. Unless potential participants in a relationship have appropriately matched permissions, or can arrogate these permissions to themselves, the generative potential of the relationship is blocked.

- action opportunities - Discourse is important, but relationships built only around talk do not usually last long or affect deeply the identities of participating agents. Engaging in joint action focuses talk on the issues and entities of greatest interest -- those around which the action is organized. And action itself reveals the identities of those engaged in it. In addition, new competences emerge out of joint action, and these competences can change agents' functionality and hence identity -- even leading to the formation of a new agent arising from the relationship itself.

Generative relationships are particularly important when a market system is in a phase of rapid cascading transformations. In such cases, it is impossible for agents to foresee all the consequence of their actions, because some of the relevant entities - artefacts, agents, or more general structures or patterns of interaction in agent-artefact space - that will determine these consequences may not be visible to the agents (or even exist!) at the moment in which the agents must act. In situations of this kind, which Maxfield and I call *complex foresight horizons*, it may be much more important for agents to decide *with whom* to interact, rather than to attempt to figure out exactly what will come from the interactions in which they engage. Clearly, the more potentially generative a relationship among agents is, the more incentive there is for these agents to interact, to explore the new attributions and interaction patterns that may emerge from their interactions. Moreover, when foresight horizons are complex, agents may act to enhance the generative potential of some of their relationships, for example by changing permissions of subordinates to encourage cross-boundary discourse and by creating action opportunities even when it is unclear exactly what outcome (and profit) might be expected to result from a joint project.

Industrial districts are often characterized by complex foresight horizons. This is obviously the case for the information technology and biotechnology clusters in Silicon Valley and the Washington Beltway, as well as the internet districts of Silicon Alley and San Francisco. Less obviously, it also holds for most of the "made in Italy" districts, where the rapid changes in artefact style that characterizes fashion industries make it difficult to foresee not only what designs will be selling in the next months, but even what materials will be used in the fabrication of products - and hence in which kinds of production processes a firm must engage. Thus, generative relationships play a key role in district phenomenology, and we should not be surprised to find that most successful districts have scaffolding structures that help to promote the formation of generative relationships among district participants.

6. Locality and Generative Potential

Generative potential may be greatly enhanced by the co-locality of agents in a relationship. To understand why, consider the following three possible interpretations of *locality*. First, local interactions might be defined as those in which the participants meet face-to-face. Such interactions differ from those in which the participants are physically remote in two important ways. First, face-to-face discourse has a broader bandwidth than remote speech, which in turn has broader bandwidth than written communication. Thus, face-to-face communication can be more nuanced, providing more space for discussions to go in unexpected, potentially fruitful directions. Second, individuals can learn much about and from one another by observing how each interacts both with people and with artefacts; face-to-face interactions provide the richest context in which this kind of learning can take place.

Locality could refer to social rather than physical space. For example, we could describe an interaction as local if all the participants belong to the same *community of practice*, even if they are physically remote. For example, the programmers who work together on freeware projects share a language, a culture and a project orientation that allows them to collaborate even if they live on different continents and have never met one another. Of course, the interactions between these programmers are local in this sense only if they stay within the boundaries imposed by the implicit rules of their community of practice.

Similarly, we might consider any interaction local that involves participants that inhabit the same geographical-social community, wherever and however the actual interactions themselves take place. As with a community of practice, co-membership in a geographical-social community implies that many kinds of interaction modes are jointly understood and accepted and may be simply enacted rather than negotiated. Of course, just what counts as a geographical-social community, and which kinds of interaction modes qualify as local for a given such community, are not obvious. Nonetheless, much of the discussion of Italian districts like Prato or the districts of the Modenese are premised on the assumption that the participants belong to such a community, and their co-locality underlies the possibility and form of many of their essential interactions - face-to-face or not. Indeed, much of the literature on Silicon Valley is based on a similar assumption.

What all three of these interpretations of locality have in common is the idea that participants, because of their co-locality, have a higher degree of shared understanding about what the interaction in which they engage means, so that there is much less need to negotiate either the context or the consequences of the interaction event itself. Shared physical space brings this about by increasing communication bandwidth and hence the direction and visibility of agreement, while shared social space does it by channelling the interaction into mutually comprehensible normative forms.

What effects does locality have on the determinants of generative potential? Certainly locality conduces to mutual directedness. A shared social space affords an *a priori* basis for mutual directedness, at least within the context of the interaction modes supported by the community, while shared physical space provides an opportunity to form either positive or negative attributions of others' modes of operating, and so either enhances mutual directedness or brings an unsatisfactory relationship to rapid closure.

Similarly, community values can ensure aligned directness for interactions directed towards transformations consistent with those values, while face-to-face communication encourages the kind of discursive exploration that can lead to the discovery of a basis for alignment that may not have been evident *a priori*. However, there is always an inherent tension between aligned directedness and attributional heterogeneity: agents that are completely aligned may come to share a relevant set of attributions to such an extent that the generative potential of their relationship may decline or even disappear; and if any relationship becomes closed to outside perturbations from other, cross-cutting networks of relations, local interactions among participants in the relationship almost certainly will hasten this process along. This risk is greatest when the participants are strongly co-located in social space, and hence begin with similar attributional systems, which the relationship may fine-tune to homogeneity. On the other hand, extended face-to-face discourse may reveal subtle unsuspected attributional differences, the exploration of which may provide space for generation of new attributions. Much depends on the way in which relationship participants regard attributional heterogeneity, whether as a threat or an opportunity for the future of the relationship. Overall, if participants in a relationship are open to the outside and to the potential fruitfulness of disagreement and misunderstanding, locality in all its forms is a potent contributor to generative potential.

For this reason, every market system provides scaffolds that make possible certain kinds of local interactions. Some of these kinds of interactions need not recur frequently: trade fairs provide an opportunity for many market system participants to keep up with general trends in artefact development and use through meetings and discussions with one another, but for most market systems one or two trade fairs a year is enough. But other kinds of scaffolded local interactions require frequent recurrence, and these are generally possible only when the networks that the scaffolds construct and maintain are composed of agents that are either spatially co-located or sufficiently socially co-located that they can maintain local interactions at physically remote sites. Scaffolds of this type include universities and community colleges that support research collaboration and training for system participants, users' groups, or meeting sites for various kinds of informal face-to-face interactions like Happy Hour at the Wagon Wheel in Silicon Valley or Silicon Alley's somewhat more formalized cyber parties. The possibility to provide scaffolding structures of this type is a strong incentive for geographic concentration of market system participants – and lies at the heart of the industrial district.

7. Industrial Districts

As defined in the introduction, an industrial district is a production subsystem of a market system, characterized by predominantly small and medium firms concentrated geographically, with a distributed or decentralized network organization. A theory of districts ought to provide an account for how a district's network structure arises and is maintained, explain how this structure carries the district's production functionality and how the district's links with the rest of the market system are organized and maintained, and characterize which market system - locality couplings are likely to give rise to districts.

From the arguments in the previous five sections, it is clear that I would expect to find districts forming in market systems characterized by persistent complex foresight horizons, where generative relationships and the capacity of firms to react rapidly to changes in the structure of the market system are much more important than economies of scale achievable by vertical integration or the kind of strategic calculation that are advantageous in markets where impersonal transactions for artefacts with stable and widely shared attributions of functionality take place. In addition, entry costs for new firms in a district production subsystem would have to be relatively low, and the value of artefacts that the subsystem produces should be socially constructed, relatively unconstrained by the laws of physics - as, for example, fashion and information technology. In such market systems, local interactions are particularly important, and if change in the agent-artefact structure of the market system occurs rapidly enough, the advantages of both spatial and social locality should be particularly pronounced.

Exactly where the resulting concentrations of small and medium firms would be located depends on many factors and historical contingencies, but one obvious prerequisite is that the geographic regions in which districts emerge must provide scaffolding structures that support the construction and re-production of the rapidly changing networks that must carry out the various competences the district's activities require. At least four kinds of competence networks are particularly important for districts as here defined: networks of *information*, through which agents learn about new developments in the organization of agents and artefacts in their market system throughout the world; networks of *interpretation*, through which agents make sense out of the information they obtain to determine the directedness of the interactions in which they will engage; networks of *production*, in which they collaborate with other district agents to generate new artefact types and produce artefacts for the market; and networks of *marketing*, through which their products find buyers in markets around the world.

Two kinds of scaffolds are particularly important for creating and maintaining these sorts of networks. The first are scaffolds that promote social cohesion within the district, while at the same time mixing heterogeneous identities among district agents. As Annalee Saxenian (1996) has pointed out, Silicon Valley developed a coherent set of practices that encouraged engineers to "talk shop" with their peers in other companies, thus providing the basis for the formation of individual-level generative relationships the benefits of which redounded to the firms for which these individuals worked (and frequently the firms they founded as the result of some of the new attributions they generated). These practices included such cognitive scaffolds as rules that do not penalize employees from switching jobs from one firm to another, and physical scaffolds ranging from Happy Hour at the Wagon Wheel to research seminars at Stanford, Santa Clara and various company labs. In Modena, the CNA (Confederazione Nazionale dell'Artigianato) holds frequent meetings in which entrepreneurs may learn about new markets and new product opportunities from one another, and as the organization grew and the production competences represented in the Modena area multiplied, its administrators came to play a brokering role, bringing together entrepreneurs whose firm's competences might be combined to produce an artefact filling a newly recognized potential market niche, discovered by another Modena entrepreneur doing his own business in, say, China or Eastern Europe.

The second kind of scaffold provides for the delivery of local services, which make it easy to start up a new firm and enable entrepreneurs to concentrate their resources on honing their particular design or production competences. In Modena, both local and provincial governments and CNA provide such services, from arranging financing to outsourcing book-keeping and legal assistance, to supporting trade fairs at which local firms may display their products to potential buyers recruited from around the world, to providing space for new firms to establish their operations, to research and educational projects like PROMO, designed to introduce local firms to best international practice. In Silicon Valley, many of the same scaffolding services have emerged privately, with little government or district-level organization. For example, venture capitalists, concentrated in several blocks of Sand Hill Road, provide not only seed financing but networking and recruiting services to prospective entrepreneurs. Local law firms specialize in legal services for start-up companies; some of them now offer start-up packages for as little as $5,000 per year (or shares in the new companies), handling all the legal problems new firms encounter in their first two or three years of business. Real estate developers build industrial parks where start-up or expanding firms can locate. The architectural design of many of these parks provide open space, inside and out, that facilitate the formation of discursive relationships within and between firms located in the parks. Probably the most important scaffold for Silicon Valley, at least in its formative years, was Stanford University, whose engineering graduates were encouraged to start companies in the area and whose professors were permitted to engage actively in consulting and entrepreneurial activities of their own. Stanford, San Jose State, and Santa Clara now all play important roles in promoting knowledge development and exchange, through seminars, research programs, and teaching (often with instructors recruited from local businesses) for Silicon Valley firms and professionals. While local governments around Modena were often in the forefront in providing infrastructural support to local districts, this was not the case in Silicon Valley, until some of the "senior citizens" from successful firms in the district established a lobbying group, the Santa Clara County Manufacturing Group, which became an important scaffolding institution for the mature district.

An interesting feature of many of the district scaffolds described above is that they are not inherently restricted to one particular market system. Thus, once a cluster of new firms is established in a region and scaffolds such as these are constructed to maintain it, the same scaffolds make it easy for other clusters operating in different market systems to develop as well. Of course, there must be some other, market system specific scaffolds around which the new cluster may form; but leveraging on the pre-existing scaffolds, it is much easier for second and successive clusters to arise in the same region. This is exactly what we find in the Modena area, as well as in Silicon Valley: the same scaffolding structures serve more than one cluster in the region. Sometimes, these scaffolds may even provide the means to initiate generative relationships between agents acting in different market systems, that can give rise to hybrid artefacts and even, at least in principle, to new districts or even new market systems.

Let us now return to the three questions I posed in the introduction. My suggested answers are as follows: district organization can be described in terms of networks of competence networks and scaffolds; which production activities might benefit from district organization may be characterized how complex are foresight horizons associated with the value and functionality of the artefact family around which the market system is organized; and how successful a district will be in organizing and maintaining its productive and innovative activities may be determined by analysing the scaffolding structure it provides to promote generative relationships and by examining how generative are key relationships among district agents and those that connect these agents to the distant markets in which the artefacts they produce are sold.

With these ideas in hand, we can now return to the two perspectives on complexity introduced in section 2, and ask whether an industrial district qualifies as a complex system. Certainly, districts as defined here satisfy the phenomenological criteria for complexity: many heterogeneous agents whose interactions give rise to interesting emergent structures (the networks that carry the district's competences and the scaffolds that construct and maintain these networks); in addition, these structures permit the district to adapt to both external and internal perturbations to the market system of which it is a part. To see whether districts satisfy the theoretical criterion for complexity proposed by Gell-Mann, we must decide what constitute a district's "regularities". It seems natural in this context to consider the processes through which the district's competence networks are constructed, maintained and enact and interlink their respective competences as the district's regularities, and it seems plausible that the minimal length of this description would be quite long - but this is a question that will have to wait until a full theory along the lines suggested here have been developed, and methods for describing and empirically identifying these processes established.

I conclude this section with a proposed, if premature, answer to another question linking districts and complexity: can a district be a complex adaptive system in the sense proposed by Gell-Mann - that is, can a district be said to possess a "schema" or compressed description of the world in which it operates, according to which it generates actions in that world, and which it may modify on the basis of its experience? I think the answer is yes, but to describe such a schema and even to talk about the district acting require further theoretical development much beyond the hints supplied here. Briefly, through its networks of communication and interpretation, the district may build up a set of partially shared attributions about the relevant agents and artefacts in its market system. Of course, this set is actually distributed among district agents, but because its elements are generally shared among these agents, they will produce some regularities in the way in which district agents interact with each other and outside agents, and as a result of these regularities, attributions that other agents make about the identity of district agents may share some elements. In this way, attributions can arise for "a way of acting like a Silicon Valley start-up company," and, at least absent strong specific contra-indications, these attributions may be applied to any "Silicon Valley start-up", thus reinforcing the behaviour that such an attribution predicts. In addition, through its scaffolding structures, a district may provide certain agents with permissions that allow them to speak or act for the district in certain kinds of situations - for example, in testimony before Congress or regulatory agencies, or in interviews with journalists who ask what "Silicon Valley" thinks about some governmental policy or macro-economic context. Both the distributed set of partially shared attributions and the set of permissions that endow district-level agency upon particular district agents may be legitimately interpreted as components of a district schema, and so we may regard the district itself - and not merely the agents of which it is composed - as a complex adaptive system in Gell-Mann's sense.

8. Methodology for a Complexity Perspective on Districts

What I have sketched above is only an approach to a theory of districts. To develop a rich theory that identifies, explains and even predicts interesting district phenomenology would require a considerable supplemental research effort. This research will have to move in two directions. The first is empirical: we need to learn how to recognize, describe and classify the network-scaffold structure of districts, and then to examine the kinds of processes that these structures support.

There are already promising lines of work in this direction. Margherita Russo has carried out several studies in the tile manufacturing district of Sassuolo, in which she applies ethnographic interviewing techniques to determine how the formation or absence of generative relationships affect the patterns of innovation in the district. In one study, she showed how a generative relationship between the owner of a tile manufacturing firm and a chemistry professor gave rise to a new technique of tile production which, however, failed to be widely adopted by other firms, a fact that she explained in terms of the network structure of the district, in which the innovating firm was isolated from the information and interpretation networks that linked other district agents. In another study, she showed how the generative relationships between tile manufacturers and the producers of tile manufacturing machinery declined, as the latter changed their networks of relations when they began to produce turnkey plants for foreign customers wishing to emulate Sassuolo production techniques. These stories would have been difficult to tell absent concepts corresponding to network dynamics and generative relationships, and impossible to uncover with usual aggregate or even firm-level data.

David Stark is leading an ambitious project to study the intra- and inter-firm structure of Silicon Alley, the district of internet firms in New York City. He has been fortunate in his project choice, since the group has been able to track both the heady emergence of Silicon Alley and its spasmodic response to the crash in internet stock prices. Using an arsenal of ethnographic techniques, including detailed observation and recording of individual interaction events, plus the tools of social network analysis, his group is developing a rich database on network formation and functioning and agent-artefact and agent-agent relationships, as well as emergent district phenomenology, including the construction and mode of operation of scaffolding structures. As the results of the group are published, it is to be hoped that researchers studying other districts will adopt the methods drawn from sociology and anthropology that Stark's group uses to supplement the economics methods of data collection and analysis that dominate most current district empirical research.

Modelling is the second important research direction necessary to develop a rich theory of districts from a complexity perspective. Through the use of models that share the network-scaffold ontology it is possible to carry out observations that may suggest interesting phenomena that emerge in such settings, which may then be sought with appropriate empirical tools in districts themselves. It will also be possible to use these models to carry out simulation experiments to evaluate causal hypotheses and predictions that the developing theory generates. To implement the ontology of entities and relations sketched in this paper, it is necessary to develop agent-based models for districts. Agent-based modelling has become in the last decade a powerful tool in social science research[5]. At least three research groups have already applied agent-based modelling to study aspects of district phenomenology. Tommaso Minerva, Irene Poli and Sebastiano Brusco constructed a cellular automaton model designed to simulate the process whereby district agents, via their communication networks, may collectively exploit a new market niche discovered by one of them. Through parameters that tune the rate of new market discovery and the density of network interactions, they use the model to predict such aggregate district features as firm-size distribution. Riccardo Boero and Flaminio Squazzoni have developed an agent-based model[6] of districts through which they can compare how effectively production networks are formed in three different environments. In the first two, the only mechanism for network formation is direct interaction between firms; in one, firms are modelled as simple myopic optimisers, while in the other, they seek long-term relationships with other firms. In the third, the district has two scaffolding structures that promote the

---

[5]The literature is already vast, as a web search on agent-based modelling will reveal. For a recent collection of interesting essays on the subject, with many references, consult Kohler and Gumerman (2000).

[6]Using the Swarm platform, developed at the Santa Fe Institute.

formation of interfirm networks. With this model, Boero and Squazzoni investigate how effective scaffolding structures are as a function of a variety of tunable model parameters that describe the market system in which the district operates. Finally, Guido Fioretti has developed an interesting agent-based model[7] of the Prato district, through which he tracks the persistence of the district throughout a series of changes in the market systems in which it operates and in the internal organization of the district itself.

Another kind of modelling enterprise may also contribute to the development of the theory sketched in this paper. There is a virtual explosion of recent work on the structure of networks, sparked in large part by Duncan Watts' work on small world models[8], and on the dynamics of network formation and growth. Using this work, it will be possible to give a much more incisive statistical description of competence networks and the ways in which they interrelate to one another, which will allow a more refined presentation of district dynamics than would otherwise be possible. In addition, we might expect in the future that district agent-based models will incorporate some of the results coming out of the work on how networks form and grow and what kinds of processes networks with different structures can support. For researchers interested in developing a theory that can explain how districts are organized and how they function, the opportunity to assimilate this emerging research on network dynamics is very exciting.

References

Kohler, T. and Gumerman, G. (2000), *Dynamics in Human and Primate Societies: Agent-Based Modelling of Social and Spatial Processes*, Santa Fe Institute Studies in the Sciences of Complexity Series, Oxford University Press.

Maxfield, R. and Lane, D. A. (1997), "Foresight, Complexity and Strategy", in: *Economy as a Complex, Evolving System II*, Arthur, W. B., Durlauf, S. and Lane, D. (eds), Redwood City CA, Addison-Wesley, Chap. 4.

Saxenian, A. (1996), *Regional Advantage: Culture and Competition in Silicon Valley and Route 128*, Harvard University Press.

Watts, D. (1999), *Small Worlds*, Princeton University Press.

[7]Also Swarm-based.

[8]Watts' early work on small worlds is presented in his book (1999).

From Marshall's to the Italian "Industrial Districts". A Brief Critical Reconstruction

Giacomo Becattini[1]

Abstract. This paper begins by reconstructing the historical genesis of the notion of industrial district, focusing on the contribution of Alfred Marshall, the economist who is thought to have discovered this system. The paper also shows how the notion of industrial district stems from the successful wedding of Marshall's keen observations of contemporary industrial reality, mainly in Great Britain, with the desire to provide a way out of the difficulties inherent in classical economic theory and answer the challenges posed by the class struggle (e.g., economic nations).

The economic conditions that fostered the development of industrial districts in Great Britain recurred in roughly the same manner in Italy after the Second World War. A few scholars studied, with an open mind, the light industrialization that was occurring in certain regions of Italy and, after taking a fresh look at Marshall's theories, rediscovered the phenomenon of industrial districts. The wealth of literature they produced soon gained attention outside of Italy. This paper, however, does not dwell on what occurred outside of Italy; it focuses instead on the development of the industrial district phenomenon and its theoretical underpinnings in the Italian economy. In particular, it studies the link between the industrial district system and the manufacture of typical made in Italy products as it seeks to explain the competitive advantage that Italian manufacturers enjoy in many industries (for example, textiles and garments, footwear, furniture, etc.). The third part of the paper presents some of the most recent developments in the study of Italian industrial districts.

The main point of the paper is that, for certain types of products, industrial districts can provide a viable alternative to a system based on a single company (with one or more plants) or a network of companies.

> We shall never fully appreciate the importance of what Adam Smith wrote unless we understand the importance it had for him. For him it was a matter of living truths, strengthening and arousing hope, that encourage, almost force us, to search for further truths. And if we wish them to be useful they must be living truths for us too.
> Alfred and Mary Marshall, 1879.

[1]I am very grateful to several colleagues and friends, most of all to Bellanca, Bellandi, Groenewegen, Raffaelli, Dei Ottati and Sforzi for useful comments, but, as usual, I do not want to imply that any one of them shares all my views.

1 The Marshallian Industrial District

1.1. It All Started with Marshall

Marshall's writings around 1870[2] contain several statements that show clearly how his position differed from the prevailing one among economists on certain crucial aspects of the theory of production[3]. In particular, Marshall disputed the standard view that the factory system, in which all manufacturing processes are concentrated under one roof with a high degree of vertical integration, was necessarily better than production systems that were technically less integrated but concentrated geographically[4].

Reading the descriptions of English industry produced by his contemporaries,[5] analyzing the most popular works on economics[6] and observing the world around him with a keen eye[7] (for example, Sheffield's cutlery works and Birmingham's metal trades), Marshall quickly came to the conclusion that, *at least for certain types of production*, there were two efficient manufacturing systems: the established method, based on large, vertically integrated production units, and a second one based on the concentration of many small factories specializing in different phases of the same production process and operated in one location or in a cluster of locations.

"We shall find", wrote Alfred and Mary Marshall in their 1879 economic manual[8], "that some of the advantages of division of labour can be obtained only in very large factories, but that many of them, more than at first sight appears, can

[2]See Whitaker (1975).

[3]The entire life and work of Alfred Marshall has been extensively and masterly explored by Peter Groenewegen. See Groenewegen (1995).

[4]It should be noted that Marshall's argument is also valid when vertical integration occurs in a small, non-specialized business. Marshall is an early exponent of the view that at every separable phase of the manufacturing process, a firm should assess whether it would be better off retaining that phase within the internal manufacturing system or outsourcing it in the marketplace.

[5]For example: Cooke Taylor (1841) and Sargant (1857), which was influenced greatly by *Les Ouvriers Européens* by F. Le Play, whose analytical method was carefully studied by Marshall.

[6]Chief among them Mill (1848); Cairnes (1874); Cliffe Leslie (1888), but also Hearn (1863). Hearn devoted some attention to the phenomenon of the industrial districts, to some extent anticipating Marshall. See Pesciarelli (1999). The main work targeted by Marshall was probably Fawcett (1863), commonly known as "Mill and water", because it was considered a mere summary of Mill (1848).

[7]Keep in mind that Marshall was not an "armchair economist". From an early age, he showed great interest in the technical and organizational details of the manufacturing process. Consider, for example, the sketches of industrial facilities and equipment he used to illustrate his survey of U.S. factories. See Whitaker (1996), vol. 1, pp. 51-59 and 80-81.

[8]See A. and M. P. Marshall, (1881), 1st ed. 1879; from the 2nd ed. pp. 52.

be secured by small factories and workshops, provided there are a very great number of them in the same trade.

The manufacture of a commodity often consists of several distinct stages, to each of which a separate room in the factory is devoted. But if the total amount of the commodity produced is very large, it may be profitable to devote separate small factories to each of these steps. If there are many factories, large or small, all engaged in the same process, Subsidiary Industries will grow up to meet their special wants[9].

But small factories, whatever their number, will be at a great disadvantage relative to large unless many of them are collected together in the same district[10].

Thus both large and small factories are benefited (...) (by the localization of industry). But these benefits are most important to the small factories, and free them from many of the disadvantages under which they would otherwise labour in competition with large factories.

And finally: "in these districts a further division of specialisation has grown up, and separate trades have sought separate localities (...). Those that work in wool do not generally live among the Lancashire cotton workers, but are collected together in Yorkshire; and they themselves are divided into the "wollen trade" and the "worsted trade", and these again spread out into various branches, each of which has a favourite district of its own"[11].

Here Marshall, in addition to offering a phenomenological description of the manufacturing district, provides some hints for a first, clear theoretical frame of reference based on "benefits", which he later called "external economies"[12].

1.2. The District Concept in Marshall's Thinking: a) The Problem of the "Economic Nations"

The most logical starting point for exploring the birth of the notion of industrial districts in Marshall's early writings can be found in the economic debates that raged when he made his debut as an economist (1871-1873)[13], especially those that arose after the publication of J. S. Mill's *Principles of Political Economy* and involved primarily J. E. Cairnes[14] and T. Cliffe Leslie. These authors had

[9]*Ibid.*, p. 52. This is the origin of the conceptual bifurcation leading on the one hand to the manufacturing district and on the other to the industrial sector or to the industrial *filière*.

[10]Here the territory is seen as a conduit for the creation and the dissemination of the "advantages".

[11]*Ibid*, p. 47.

[12]Notable studies that provide a historical framework for the development of Marshall's concept of district include those by Bellandi (1982); Loasby (1998); Loasby, (1998); and Raffaelli (forthcoming). See also, for a good background Loasby (1989).

[13]Marshall's first theoretical economic work was an 1872 review of The Theory of Political Economy (1871) by S. Jevons. See Pigou (1925), pp. 93-100.

[14]The essay containing Marshall's first serious theoretical work is a critique of Cairnes's critique of J. S. Mill, See Pigou (1925), pp. 119-133.

discovered that the professional and territorial geographical mobility of labour and capital postulated by classical economic theory did not exist - not even in England, the promised land of classical economic laws. In practice, British social reality was fragmented into regional, sectorial and social compartments, which hindered the free circulation of capital and labour that was supposed to exist in theory. The problem was then how to detect (or circumscribe) areas or groups of agents so as to allow the to and fro between phenomena and theory required by an empirical science like economics. Lacking such a methodological exploration, classical economic theory would be left floating in mid-air. This explains the cruciality of the theme of the "economic nation", which the young Marshall derived from the theory of "non-competing groups" and developed into two separate versions[15].

In the first version, Marshall's "economic nation" is a place, or a system of places, characterized by such a high degree of cultural homogeneity (values and institutions), free circulation of information and territorial continuity that the movement of capital and labour can quickly equalize profit and wage levels. One could then say that any geographical area of this type constitutes a sort of "economic nation", even when it does not become fully conscious of its autonomy and does not develop a unified system of government. A "political" nation (the usual nation-state) consists of a series of "economic" nations, and it is entirely possible for an economic nation to straddle the borders of several political nations[16].

In the second version, an economic nation is a "block of subjects" within a nation-state (e.g., miners and owners of coal mines, the farm lobby, etc.) or encompassing several nation-states (e.g., the working class) who recognize that their fundamental economic interests (or an important part of those interests) coincide and are at variance with the interests of similar blocks in the same nation. The knowledge awareness of a commonalty of economic interests interacts over time with other non-economic aggregants to make these social bodies coalesce into different forms and change over time the degree of their social cohesion.

This second type of "economic nation" also differs from the first in that it always has a more or less explicit form of government and a more or less complete and coherent "foreign policy", so to speak, on the basis of which it forms alliances or engages in confrontations with similar groups[17].

This also means that a dialectical interaction, the nature of which has yet to be fully understood, entails the existence in every "economic nation" of the second type not only of a contrast between the interests of masters and men but also of a common aggregant that enables them to stand together against the rest of society.

[15]On the young Marshall, see: Dardi (1984).

[16]I discussed more fully the notion of "economic nation" in my essay "Nazione economica e nazione politica nel pensiero di Alfred Marshall" in Roggi (1994). See also, in the same book, some brief remarks by Marco Dardi on this subject.

[17]At the time, this construct, which may seem bewildering to contemporary common sense, was not just the artifice of a scholar in his ivory tower. To understand this, all one has to do is think of the "guild socialism" political movement, which, not accidentally, attracted Marshall's attention. See his book (1919), Book III, par. XIV, 7.

This aggregant is not necessarily the same as the one that binds the inhabitants of a certain area or the one that is presumed to hold together the international working class, but is probably of the same type[18].
What is the general thrust of Marshall's effort to develop the notion of an economic nation separate from the nation-state? In my opinion, it expresses Marshall's refusal to derive the basic unit of economic research from real world entities such as nation-states and provinces. In this sense, it is difficult to define a nation as an empirical entity. Even the use of history or geography as a means of classification entails an effort at conceptualization or abstraction that can be conditioned by countless and undetermined historical and geographical influences. The unit of research must be such as to facilitate an abstract-concrete-abstract flow. The overall action of socioeconomic forces constantly defines and redefines the unit of research, which is at any rate no less than some sort of "economic nation", i.e., a plurality of individuals who feel more closely linked than is generally the case, in a grouping that can fall under either the first or the second definition, or both[19]. I believe that one of the traits that most distinguishes Marshall, even in the classification process, is the importance he attaches to personal empowerment, i.e. his concern for the individual and his thought process, without making claims of absolute objectivity. It may be worthwhile to emphasize this point.

This point of view hints at the possibility of shifting the analytical approach typically used by economists from the study of the behaviour of individuals in the commodities markets[20], the identification of which heavily conditions the conclusions drawn from economic theory, to the study of the behaviours that are typical of representative subjects. Obviously, this approach leads to human groupings that are represented by a limited number of agents (e.g., the inhabitants of a certain town) and/or enterprises (e.g., a "population" of firms). In my opinion, if this different way of "slicing" or approaching social phenomena contemplated by Marshall in his early writings had become established[21], it would have yielded

[18]The only contemporary economist who has, to my knowledge, addressed this issue is Jan Steedman, who also makes reference to Marshall in his essay, 1986. However, he does not seem to have followed up on the new ideas presented in his essay.

[19]The modernity of this notion is clearly shown in the following passages by C. P. Kindleberger: "The optimum social area is a function of the average citizen's sense of participation, of having a share in decisions, of counting (...). While the optimum economic area may be large, the optimum social area is clearly much smaller (...). Moreover while the optimum scale of economic activity is getting larger and larger, the optimum social scale appears to be shrinking". Kindleberger (1989), pp. 87-88.

[20]In this area, see Lange (1952), Appendix, who writes: "In economic science, however, the classification of goods cannot be made on a purely arbitrary basis, because the laws of economics would then be dependent on the particular classification adopted. This would restrict the significance of the propositions of economics to a degree that would make them useless".

[21]Marshall, pressured by a mainstream economics that was moving in a totally different direction, did not have the will, or the strength, to follow this approach to its ultimate

important consequences, in terms both of the division of scientific labour in the study of social phenomena[22] and the ongoing development of strictly economic thinking.

1.3. The District Concept in Marshall's Thinking: b) Humans as Variable Entities

Alfred Marshall's ability to see districts where everyone else saw mere industrial agglomerations, or, in other words, his willingness to apply Smith's notion of the division of labour to entities other than the world, a nation-state or a factory (for example, Smith's pin factory), is deeply rooted in the classifications of capitalism developed while he was still young.

While this is not the place to discuss this point in detail, a few brief remarks and references seem to be in order.

There are at least two core theoretical problems - we understood this only recently[23] - with the notion of district: the first, already mentioned, concerns economic nations and the second, perhaps more profound, is rooted in the studies of the human mind carried out by Marshall during his younger years. The "district anomaly", as I like to call the complex of ideas that gravitate around the notion of industrial district[24], contains a transparent reference to the social and dynamic concept wherein the process of growth of human intellectual abilities is the specific engine that drives the human aspects of natural history. Marshall believed that humans evolved mainly through work and, therefore, evolved differently in different organizational and social contexts[25]. His industrial atmosphere, for instance, is more than just the product of an uninhibited observation of the facts. It is also a symbol of Marshall's notion of natural sociability of man and of the historical roots of a social process that is simultaneously a process of production of commodities and of the associated social change. In the final analysis, this localized process is nothing more than a "fragment" of social life, taken at a given point in time and space and viewed from the standpoint of the need to meet other historically perceived human needs.

---

consequences. However, it: a) continued to have a "subterranean" influence on the way Marshall approached the study of economic phenomena and b) caused profound incompatibilities with mainstream economics.

[22]The division of the study of social phenomena into economics, sociology, anthropology, social psychology etc., which became established during the second half of the nineteenth century, would have been unthinkable.

[23]Especially after the studies by Raffaelli. See Raffaelli (1994).

[24]For a more general understanding of the anomalies of Marshall's thinking compared with mainstream economics, see my essay "Anomalie marshalliane", 2000a. For an English version, see: "Marshallian Anomalies", 2000, new series.
http://www.cce.unifi.it/dse/marshall/welcome.htm.

[25]There are numerous passages, especially in his earlier works, where Marshall clearly postulates this type of environmental and cultural conditioning. In particular, see the text of the lecture entitled "Some Features of American Industry" in Whitaker (1975), pp. 355-376.

As opposed to dogmatic economists, who on the one hand refused to give up the "wage fund" theory and on the other were blinded by the economies of scale and the factory system[26], Marshall showed with a concrete example (the district formula) that there was already, in practice at least, one necessary alternative or perhaps a complement to the esosomatic increase in the productivity[27] (more and better machines *per capita*) of individual workers, i.e., an "endosomatic" increase in productivity achieved by paying attention to values and knowledge that enhance the compositeness of the place with which the worker himself "identifies".

Another characteristic of Marshall's analysis that dates to his early work is a dialectic juxtaposition of routines and organizational innovations in the broad sense, duplicating a framework the roots of which can be traced to Marshall's early reflections on the workings of the human mind. Society, in a manner not significantly different from the human mind, alternates routine behaviour with innovative behaviour, which, if successful, becomes in turn routine, freeing new energies for innovation[28].

In this way, Marshall shifts the focus of economic analysis from the esosomatic means of production (equipment, facilities, canals, railroads, etc.), which incorporate routines and are acquirable and cumulable, to individual, endosomatic tools, such as the intellectual potential of the human mind, which are constantly open to innovation and are inseparable from their carrier, and to collective assets, such as social, local or sectorial capital, which by their very nature cannot be appropriated because they consist of delicate and shifting networks of interpersonal relationships[29]. This position inadvertently created a wide chasm between Marshall and mainstream economists, both classical and marginalists. It also caused Marshall's thinking to move further away from Marx's notion of the capitalist production system, which placed excessive emphasis on the separation between workers and the "material" means of production[30].

---

[26]Conservative and a few "progressive" economists bow reverently before the so called wages fund theory. Marx, who is not fooled by this trick, accepts acritically - as history later demonstrated - the claims of those who sang the praise of the factory system and the economies of scale. In my opinion, this view of Marx's, which anticipates Fordism, harbors the early roots of the failure of centralized economic planning.

[27]This distinction comes from the writings of Georgescu Roegen. See Georgescu Roegen (1966), pp. 98-99.

[28]See Raffaelli (1994). But also his forthcoming (Routledge 2002) book on Marshall.

[29]Marshall speaks of "social credit". See Marshall (1919), Chap. IX, 1.

[30]It is interesting how today's crisis of Fordist capitalism, which lent itself better to a Marxian analysis, gives new significance to the viewpoint proposed by Marshall in this area.

1.4. Semi-Automatic and Man-Made Districts

In his work, Marshall differentiates between two types of districts: one that represents an evolution of the "centres of specialized skills" from the pre-industrial age and another created intentionally and in an organized fashion as the result of the spillover of some manufacturing and craft activities from the industrial cities typical of English capitalism at the end of the nineteenth century.

In the first case, of which Sheffield, with its long iron-working tradition, is a perfect example, the key feature is the interdependence between technical-economic issues and the underlying sociocultural relationships. It is a long and complex process in which a steady flow of socioeconomic innovations leads to a process whereby the most efficient ones are consolidated into institutions[31].

The second case foreshadows, in embryonic form, the role of public policy in the formation of manufacturing districts. In a paper that became the manifesto of the *Society for Promoting Industrial Villages*, which was close to the *Garden City Movement,* of which, significantly, Marshall was a member, he supported "the formation of a colony in some place well beyond the range of London smoke (...). The work of several firms, not always in the same business, might in some cases be sent together. Gradually a prosperous industrial district would grow up, and then self-interest would induce employers to bring down their main workshops, and even to start factories in the colony (...). It is only the first step that costs; every succeeding step would be easier"[32].

In conclusion, Marshall's vision of the manufacturing district is not just the concentration of manufacturing activity in a given territory (the *clustering* or *networking* of businesses referred to by many economists today) that results from the "natural" development of market automatisms[33]. In Marshall's vision, economic and socio-cultural phenomena are fused into a single conceptual block, paving the way for the implementation of cogent public policies[34].

1.5. Paths of Industrial Development

To better understand the "district anomaly", we must consider briefly some of its implications in terms of the understanding of the concept of industrialization[35]. If the "district anomaly" had succeeded, industrial capitalism would have been

---

[31]Several fragments of a historical-conceptual reconstruction of the path leading from the pre-industrial centres of specialized skill to the industrial district are scattered among various chapters and appendices of Marshall (1919).

[32]See Pigou (1925), pp. 149-150.

[33]Marshall likes to speak instead of "semi-automatisms". E.g. Marshall (1919) p. 599.

[34]This clearly differentiates Marshall's districts from any involuntarily created locals systems, such as those obtained through an extension of Hayek's models. For example, Parri (1997), pp. 175-190.

[35]For a more detailed analysis, see Becattini (2000b), p. 23. An English version of this book is forthcoming.

understood as having not just one development path based on accumulation and technical progress, with firms of increasingly large average size combined in increasingly concentrated industries (as most of the economists of the time thought and almost all later economists would continue to believe for a long time), but of multiple paths (at least two), one of which would have produced territorial agglomerations of medium-size and small firms.

An early understanding of the existence of "multiple paths to industrialization", which, by crisscrossing and cross-pollinating each other, coexist over time, could have had far-reaching consequences. I will just mention a few of them: a) an understanding of the persistence and success of independent forms of employment and small businesses[36] in certain industries would have refuted the mechanistic extrapolation that industrial capitalism produces a social polarization between the two classes of owners and proletarians[37]; b) a denial of the view that capitalism leads to the creation of a few megalopolises would have helped focus territorial studies toward a plurality of alternative models of territorial agglomeration; c) the assumption that there is a plurality of interactive commonalities between manufacturing activity and everyday human life would have refuted the vision of a manufacturing system increasingly enclosed inside the factory and/or secluded in "industrial quarters" that exploits and upsets the balance of surrounding socio-cultural and natural resources[38]; and d) a refusal of the "easy" view, supported by socialist critics of capitalism, that the trend toward the technical, economic and financial concentration of the manufacturing sector paves the way toward a generalized shift from private to public property, would have spared humanity some painful disappointments[39].

The fundamental question raised by Marshall in his early writings about which basic unit is best suited for a correct economic analysis casts doubts on the acritical acceptance of *de facto* empirical compartmentalisations - such as the concept of nation-state at a given time in history - and paves the way for the

---

[36]This, which is a recurring theme in Marshall's work, found a non-receptive cultural environment and caused Marshall to be seen as a *laudator temporis acti*.

[37]Marshall does not accept the idea that the concentration of wealth and poverty at the two extremes of society is inevitable. He recognized that capitalism contains strong tendencies toward that outcome, but also recognized the existence of counterbalancing tendencies, especially in the economic domain, (for example, the tendency toward bureaucratization among large public and private entities is contrasted by an endless supply of new initiatives from small entities striving to succeed) but also in the moral and intellectual domains (the trend toward more civilized behaviour in all social strata).

[38]It is not by accident that Marshall always viewed as "problematic" the so-called *sweating system*, because, in his mind, this system was one of the manifestations of the embedment of industry into society and could also be a cause of growth. This position made him highly unpopular among left-wing politicians and union representatives. See, for example, Pigou (1925), p. 225.

[39]See Marshall (1919).

current phase of territorial fragmentation-recomposition (international integration and sub-national segmentation)[40].

The economic, cultural and political consequences of the "district anomaly", widely interpreted, had it been accepted, could have been significant. Unfortunately, in his later years, when Fordism was triumphant, Marshall, while remaining critical[41], flirted with the dominant theory and shifted his external/internal economies from the industrial "district", where they had been born, to the industrial "sector", where, as it is well known, they do not fit very well[42].

2 The Italian Industrial District

2.1. The Unnoticed Decline of Fordism

Toward the end of the 1960s, many years after Marshall's death in 1924, a few Italian economists detected some interesting developments. In particular: a) in certain regions of Italy (Tuscany, for example)[43] where large private and public-sector companies in capital-intensive and/or high-tech industries were showing clear signs of weakness, there was a "strange" bloom of small manufacturing businesses, which produced increases in local income, jobs and exports[44]; and b) the small companies included in these clusters were adequately equipped on a technical level - at least as well as their large competitors were - to do the things they were doing[45]. These two observations were "scandalous" for those who believed in the then dominant economic theories. As a result, for several years the scientific and political communities dismissed these businesses as precarious and

---

[40]Kindleberger writes: "social and political disintegration (...) illustrates a tendency towards smaller units. The reaction against the multinational corporation (...) is a response to intrusion by foreign elements and the fear that no-members of the group are making decisions affecting it". See Kindleberger (1989), p. 87.

[41]In line with his unpublished manuscript (*Ye Machine*), Marshall levels important, criticisms at the notion of scientific organization of labour, which was accepted acritically by most mainstream economists and was incorporated illogically and with well-known results in the socialist ideology of centralized economic planning. See Marshall (1919).

[42] I allude to the criticisms by Sraffa. A discussion of the reasons for this shift would be too complex to tackle here. In his first criticism of Marshall, Sraffa grasps perfectly well the alternative, but fails to explore its territorial ramifications. See Sraffa (1998), p. 346.

[43]See Becattini (1978), Sforzi (1994).

[44]The early classifications, still based on first impressions, identify a Third Italy (Bagnasco) and a North-East-Centre, NEC region (Fuà). However, economists sensed from the start that the reason for the efficiency of clusters of small businesses must be some factor that links the production system with its social and cultural hinterland context. On this topic, please see my work (1978), pp. 107-123.

[45]See Brusco (1989), pp. 59-154.

marginal and as the alleged harbingers of an inevitable deindustrialization, leading to the social and civil decay of those regions that were being squeezed between industrially advanced countries with their wealth of capital, manufacturing infrastructures and technical expertise, and underdeveloped countries with much lower labour costs.

Despite these views, the phenomena described above continued, developing a new and equally "strange" characteristic: the new companies were not being created inside the industrial cities and across the full range of industrial sectors. Instead, they were being established across a vast geographical territory - somewhere between the regions of traditional capitalistic development and the economically depressed areas of Southern Italy - where they were concentrated in relatively small areas that, in most cases, should not have been attractive (because insufficiently infrastructured or not easily accessible, for example) according to the theories of industrial localization prevalent at that time and in industries that most industrial economists considered "mature" and lacking in growth opportunities (textiles, garments, footwear and leather goods, wood furniture, etc.), using forms of business organization (small companies, primarily family owned) considered obsolete. What was the reason behind this vigorous resurgence of archaic and irrational forms of manufacturing in washed-up industries? Before attempting to give a direct answer to this question, I would like to make some parenthetical observations.

2.2. General Conditions for the Rise of Industrial Districts

Let us start with a general question: under what conditions can industrial districts flourish and when do these conditions occur? These conditions can be divided into two groups: "local supply" conditions and "general demand" conditions.

The local supply conditions include "the existence of countries that, throughout the classical period of industrialization and the First and Second World War, retained in some parts of their territory: a) sufficient cultural complexity, made up of values, knowledge, institutions and behaviours (...) that elsewhere had been pushed aside by a generally industrial and massifying culture, and a manufacturing system that included factories, craft or artisan workshops, home workers and family production units; and b) a credit system (...) ready to finance small but promising businesses and help people use their increasing amount of spare time to produce goods that could be sold in the marketplace"[46].

The general demand conditions include the progress, in many countries, of large segments of the middle class beyond the normal *standard of comfort* found in each of these countries[47]. Progress beyond this standard creates the conditions for the

---

[46]See Becattini (2000b), p. 23.

[47]For the concept of *standard of comfort*, as opposed to that of standard of living, one can refer to the author who first proposed this distinction, i.e., Alfred Marshall. See Marshall (1961), Book 6.

emergence of new sets of needs with a high social and "qualitative" content that produce a highly variable demand for differentiated and personalized goods.

The combined impact of this twin set of conditions undermined the advantages of large factories, consisting primarily of internal economies of scale in the production of standardized goods and continuous production processes (i.e., processes that could not be broken up into phases), favoring instead smaller factories with stronger ties to the surrounding community, where people worked separately in a limited number of phases of a common (to the industrial district) complex process. When there is an adequate "industrial atmosphere", that includes both technical knowledge and "commercial morality", the shift from an environment characterized mainly by hierarchical relationships between departments - as they exist within a large enterprise - to one dominated by transactions between independent firms (where the existing inequalities tend to diminish and eventually, for the successful ones, to disappear) creates a positive trust differential[48] between the participants that lowers the overall cost of the cluster's production[49].

Whenever these two sets of conditions come into existence in a situation dominated by manufacturing in its industrial or preindustrial state, the district form of production becomes relatively stronger[50].

In a nutshell, my conclusion is this: Italy, during its "economic miracle" phase, when it was part of a rapidly expanding Western world, saw the development of conditions similar to those experienced by many British cities and towns during the first great expansion spawned by the industrial revolution. In both cases, a long and socially unequal income expansion concentrated wealth in the pockets of a large segment of the middle class, who, having achieved the current *standard of comfort*[51], sought increasingly differentiated and personalized goods and services as a means of experiencing "new sensations" and achieving social prestige. Endless series of standardized objects became increasingly less attractive to this new, affluent class, whose demand was increasingly fragmented and variable. These developments on the "real" side marked the birth, or rebirth, of a "new" theory of industrial districts.

---

[48]Something of the same kind of the already mentioned marshallian "social credit".

[49]Marshall pointed out that the "scholastic" concept of production costs for a single product was not realistic. A study of the production process as part of the industrialization process inevitably leads to the conclusion that an analysis of intermediate and long-term development must take into account the web of cost conjunctions and connections. See Marshall (1919), Book II, Chap. 1.

[50]Naturally, this can be understood only if there is an economic theory that identifies this phenomenon. Districts can be flourishing in the real world, but the prevailing theory can fail to recognize them. It is here, at the dialectic intersection between phenomena and theory, that the decisive importance of Marshall's break with classical orthodoxy and neoclassical economic theory becomes fully apparent.

[51]In Victorian times it meant: home with garden, nice living room possibly with Morris's wall paper, sons at the public school, etc.; in post-war Italy: town flat, home appliances, car, vacations, etc.

2.3. The Italian Way to Industrial Districts

An explanation of these phenomena, which came naturally to some Italian researchers as a way to solve the *impasse* they were facing when applying the then dominant theory, was the existence of a factor that increased productivity (and innovation potential) and which had more to do with physical contiguity than with investments in means of production. This factor of higher productivity - it should be noted - has nothing to do with the exploitation of the inferior position of employees, as often occurs in areas where many small firms gravitate around just one or few large companies[52].

A factor of this type already existed in economic theory: economies that were external to any one firm, but internal to an industrial sector or territorial group of firms. Driven from the context of the industrial sector by Sraffa's criticism[53] and little loved by mainstream economists because they provided a theoretical justification for government intervention in economic matters[54], external economies, in a backward leap to Marshall's thinking in 1879, reappeared in Italian economic thinking in a territorial context[55].

However, this was not the only path[56] followed by Italian scholars to develop the concept of district. Sebastiano Brusco, for example, arrived at the notion of industrial district from a theoretical basis that has nothing to do with Marshall. By

---

[52]F. Pyke and W. Sengenberger called this the "high road" to economic development, as opposed to the "low road," which is based on exploitation. See Pyke and Sengenberger (1992). In view of the decisive role it played in going to the roots of the industrial district issue and publicizing it internationally, I must also mention the International Labour Organisation (ILO) research program, *The New Industrial Organization*, which was launched in 1986 and coordinated by W. Sengenberger, with the initial support of M. J. Piore. It produced several books: from: Pyke, Becattini and Sengenberger (1990), Cossentino, Pyke and Sengenberger (1996).

[53]I am referring to the dispute about costs originated from a famous Clapham's article. See Clapham (1922a). See also Clapham (1922b).

[54]I am referring to the area of research stemming from Pigou (1920).

[55]The first work where, to my knowledge, internal and external economies started again to perform the function envisioned by Marshall in his early works is: Istituto Regionale Programmazione Economica della Toscana - IRPET (1969). Republished in: Becattini (1999), pp. 116-146.

[56]I am afraid I may have overlooked some of the other paths that led to the notion of industrial clusters, and I apologize to the colleagues whose names I did not mention. I will provide a more complete analysis on another occasion. However, I would like to mention several books on this subject that will give the reader a fuller picture of this issue: Garofoli (1983); Becattini (1987); Brusco (1989); Gobbo (1989); Pyke, Becattini, and Sengenberger (1991); Nuti (1992), 2 voll; Onida, Viesti and Falzoni (1992); Bellandi and Russo (1994); Dei Ottati (1995); Brusco and Paba (1997); Varaldo and Ferrucci (1997); Belfanti and Maccabelli (1997); Corò and Rullani (1998); Becchetti, Belussi and Gottardi (2000); Tattara (2001); Becattini, Bellandi, Dei Ottati and Sforzi (2001), which includes a rich bibliography. The literature is much vaster but I think that the works mentioned above point the way to most of the Italian and foreign work on industrial districts.

applying his analysis to the industrial phenomena of the Emilia Romagna Region, Brusco developed the concept of industrial district without the aid of external-internal economies and despite what I like to call the handicap of his Sraffian background[57].

2.4. Made in Italy Products

At the beginning of the 1990s, a major research project carried out by Michael Porter and a group of associates[58] opened a new approach to understanding the issues of manufacturing systems. Drawing on an in-depth analysis of world commerce, Porter discovered, (especially in Italy but in other countries as well), the existence of territorial clusters of small businesses that had gained significant competitive advantages in relatively important niches in the world trade for manufactured goods (textiles, garments, footwear, leather goods, furniture, decorative tiles, etc.). These advantages were achieved despite the initial technological superiority of industrial giants and the lower relative labour costs of other competitors. How could this development be explained? In a series of papers[59], Porter and his collaborators refined the concept of business cluster, giving it a form that closely resembled that of Marshall's industrial district. The prestige of the center of learning where these analyses were carried out - the Harvard Business School - and the fact that Porter's conclusions were in tune with the views prevailing at the time - i.e., that business competitiveness is at the centre of everything - paved the way for acceptance of the concept of business clusters in mainstream economics and forced average economists to take a new look at the notion of industrial districts, which until then they had found methodologically troubling.

In Italy, this convergence of conclusions - the efficiency of local systems claimed by the supporters of industrial districts and the competitive advantages pointed out by Porter - led to the development of a new area of research that focused on those products that are emblematic of Italian exports. A number of Italian scholars (Marco Fortis, Giuliano Conti, Stefano Menghinello and others)[60] sought to verify, expand and consolidate the theme, which Porter had touched on only briefly, of the relationship between industrial districts and the success of Italian exports. Fortis's work helped bring into focus certain industries that

[57]Why do I use the word "handicap"? Because the entire edifice of Sraffa's theories is designed to prevent the entry of the entrepreneurial contribution in the core of economic theory. It is truly remarkable how Brusco, starting from these premises, developed the concept of industrial clusters, which are genuine incubators of the entrepreneurial spirit. Brusco's approach to the problem is described in some detail in Brusco (1989). Peter Groenewegen remindered to me that Maurice Dobb read a paper, in 1923, on "The entrepreneur myth", to the economic Section of the Cambridge Club *The Heretics*.

[58]See Porter (1989).

[59]See Porter (1999).

[60]See e.g. Bagella and Becchetti (2000).

manufacture typical Italian exports, concluding that they represent a true "system of manufacturing sectors and sub-sectors" that are interdependent and specialize in the production of household and personal goods[61]. Conti and Menghinello studied the relationship between export performance and territorial concentration in detail, consolidating and expanding Porter's conclusions[62]. Becattini and Menghinello sought to identify, within the general universe of "made in Italy" products classified by industrial sector), a core group of products manufactured in industrial districts that represent the cutting edge of Italian exports[63].

2.5. The New Panorama of Italian Research on Industrial Districts

ISTAT, the Italian statistical institute, with a vital contribution from Fabio Sforzi, divided Italy into local systems[64] that included 199 industrial districts[65]. This permitted a comparison of the many different characteristics that distinguish district areas from non-district areas. The most important product of the studies carried out in this area was an econometric study coordinated by Luigi Federico Signorini for the Research Department of the Bank of Italy that confirmed the existence of many of the relationships that are typical of the "theory of industrial districts"[66].

Two institutions also came into existence at this time: a) the *Club dei Distretti* (1996), without permanent headquarters, which sought to represent the interests of industrial districts *vis-à-vis* public authorities at all levels, from local municipal governments to the European Union and the WTO, and b) an Industrial Districts Observatory, which was established in Milan (1997) through a collaborative effort of Montedison Economic Research and Innovation Department and the Research Centre in Economic Analysis, International Economics and Economic Development (CRANEC) of Università Cattolica, joined in 1999 by Fondazione Comunità e Innovazione[67].

---

61 See Fortis's several books, but particularly, 1996, 1998.
62 See Conti and Menghinello (1996), pp. 286-303.
63 See Becattini and Menghinello (1998), vol. V, n. 9.
64 See Sforzi (1989), ISTAT (1997).
65 See Istat (1996), pp. 262-271.
66 See: Signorini (2000).
67 The work of two Rome-based research institutions, Centro Studi Investimenti Sociali (CENSIS) and Istituto Tagliacarne of the UICCIA, on districts and local issues is also worth mentioning.

2.6. Principal Characteristics of Italian Industrial Districts

The recognition of the heuristic usefulness of the concept of industrial district and, consequently, of the empirical existence of groups of phenomena that could be effectively analyzed with such a tool (this happened more or less at the beginning of the 1990s) was followed by a sort of terminological explosion: from the notion of "industrial district" to system area, to local production system, to *milieu innovateur*, to cluster and so on. This is not the place to review such immense and varied literature, nor would it be useful to lump often heterogeneous concepts. I shall just explain the path that I followed with some of my associates.

Instead of starting with a model developed by finding similarities with (and differences from) other well-known types of manufacturing system organizations (for example, industrial poles), which could have been restrictive and distortive of new emerging phenomena, our group began with the assumption that, since industrial districts follow their own peculiar productive and reproductive logic, an understanding of their development could come only from an in-depth study of an industrial district in action over a sufficiently extended period of time. Having determined that Prato was an industrial district - perhaps the archetype of all industrial districts - we concluded that a study of the Prato district would yield significant insights into the anatomy and physiology of all of the industrial districts[68].

What did we find in this industrial cluster?

First of all, a population of families and businesses interacting with each other in various ways within a well-defined territory. Looking more closely, we discovered that the businesses could be broken down into different populations working on different production phases of production (spinning, weaving, dyeing, finishing, etc.)[69] organized in flexible teams[70] normally headed by a finished goods manufacturer (e.g. wool fabric mill, or *impannatore*) that interacted with the external market. The labour and consumption behaviour of the families was divided, more or less, along lines that reflected their position within the local manufacturing process (industrial entrepreneurs and craftsmen, regular employees, "textile sharecroppers" tradesmen, etc.).

More importantly, we were surprised to discover that the prices of semifinished goods were determined by a system that on the one hand pegged prices to the same (or practically the same) goods found elsewhere in the world and on the other set prices that were strictly related to the standard of living and social status of the main categories of families living in the district. Because of their impact on the income of crucial socio-economic groups within the district, the relative prices of processing tasks played a major role, together with wages levels and the profits earned by finished-good manufacturers, in determining the system's compatibilities.

---

[68]See Dei Ottati (1995); Bellandi and Trigilia (1991), n. 70, pp. 121-152; Becattini (2000b).

[69]For an early assessment of production-phase businesses and their development after the Second World War, see: Lazzeretti and Storai (1999).

[70]See Ciappei and Mazzetti (1996).

The upshot of all this is that there must be a close, generally reproductive, overlap between a block of economic and production relationships, occurring within a spatially defined area, and another block, also spatially defined, of socio-cultural relationships[71].

The next step, which is perhaps not yet fully understood, is the definition of the synergies and antagonisms between the two engines of the Italian economy: large and medium-size businesses in leadership positions and industrial districts[72].

The research coordinated by Signorini that I mentioned earlier confirmed independently and with completely different methods the concrete existence and relevance of the relationships discussed above. It also went even further, agreeing with a conclusion of the staunchest supporters of the district theory, who say that the unit of research that is best suited to industrial economics is not the individual firm, but a group of firms, possibly located within the same territory[73], a statement that clearly entails major theoretical implications.

3 Work in Progress

3.1. From the Study of the District as Such to the Study of the Forms of the Industrialization Process

This phase in the analysis of industrial districts opens at least two alternatives: we either carry out an in-depth study of industrial districts as the basic unit of research, with the goal of isolating its component parts (for example, manufacturing system, institutional network, system of values, etc.) and seeing how they are connected within the "local system", or move to the next level of research, seeking those processes the combined operation of which determines whether a certain territorial distribution of a manufacturing system is or is not approaching the district form. I started by pursuing the first avenue[74], but, seeing that it led to a sort of frenetic "district hunt" in space and time, with a plenty of "bastard" districts and an excess of taxonomic quibbling[75], I decided to attempt dissecting the district phenomenon into several distinct processes that were convergent at birth, parallel during their lifetime and divergent when the district dissolved.

I identified the following processes: a) a gradual, targeted and self-contained subdivision of certain production processes and of other complementary and ancillary processes; b) the formation and multiplication over time of dynamic links between "complexes of specialized production skills" developed in specific

[71]These conclusions are articulated in: Becattini (1990).

[72]This topic started to be investigated by Bellandi (2001).

[73]See Signorini (2000), p. XXII.

[74]This phase of the research work conducted by our group is summarized in Pyke, Becattini and Sengenberger (1990).

[75]See Becattini (1998).

territorial settings and "nuclei of needs" that emerge from the general panorama of needs[76]; c) the sedimentation of social practices into formal and informal, tangible and immaterial institutions that at the same time protect both competitiveness and the social and naturalistic reproducibility of the local system; d) the dynamic integration of contextual production expertise[77], which is often implied, with codified scientific/technical expertise into the manufacturing process; e) the development of figures and institutions (versatile integrators) that serve as mediators between the contrasting needs of specialization and versatility; f) the formation, consolidation, decay and dissolution of the "sense of belonging" among the individual agents of social production; and, finally, g) constant support for local social and professional mobility.

3.2. Inside Change: The Driving Force

The driving force in the development of district-type organizations is the trend, triggered at any given point in time, towards a growing, focused and self-contained subdivision[78] of a production process that is "typical" of a given cluster of production processes into separate phases that are generally performed in separate facilities by different enterprises. It is the Smithian-Marshallian-Youngian principle[79] of the gradual division of labour as a wheel that, starting from within the individual[80], moves the entire social world. However, one thing should be quite clear: the notion of the division of labour is here not referred to a single company, as in Smith's example of the pin factory, nor to the market in its abstract totality, as in Young's model, but to an intermediate, meso-economic entity that perhaps lacks a legally recognized status and aggregates and disaggregates in its different manifestations - territorial and otherwise - in response to overall long run changes in socio-economic relationships at both the local and global level[81].

The continuous and growing "local articulation" of the manufacturing process that is "typical" of an industrial district produces two types of effects. First of all, it is a natural vehicle for increasing the productivity of labour within the district,

---

[76] What I am alluding to is an attempt, perhaps still relatively raw, to formulate a theory that could link in organic dualities nuclei of needs that are perceived as joined (for example, clothing or, as part of it, footwear) but are constantly modified by disaggregations and reaggregations of perceived needs on a global scale, with territory-based nuclei of manufacturing expertise that have roots in the history and geography of the respective territories.

[77] For a wider development of the theme, see Becattini and Rullani (1993), and Nonaka and Takeuchi (1997).

[78] The most difficult, and as yet unsolved, problem is how to study the internal and external forces that make self-containment possible.

[79] The key, and perhaps sole, reference point is found in Young (1928).

[80] Raffaelli's work, quoted earlier, is essential to understanding this connection.

[81] See Bellandi (2001), pp. 189-210. But also, for different aspects, Solinas (1993).

which is obtained by an intensive use of the existing know-how and the ability to acquire new know-how. However, this characteristic is also the source of the district's biggest problem: the need to find increasingly large outlets for its output. Like capital, the industrial district either increases or perishes.

The increasing "articulation" of the main process is also important because it permits an inter-penetration of the production process and of the daily life of the reference community, which constitutes one of the most distinctive peculiarities of the "district-type method of production". It is because of this peculiarity that the "surplus" time, know-how and work energy that the local production process generates "as it runs" find within the same process new economic spaces and relative competition gaps, which they can fill. By constantly evolving internally, the production process of an expanding district creates the means of monetizing many kinds of resources by transforming them into goods and services that can be sold at a profit. At the same time, it embeds in the individual and collective know-how, which can be put to new uses, the manufacturing and marketing experiences that proved to be successful[82].

Viewed from this perspective, the increasing subdivision into phases, the induction of collateral processes, the formation of open (but with a district bias) teams of specialized businesses, the birth of markets within the cluster and of an embryonic "local price system" are different aspects of the same process: the gradual formation - concurrently with the unfolding of a "typical" multiphase production process that involves a large number of autonomous, spatially contiguous agents - of an organic plurality of local institutions centered on a grid of local markets for semifinished goods (e.g., yarns, fabrics, etc.), new and used equipment, and manufacturing processes (e.g., dyeing, finishing, etc.).

However, it would be impossible to understand the formation of this plexus of internal markets, pivoting around a unique "district market" for human resources, without also taking into account the coalescing around these markets of formal (e.g., craftsmen's associations, unions, interest groups, technical schools, institutions, the ultimate function of which is building in every potential specialized contributor the confidence that he will be able to sell his specialty easily and at a remunerative price. This local confidence-building process is just the other side of the formation of a web of local and external markets.

3.3. Inside Change: Social Reproduction and District Governance

There is another dimension to the district mechanism that is often overlooked: with the end of Fordism, efficiency and the ability to innovate and, consequently, business competitiveness became increasingly dependent on the productive performance, commitment and flexibility of the employees of those companies that were part of the "team". For example, if a subcontractor feels bullied, he will never perform beyond his contract minimum, and may even cheat a little, feeling

[82] See, e.g. Bellandi (1989), Dei Ottati (1994), Sforzi (2000).

free to find new customers for whom to work whenever conditions allow it and make it convenient.

In other words, an entrepreneur cannot ask his partners (workers and subcontractors) to exert themselves excessively to help grow, or rescue, his business when it is clear that at the first opportunity he would fire an employee and/or change a supplier. In an "ideal-type district", the situation is different because: a) the production cost of a "representative team" of firms depends "intelligibly" on the overall functioning of the local society, which is affected to a significant degree by the decisions (zoning, waste treatment facilities, highways, etc.) of democratically elected officials; and b) both the "representative" worker and subcontractor have a good understanding of the overall economic conjuncture of the "district system". In other words, district workers and subcontractors are well aware that they are not powerless *vis-à-vis* their terminal businesses, whose quest for cheaper labour is limited, in the environment outside the district, only by competition. These conditions provide a basis for a genuine and realistic negotiating process[83]. It is as if inside the heads of the representatives of the different parties there was a common simplified model of how the cluster operates that defines, for every degree of integration, the allowed margin of fluctuation for the average wage and price of each phase operation. This margin is relatively narrow and each of the parties will be careful not to destroy, together with the opposing party, their golden opportunity (provided that this is their perception), i.e., the opportunity to live in a "district-type local system" where a widespread competitive attitude is mitigated by the presence of relationships based on interfirm cooperation, good information and reciprocal dependence among operators[84].

3.4. Summing It Up

Going back to our starting point, we can say that Marshall, in his early works, was right to question the compatibility of increasing returns, which in turn engender cumulative processes, with an economic theory centered on a model of static equilibrium. We can also say that Marshall was perhaps wrong when, later in life, he failed to bring his basic intuitions to their ultimate conclusions[85].

The overall and, obviously, remote goal of the approach to industrial phenomena here suggested is to look for self-reproducing organizational forms of the societal production process that are defined, in each case, by a combination of processes as specified in 3.1. The industrial district is the self-reproducing form that Italian economic reality has placed before our eyes, but nothing prevents us from thinking that others may exist.

---

[83]Gabi Dei Ottati started to reseatch this issue in Dei Ottati (2001).

[84]This reminds much some expressions of Kindleberger's analysis.

[85]The position of Marshall on the issue in 1906 can be seen in: Raffaelli (1996) n. 1. I examine more extensively the issue in a forthcoming paper: Becattini (2001).

In other words, the ultimate unit of analysis for the study of change is not necessarily the industrial district, but an entity that changes with the overall change, yet, up to a certain point, remains the same. If we pursue this path of research to its conclusion, we will end up, I guess, not with a mosaic of markets for homogeneous and clearly defined commodities, searching for irrelevant, or impossible, equilibria, between supply and demand which change of "meaning" with the qualitative change of the commodity, but with areas of convergence (temporary but relatively stable) between economic and socio-cultural processes. These are the meso-socioeconomic units that we need.

References

Bagella, M. and Becchetti, L. (2000), *The Competitive Advantage of Italian District - Theoretical and Empirical Analyses*, Berlin, Physica Verlag.

Becattini, G. (1978), "The Development of Light Industry in Tuscany: an Interpretation", *Economic Notes*, n. 2-3, pp. 107-123.

Becattini, G. (1987), *Mercato e forze locali - il distretto industriale*, Bologna, Il Mulino.

Becattini, G. (1990), "The Marshallian Industrial District as a Socio-Economic Concept", in: *Industrial Districts and Inter-firm Cooperation in Italy*, Pyke, F., Becattini, G. and Sengenberger, W. (eds), Geneva, IILS.

Becattini, G. (1994), "Nazione economica e nazione politica nel pensiero di Alfred Marshall", in: Roggi, P. (ed.), *Quale mercato per quale Europa*, Milano, Franco Angeli.

Becattini, G. (1998), "Dal distretto industriale al processo di distrettualizzazione", in: *Le vie dell'industrializzazione europea*, Fontana, G. L. (ed.), Bologna, il Mulino.

Becattini, G. (1999), *L'industrializzazione leggera della Toscana. Ricerca sul campo e confronto delle idee*, Milano, Franco Angeli.

Becattini, G. (2000a), "Anomalie marshalliane", *Rivista Italiana degli Economisti*, No. 1. English translation: "Marshallian Anomalies", *Marshall Studies Bulletin*, 2000, n. 7, new series, http://www.cce.unifi.it/dse/marshall/welcome.htm.

Becattini, G. (2000b), *Il bruco e la farfalla. Prato nel mondo che cambia (1954-1993)*, Firenze, Le Monnier and The City of Prato.

Becattini, G. (2001), "Alfred Marshall fra analisi e storia economica", *Rivista di Storia Economica*, Dicembre.

Becattini, G. and Menghinello, S. (1998), "Contributo e ruolo del made in Italy "distrettuale" nelle esportazioni italiane", *Sviluppo locale*, vol. V, n. 9.

Becattini, G. and Rullani, E. (1993), "Sviluppo locale e mercato globale", *Economia e Politica industriale.*

Becattini, G., Bellandi, M., Dei Ottati, G. and Sforzi, F. (2001), *Il caleidoscopio dello sviluppo locale. Trasformazioni economiche nell'Italia contemporanea,* Torino, Rosenberg & Sellier.

Belfanti, C. M. and Maccabelli, T. (1999), *Un Paradigma per i distretti industriali*, Brescia, Grafo (ed.).

Bellandi, M. (1982), "Il distretto industriale in A. Marshall", *L'industria*, n. 3.

Bellandi M. (1989), "Capacità innovativa diffusa e sistemi locali di imprese", in: Becattini, G. (ed.), *Modelli locali di sviluppo*, Bologna, Il Mulino, pp. 149-172.

Bellandi, M. and Russo, M. (1994), *Distretti industriali e cambiamento economico locale*, Torino, Rosenberg and Sellier.

Bellandi, M. and Trigilia, C. (1991), "Come cambia un distretto industriale. Strategie di riaggiustamento e tecnologie informatiche nell'industria tessile pratese", *Economia e politica industriale*, n.70, pp. 121-152.
Bellandi, M. (2001), "Local Development and Embedded Large Firms", *Entrepreneurship & Regional Development*, 13.
Belussi, F. and Gottardi, G. (2000), *Evolutionary Patterns of Local Industrial Systems. Towards a Cognitive Approach to the Industrial District*, Ashgate, Aldershot.
Belussi, F., Gottardi, G. and Rullani, E. (eds) (2002, forthcoming), *The Net-Evolution of Local Systems. Knowledge Creation, Collective Learning and Variety of Institutional Arrangements*, Kluwer, Dordrecht.
Brusco, S. (1986), "Small Firms and Industrial Districts: The Experience of Italy", *Economia Internazionale*, Vol. XXXIX, n. 2-3-4.
Brusco, S. (1989), *Piccole imprese e distretti industriali*, Torino, Rosenberg and Sellier.
Brusco, S. and Paba, S. (1997), "Per una storia dei distretti industriali italiani dal secondo dopoguerra agli anni novanta", in: *Storia del capitalismo italiano dal dopoguerra ad oggi*, Barca, F. (ed.), Roma, Donzelli.
Cairnes, J. E. (1874), *Some Leading Principles of Political Economy*, London.
Ciappei C. and Mazzetti G. (1996), *Lo sviluppo produttivo a rete nel distretto tessile pratese*, Firenze, Regione Toscana.
Clapham, J. H.(1922a), "Of Empty Economic Boxes", *The Economic Journal*, sept.
Clapham, J. H. (1922b), "The Economic Boxes: A Rejoinder", *The Economic Journal*, dec.
Cliffe Leslie, T. (1888), *Essays in Political Economy*, London, Longmans, II ed.
Conti, G. and Menghinello, S. (1996), "Territorio e competitività: l'importanza dei sistemi locali per le esportazioni italiane di manufatti. Un'analisi per provincie (1985-94)", *Rapporto sul Commercio Estero, 1995*, Roma, ICE, pp. 286-303.
Cooke Taylor, W. (1841), *Notes of a Tour in the Manufacturing Districts of Lancashire*, London, Duncan and Malcolm.
Corò, G. and Rullani, E. (eds) (1998), Percorsi locali d'internazionalizzazione. Competenze e auto-organizzazione nei distretti industriali del Nord-Est, Milano, Franco Angeli.
Cossentino, F., Pyke, F. and Sengenberger, W. (1996), *Local and Global Response to Global Pressure: the Case of Italy and Its Industrial Districts*, Geneva, IILS.
Dardi, M. (1984), *Il giovane Marshall. Accumulazione e mercato*, Bologna, Il Mulino.
Dei Ottati, G. (1994), "Trust, Interlinking Transactions and Credit in the Industrial District", Cambridge Journal of Economics, Vol. 18, n.6, dec., pp. 529-546.
Dei Ottati, G. (1995), *Tra mercato e comunità: aspetti concettuali e ricerche empiriche sul distretto industriale*, Milano, Franco Angeli.
Dei Ottati, G. (2001), "Concertazione e sviluppo nei distretti industriali", in: *Evoluzione del lavoro, crisi del sindacato e sviluppo del paese*, Ninni, A., Silva, F. and Vaccà, S. (eds), Milano, Franco Angeli, pp. 163-195.
Dobb, M. (1923), *The entrepreneur myt*, paper read in Cambridge Club The Heretics.
Fawcett, H. (1863), *Manual of Political Economy*, Cambridge, Macmillan.
Fortis, M. (1996), *Crescita economica e specializzazioni produttive. Sistemi locali e imprese del "made in Italy"*, Milano, Vita e Pensiero.
Fortis, M. (1998), *Il "made in Italy" - Quando stile e creatività non sono solo moda*, Bologna, Il Mulino.
Garofoli, G. (1983), *Industrializzazione diffusa in Lombardia: sviluppo territoriale e sistemi produttivi locali*, 1st ed., Milano, Franco Angeli.
Georgescu Roegen, N. (1966), *Analytical Economics: Issues and Problems*, Cambridge, Mass, Harvard University Press.
Gobbo, F. (1989), *Distretti e sistemi produttivi alla soglia degli anni 90*, Milano, Franco Angeli.

Groenewegen, P.(1995), *A Soaring Eagle. Alfred Marshall, 1842-1924*, Brookfield, Edward Elgar.
Hearn, W. E. (1863), *Plutology, or the Theory of the Efforts to Satisfy Human Wants*, Melbourne, G. Robertson and Co.
ILO (1985), *The New Industrial Organization*, coordinated by W. Sengenberger, with the initial support of M. J. Piore.
IRPET (1969), "Lo sviluppo economico della Toscana: un'ipotesi di lavoro", *Il Ponte*, No. 11.
ISTAT (1996), *Rapporto Annuale. La situazione del paese nel 1995*, Roma, ISTAT, pp. 262-271.
ISTAT (1997), *I sistemi locali del lavoro 1991*, Roma, ISTAT, Argomenti n. 10.
Jevons, S. (1871) *The Theory of Political Economy*, London, Macmillan.
Kindleberger, C. P. (1989), *Economic Laws and Economic History*, Cambridge, Cambridge University Press.
Lange, O. (1952), *Price Flexibility and Employment*, Chicago, Cowles Commission.
Lazzeretti, L. and Storai, D. (1999), *Il distretto come comunità di popolazioni organizzative. Il caso Prato*, Prato, Quaderni dell'IRIS, n. 6.
Le Play *Les Ouvriers Européens*.
Loasby, B. J. (1989), *The Mind and the Method of the Economist*, Aldershot, Edward Elgar.
Loasby, B. J. (1998), "Industrial Districts as Knowledge Communities", in: *Industry, Space and Competition. The Contributions of Economists of the Past*, Bellet, M. and l'Harmet, C. (eds), Cheltentham, Edward Elgar.
Marshall, A. (1919), *Industry and Trade, A study of industrial technique and business organization; and of their influences on the conditions of various classes and nations*, London, Macmillan, First Edition.
Marshall, A. (1961), *Principles of Economics*, Variorum edition overseen by C. Guillebaud, London, Macmillan.
Marshall, A. and Marshall, M. P. (1881), *The Economics of Industry*, London, Macmillan, 2nd ed.
Marshall, A. *Ye Machine*, early manuscript published and edited in Raffaelli, T. 1994.
Mill, J. S. (1848), *The Principles of Political Economy*, 1st ed.
Nonaka, I. and Takeuchi, H. (1997), *The Knowledge Creating Company*, Milano, Guerini e Associati.
Nuti, F. (1992), *I distretti dell'industria manifatturiera in Italia*, 2 voll., Milano, Franco Angeli.
Onida, F., Viesti, G. and Falzoni, A. M. (1992), *I distretti industriali, crisi o evoluzione?*, Milano, Egea.
Parri, L. (1997), "Risultati di azione umana ma non di progetto umano. I distretti industriali per Hayek e la Scuola austriaca", in: *Un Paradigma per i distretti industriali*, Belfanti, C. M. and Maccabelli, T. (eds), Brescia, Grafo, pp. 175-190.
Pesciarelli, E. (1999), "W.E. Hearn on the industrial organization of society", in: *Industry, Space and Competition*, Bellet, M. and l'Harmet, C. (eds), Cheltentham, Edward Elgar.
Pigou, A. C. (1925), "Where to House the London Poor", in: *Memorials of Alfred Marshall*, Macmillan, London, pp. 149-150.
Pigou, A. C. (1925), *Memorials of Alfred Marshall*, London, Macmillan.
Pigou, A. C. (1914), *Wealth and Welfare*.
Porter, M. (1989), *The Competitive Advantage of Nations*, New York, Macmillan.
Porter, M. (1999), *On Competition*, Boston, Harvard Business School Press.
Pyke, F. and Sengenberger, W. (1992), *Industrial Districts and Local Economic Regeneration*, Geneve, IILS.

Pyke, F., Becattini, G. and Sengenberger, W. (1990), *Industrial Districts and Inter-firm Cooperation in Italy*, Geneva, IILS.
Pyke, F., Becattini, G. and Sengenberger, W. (1991), *Distretti industriali e cooperazione fra imprese in Italia*, Firenze, Banca Toscana.
Raffaelli, T. (1994), "Alfred Marshall's Early Philosophical Writings", in: *Research in the History of Economic Thought and Methodology*, Samuels, W. (ed.) London, JAI Press.
Raffaelli, T. (1994), "Alfred Marshall's Early Philosophical Writings", in: *Research in the History of Economic Thought and Methodology*, Samuels, W. (ed.), London, JAI Press. forthcoming (Routledge, 2002) book on Marshall
Raffaelli, T. (1996), "La formazione di Keynes nel laboratorio marshalliano", *Rivista Italiana degli Economisti*, n. 1.
Raffaelli, T. (forthcoming), Il concetto di distretto industriale nel sistema teorico di Alfred Marshall, *Sviluppo Locale*.
Sargant, W. L. (1857), *Economy of the Labouring Classes*, London, Simpkin, Marshall and Co.
Sforzi, F.(1989), "The Geography of Industrial Districts in Italy", in: *Small Firms and Industrial Districts in Italy*, Goodman, E.and Bamford, J. with Saynor, P. (eds), London, Routledge, pp. 153-173.
Sforzi, F.(1994), "The Tuscan Model: an Interpretation in Light of Recent Trends", in: *Regional Development in a Modern European Economy: The Case of Tuscany*, Leonardi, R and Nanetti, R. Y. (eds), London, Pinter.
Sforzi, F. (2000), "Local Development in the Experience of Italian Industrial Districts", in: *Geographies of Diversity. Italian Perspectives,* Cori, B. et al. (eds), Roma, CNR, pp. 133-156.
Signorini, L. F. (2000), *Lo sviluppo locale. Un'indagine della Banca d'Italia sui distretti industriali*, Roma, Meridiana Libri.
Solinas, G. (1993), "Competenze, grandi imprese e distretti industriali. Il caso Magneti Marelli", *Rivista di Storia Economica*, n. 1, pp. 79-111.
Sraffa, P. (1998), "On the Relations Between Cost and Quantity Produced", in: *Italian Economic Papers*, Pasinetti, L. L. (ed.), vol. III, Bologna, Il Mulino.
Steedman, J. (1986), "Trade Interest versus Class Interest", *Economia Politica*, August.
Tattara, G. (ed.) (2001), *Il piccolo che nasce dal grande. Le molteplici facce dei distretti industriali veneti*, Milano, Franco Angeli.
Varaldo, R. and Ferrucci, L. (eds) (1997), *Il distretto industriale tra logiche di impresa e logiche di sistema*, Milano, Franco Angeli.
Whitaker, J. (1975), "Some Features of American Industry" in: *Early Economic Writings of Alfred Marshall,* London and New York, Macmillan, 2nd vol., pp. 355-376
Whitaker, J. (1975), *Early Ecomonic Writings of Alfred Marshall*, London, Macmillan, 2 vols.
Young, A. (1928), "Increasing Returns and Economic Progress", *Economic Journal*, pp.527-542.

Diversity, Not Specialization: The Ties That Bind the (New) Industrial District

Charles F. Sabel

Abstract. Whatever they are exactly, industrial districts are also a worldly success and a conceptual innovation. In Italy, the map of district agglomerations of small and medium-sized firms specializing in particular branches of light industry is no longer limited to the triangle marked by Venice, Florence, and Ancona in the Adriatic Marches. New agglomerations are spreading, as though by contagion, down the shores of the Adriatic and the Mediterranean, now and again obstructed in their march south by the detritus of failed, heavy industrial complexes. Quite apart from this expansion of the districts' homeland, district-like agglomerations are emerging in some patches of the South of Italy. Looking at the economic map of Europe, the economic agronomist sees a "hot banana" of thriving districts curving from London, through Switzerland and the Southwest of Germany into Northern Italy. In the developing economies districts are identified as a motor of growth in countries as different as Chile, Brazil, and India.

The advance of the industrial district as a concept has at least kept pace with its progress in fact. Discussion of economic growth in both the advanced and developing countries is dominated by what is often called the Washington consensus - concern for opening economies to free trade, getting exchange rates and other prices right, and (lately) building the institutions needed to do this. But insofar as it is not, debate gravitates toward the creation and expansion of "clusters" - the business name for districts. In the European Union in particular fostering clusters is often seen as a way of encouraging economic competitiveness without giving in to US pressure to give markets free reign. In theoretical debates about economic growth and international trade too agglomerations, often explicitly identified with districts, have come to play a central role (although, as we shall see, with surprisingly subversive effects).

Why, then, are the districts prospering and diffusing? What explains their success in today's volatile and rapidly changing markets? Surprisingly - or not, you be the judge - there is no compelling answer to such brusque and elementary questions. Not that familiarity with the subject has dulled curiosity about its causes. Rather, I suspect the fruitfulness of the district idea as a research agenda and as a policy tool has dulled incentives to inquire too closely into the consistency and generalizability of the sustaining ideas. Concepts that both allow exploration of economic forms uncognizable in the light of standard theories of the firm and discussion of public engagement in economic development otherwise taboo seem somehow self justifying if not self evident. Put another way, you don't look a gift horse in the mouth, especially when you are bestride it.

The result is a growing gap between a venerable and originally plausible explanation of district success and the sources of competitive flexibility generally, and a discrepant mass of material regarding developments in the world, research findings and alternative views of innovation. The familiar explanation focuses on the specialization of activity and the tacit or lived quality of the knowledge that such specialization produces. The discrepant material, on the contrary, links the innovative capacities of agglomerations to diversity and the (partial) formalization of the knowledge that coordination across difference demands. The districts initial success was mysterious. How could small, traditional artisan firms outperform giant, technologically sophisticated corporations in the application of new technologies in industries as diverse automotive components and textiles? Now the deep changes wrought by the districts' capacities for transformation make the grounds of their continuing succcss mysterious again.

In this essay I want to argue for the discrepant view. More exactly, I want to argue that under current conditions innovation, and problem solving generally, depend on disciplined comparisons of alternative solutions, and these in turn require transforming tacit knowledge into what might be called pidgin formalizations: accounts sufficiently detailed to be recognizable to those who know the situations to which they refer first hand, but sufficiently abstracted from them to accessible to outsiders, from various disciplines.

Such pidgins are more than mere descriptions yet less than fully developed causal theories. As the work of Galison (1997) shows, they are familiar to the history of science as the working languages of collaborations between, say, experimentalist instrument makers or engineers on the one hand and theoreticians on the other. From these collaborations radically innovative theories sometimes emerge. For our purposes, though, these formalizations are less interesting for their generative potential than for the working exchanges they enable. Their availability, we will see, breaks down or at least attenuates the distinction between insiders and outsiders, intimates and strangers upon which the success of districts, and related organizational forms, was premised. Conversely, organizations designed to generate such formalizations, and collaborate with others doing the same, have internal structures and build relations not contemplated in the familiar district literature or in its analogue in the discussion of large firms. These novel organizations are neither craft communities that organize themselves on the basis of common, tacit skills nor hierarchies tied together by tacit routines. Their distinctive feature, the one from which flows their peculiar openness to change within and reconnection to the world without, is precisely the ability to question their routines and the foundations of their common skills without undermining the utility of either. They represent, we will see in the end, a form of self-organization that engages the human capacities for self-reflection and deliberation without supposing that we are endowed with anything like a panoramic understanding of our situation.

To intercept misunderstanding at the outset let me underscore two important limits to the argument. First, in emphasizing diversity and formalization rather than specialization and taciturnity as conditions of problem solving I am not arguing that things have always been so. The traditional view of the districts is still a good account - the best we are likely to have - of how districts operated traditionally (until roughly the early 1990s). The discrepant material, and the reading given here of it, is less concerned with revising history than taking account of new circumstances. Second, however great the changes entailed in accommodating the changed circumstances there is no compelling reason to think that districts, starting from their traditional constitution, will be unable to make the adjustment, or will be changed beyond recognition if they do. Indeed, on the current evidence "new" districts agglomerating firms and other institutions capable of jointly formulating pidgins as and so that they can collaborate are often arising from the stepwise transformation of "old" ones.

In presenting these views I am, some readers will know, speaking out of both sides of my mouth. I was a partisan of something akin to the "traditional" view of the districts when it seemed to many outlandish. I'm reasonably sure that the concept of flexible specialization that Michael Piore and I (1984) coined to capture the innovative responsiveness of districts apparently bound by craft tradition is ambiguous enough to paper over the differences explored here (see also Sabel and Zeitlin 1997). But nothing is served by this kind of evasion. Self-reflection, and the distance on past selves it requires, is by definition not just for the others.

The conception of innovation and flexibility founded on tacit specialization finds a natural expression in the idea of a craft or artisan economy. In such a world problem solving is associated with particular tools and materials. Through apprenticeship the use of the materials and instruments characteristic of a particular trade become second nature. So too apprenticeship teaches that the exercise of individual prowess draws on, and replenishes a fund of community knowledge. Mastery comes with further experience, as the artisan gains the autonomy to apply the familiar techniques to the solution of unfamiliar, even novel problems.

To connect this form learning by socialization to an economic structure, take the simplest case, where each artisan in a given trade owns and operates a single lathe or loom. Together the ensemble of owner-operators command the skills and machines needed to whatever specialized good the market for their class of product demands. A broker (who may also be an artisan) solicits customers and contracts with the appropriate group of skilled producers to deliver the goods. Since producers and brokers can diversify the risk of any transaction by pursuing other projects with different partners simultaneously or in sequence, the artisans and brokers can jointly respond to highly volatile markets while safeguarding or advancing their separate interests. Given the manifest advantages of cooperation, moreover, the artisans can easily agree to create institutions - public or private - to provide themselves with services (bookkeeping, monitoring of technical standards, quality assurance) that none, given the small scale of operations, can provide for him or herself (Brusco 1989).

Notice that this craft version of the district economy is largely self-governing in the sense of (almost) proof against opportunism or self-seeking guile, and for related reasons expansively self-reinforcing. The governance of opportunism is a limited problem because the artisans' skills are largely complementary. Homogeneous viewed from the outside, they are highly differentiated in their own view, with micro-specializations - the use of a particular kind of loom or lathe - that make most of their colleagues into potential partners rather than rivals. To be sure, even partners are rivals when it comes to the distribution of the gains to cooperation; but such rivalry is disciplined in explicit partnerships and in the craft districts by the recognition of the mutual dependence. Socialization into the community of craft knowledge, and the sense of dignified competition for the respect of ones' peers that goes with it, echoes the conclusions of prudent calculation. So it proves relatively easy to establish institutions to police the wage cutting and substitution of shoddy materials that, in hard times, tempt strapped producers to defend their individual livelihoods at the price of jeopardizing the stability and reputation of the district ensemble.

This craft model of the district is naturally expansive: given that skills are fixed in a community and a community in a place, the more specialized the district, the more it attracts similarly specialized resources. Alfred Marshall is famous today not least for having seen this connection. In craft communities, he noted, looking to the metalworking agglomeration of the Birmingham of his day, the "mysteries" of a trade are "in the air". Youngsters learn them before they are fully conscious of learning at all, let alone investing in their human capital. Once acquired, the skills become a local resource, attracting outsiders who need in some sense to incorporate themselves into the community to benefit from what it can do only where it is. Today we would say there are positive returns to specialization. Marshall put it this way:

> When an industry has thus chosen a location for itself, it is likely to stay there long: so great are the advantages which people following the same skilled trade get from near neighbourhood to one another.... A localized industry gains a great advantage from the fact that it offers a constant market for skill.... Employers are apt to resort to any place where they are likely to find a good choice of workers with the special skill which they require; while men seeking employment naturally go to places where there are many employers who need such skills as theirs and where therefore it is likely [they will] find a good market (Marshall cited in Fujita 2000, p.10. See also of course Becattini 1990).

The craft model of the small-firm district connects to large-firm settings by two routes; and it is useful to consider these briefly to see the full reach of the assumptions of specialization and taciturnity on current discussion of innovation and flexibility. The more direct, but also more parochial connection is simply that in some cases large firms are an assembly of craft shops under one corporate or physical roof. This is notably so in Germany, where large firms did not specialize in the mass production of standard consumer goods (cigarettes, packaged cereal, cars). Instead they traditionally manufacture customized producers or investment goods: for example, the complex machine tools that mass producers use in their own factories to turn out long runs of a single make or model of product. Workers are largely autonomous in the same way, and by virtue of a similar (though more formal) experience of apprenticeship as artisans in their small firms. Because the firm is large, it can provide itself with many of the support services provided by the institutional exoskeleton in the craft district. But familiar collective action problems - especially fear that trained workers will be poached away from firms where they apprenticed by firms that provide no training at all - require that here, too, the community character of specialized skill is formally acknowledged and protected against tragedies of the commons.

The second connection to large firms is via the school of evolutionary or neo-Schumpeterian economics (for the *fons et origo* see Schumpeter 1926). It is less obvious but broader in reach than the first, and shows that the idea of specialized, tacit knowledge need not be associated with the notion of craft in any familiar sense at all. One key assumption of this school is that large firms are configured not as (nearly) self-organizing communities of producers, but rather as hierarchies. The peak of the hierarchy fixes a master plan, including rules for partitioning that plan into executable tasks. Successive tiers of superiors then actually decompose the tasks into finer and finer jobs for their respective subordinates. This is specialization by task. Because it is disconnected from any systematic exploration of tools or materials it is disconnected from, even antithetic to specialization by craft. A second assumption is that the partitioning of the rules by which complex projects are subdivided and the rules by which the outputs of the partitioned tasks are aggregated into useful wholes are never comprehensive and unambiguous enough to cover all contingencies. The written rules therefore have to be supplemented and corrected by informal, lived experience of how to actually get things done in the organization. This is tacit knowledge. But it is tacit knowledge at least as much of process and organization as of substance: of the routines that define a firm and enable it to execute as a collective tasks that employees could not manage if they needed to think both about their work and how to link it with the efforts of others.

Unlike the craft model district, the tacitly routinized large firm faces substantial governance problems and is not inherently expansive. The governance problem arises from the fact that the firm's assets, physical and human, are specialized to work most efficiently with each other as provided in the master plan. Their value in alternative arrangement - in conjunction that is, with assets not specialized according to the same plan - is close to zero. Hence the hold up problem: suppose a power-plant owner and a coal-mine owner agree respectively to build a power plant and dig a coal mine that are co-located and otherwise specialized to each other. Whoever invests first is at the mercy of the other, who can withhold the complementary investment until the dupe agrees to redistribute the benefits of the venture in favour of the knave. The solution of course is for the one owner to make both investments (with the one better positioned to use the whole effectively taking control). This vertical integration formally closes the large firm in on itself in a way that contrasts with the - formal - openness of the craft districts.

With such closure comes another kind of self-limitation. All the crucial aspects of large-firm operation have to be under exclusive control of a single owner, and the exercise of control becomes more expensive as the complexity of what is controlled increases. Firms therefore do not benefit from the positive returns to agglomeration in the way that "centreless" districts do. Or rather they do no benefit from such returns as a matter of course. Economies of scale - the larger the production run, the lower the unit costs of production - favour such expansion; the costs of extending control to every more complex situations cut against it. (We are, many readers will recognize, in a Coasian world (Coase 1937) where the decision to extend the organization, and how, or rely on the market, depends on a comparison of the relative costs of executing the marginal transaction in each of these ways).

But beyond these differences the craft district and routine large firm views are fundamentally similar in the emphasis on tacit knowledge as both a precondition but also limit to innovation and flexibility. Because craft skill and corporate routine are taken for granted and second nature, they allow for "spontaneous" or self organized co-ordination within certain boundaries. For the same reason, they make it hard to identify those boundaries, and harder still to say when they are obstacles to problem solving, and hence to be corrected.

The neo-Schumpeterians are especially clear on this point. Given the grip of routine, large firms can only innovate - break free of past assumptions to aspire to "autonomous strategic behaviour" - if they can establish "breakout" structures: "entities separate from current operations which" are used to "incubate new projects" (Teece 1998, p. 152). This recognition that the large firm can only be reformed from without is tantamount to an admission that it can not be transformed from within. A congruent view is current in network analyses of economic sociology (Granovetter 1973). Ties among neighbours - friends who are themselves likely to be friends—are robustly self-reinforcing. They are the social substratum of the reciprocity that fosters trust and tacit knowledge. These ties do the work of coordination, but, again, at the price of closing the networked organization off from the world. The information about the outside on which innovation ultimately depends comes from sparse, "weak" ties to distant actors: the cousin of a distant friend from my home village who tells me where to look for a job in the city to which we have both migrated. These ties convey formal knowledge, but neither its tacit complement nor the propensity to trust from which collaboration grows.

Accordingly, boundary blurring, "breakout" structures aside, innovation for these schools of thought occurs largely outside the firm, in university or government research labs or "distant" companies. Different from country to country, these constellations form what the neo-Schumpeterians call the "national innovation system". This system is to the large firms what the exoskeleton of service providers is to craft firms in the districts: the external, innovative corrective to the routines of self-coordination within the firms themselves. Explicitly in the case of the neo-Schumpeterian large firm, implicitly in the case of the craft district, the assumption is that competition selects for those firms or bundles of routines that make the most effective use internal and external resources. Economies of large or small firms survive in the long run in part because their constituent "organisms" are sufficiently diverse in the genetic material of their tacit knowledge to accommodate even the broadest shifts in the conditions of competition. But they survive in part too because the "we" of society is capable of renovating the "external" institutions of the district or the national economy as circumstances warrant. If these routine-revising institutions were themselves prisoner of their own incorrigible routines, only blind luck would save us from tragic attachments to solutions that no longer serve.

So in the end, and in the large - at the level of society at various scales - some kind of reflection on and amendment of routine is possible and necessary. But only outside the economy, in "society"? Or at least the firm? Is it really plausible that large firms can devise "breakout" structures and small firms district institutions, yet neither is capable of continuing self-revision in the small of its on-going activities? Surely the existence of these structures and institutions attests awareness of the problem of routine entrapment. Is that awareness fully exhausted in the split solution - routine firm, routine-jiggling frame - these theories take for granted? Or is this assumption itself an example of a (scientific) routine that impedes recognition of a novel solution? The evidence presented next strongly suggests the latter.

It is common, though perhaps not yet commonplace, to observe in the business and business-school, engineering, and organizational sociology literatures that firms are today able to adjust their operating routines and organizational structures, and that disciplined comparisons of alternatives is an indispensable aid to this. Most of the organizational innovations that permit this were pioneered in modern times by Japanese firms, but are no longer limited, or indeed most effectively developed by them. Here, sidestepping any considerations of how these changes came about, I present an eclectic review of the innovations to highlight the way the formalization of difference is key to the revision of routine. We start with the operating units of the economy: first large, then small, district firms. The distinction can be tenuous given that the latter are often suppliers to and must mesh with the structures of the former, and that decentralization on the one side and centralization on the other produces another kind of convergence. But the discussions we have been canvassing draw this line, and we respect it here if only to show that it is irrelevant to the innovations of interest. Finally we look at some recent work on economic growth and geography which, innocent of any concern for the structure of firms, comes to a complementary finding regarding the contribution of diverse settings, rather than (relatively) homogenous and specialized ones, to innovation.

Developments in large firms are reflected in the explosive diffusion of standards that create and signal the existence of the preconditions for the creation of formal pidgins or lingua francas for collaboration across firms and areas of expertise. On these are built in turn disciplines that specify processes for comparing alternative solutions to design problem or trouble shooting existing production set-ups (Helper, MacDuffie and Sabel 2000). The most common of the standards is the ISO 9000 series maintained by the International Bureau of Standards in Geneva. It requires simply that firms be able to demonstrate that they can say what they do and do what they say with whatever level of resolution is relevant to their industry. Put another way, the standards require that a firm be able to demonstrate that it can make its "tacit" routines for, say, registering an order, or designing a product, or correcting a manufacturing defect explicit to a potential customer. Certification by a third party under the standard does not, of course, ensure that any actual customer will find the explication sufficiently detailed and reliable for its own collaborative needs. But certification is widely regarded in the automobile, chemical, semi-conductor and computer industries as a precondition of a more demanding examination of a firm's capacity for making its practices transparent enough to be discussed with outsiders. And this is surely indicative of shifting assumptions regarding the "stickiness" or embeddedness of (once) firm-specific knowledge (von Hippel 1998).

The comparative disciplines, having arisen in practice, not under official aegis, do not have standard names. In design the relevant practice is often called concurrent or simultaneous engineering. A new product is defined initially and provisionally by benchmarking (itself a comparative discipline) a current model against successful competitors, relevant analogues in other fields, and adjusting these comparisons to take account of laboratory innovations likely to be ready for market in the next product cycle. Then the provisional design is decomposed into modules (engine, transmission, seats, instrument panel, heating, ventilation and air conditions system, and so on in the case of automotive vehicles) and each module is then presented to the scrutiny of one or more of the corresponding module makers. These module makers then benchmark their systems and suggest refinements in the design, including if necessary changes in the performance specifications of closely connected modules. Once the lead design group adjusts the overall plan and selects module makers from those competing, the process is iterated again, with the chosen module makers decomposing their systems into sub-modules, and revising their plans in the light of their suppliers' suggestions.

This method both shortens design times and increases the reliability of designs, while of course increasing the possibilities for incorporating innovations. This is counterintuitive. It seems reasonable, after all, that successive refinement of a single design will be either quicker, or more reliable than repeatedly correcting a vague initial idea in the light of comparisons of variants, or both. Such in any case was the view in the vertically integrated firm, where the design process was typically seen as a mirror of the process by which the company partitioned itself hierarchically. Concurrent engineering succeeds nonetheless in overcoming an apparently inevitable trade-off because careful comparison of alternatives reveals the defects and virtues of each variant, allowing for quicker, more reliable, and more inclusive choices (Actually the comparisons involve a kind of double diversification: different projects are compared by design teams whose members, drawn from different specializations, are cognitively diversified. The underlying mechanism of instigating surprise through investigation of difference remains the same).

Similar mechanisms are at work in new regimes for error detection and correction. An example is the five "why": Why is machine a broken? No preventive maintenance was performed. Why was the maintenance crew derelict?

It is always repairing machine B. Why is machine B always broken? The part it machines always jams. Why does the jam recur? The part warps from heat stress. Why does the part overheat? A design flaw.

Thus error-detection and correction, like benchmarking and simultaneous engineering, discovers unexpected (mis-) connections among the parts of complex endeavours. The cumulative effect of these results is captured in improvements in the benchmark standards for various production processes.

Apparently modest, even commonsensical institutions such as these embody a deep innovation in the cognitive capacities of (large) organizations. For benchmarking, simultaneous engineering and error-detection methods like the five "why" are procedures for doing just what the tacit-knowledge view of specialized action says cannot be done: routinely questioning the suitability of current routines. Whether in the initial specification of new designs (benchmarking), the concretisation of these approximations (simultaneous engineering), or in the course of their practical application (error detection), this disciplined inquiry of routines occurs at just those times when self-interrogation seems most valuable but most difficult.

These mechanisms oblige the actors to search for solutions in a circumscribed space of possibilities (the set of best current or potential designs, the activity chains that might have caused a particular breakdown) whose exact contours and contents they could not have anticipated. The outcome of the search is thus likely to be sufficiently unfamiliar and disconcerting to force re-evaluation of habitual responses. The new large firm is thus a member of a new class of institutions defined not by the fixed routines to which they are oblivious, but rather by the routines they use for interrogating and altering their routines. Think of the new institutions as pragmatist: they systematically provoke doubt, in the pragmatist sense of an urgent suspicion that habitual beliefs are poor guides to current problems. They create the pidgins that make collaboration between insiders and outsiders, neighbours and distant acquaintances, and specialists or professionals of radically different kinds possible (for a thoughtful account of the relation between neo-Schumpeterian theory and the new disciplines that dovetails with the discussion here, see Winter 2000, pp. 56-58).

The new pragmatist standards and disciplines suppose and foster a distinctive form of industrial organization. In contrast to the tacitly routinized large firm, the pragmatist corporation is federated, not hierarchical and centralized: decisions of higher-level entities are crucially shaped by the decisions of their constituent units. The federation is open, not vertically integrated: components or services crucial to the final product of one firm can be provided by independent companies, and the firm's internal specialized producers can provide outsiders with crucial inputs. These differences repose the question of governance; and the resulting answers prompt reconsider of, among other things, the distinction between weak and strong ties. We return these matters after completing the eclectic tour of discrepant material.

Consider next the coordinate changes in the organization of districts and district firms. Some of these changes are of course so directly tied to the pragmatist reorganization of the large firm as to count as part and parcel of the latter, not independent developments. When large firms begin to collaborate more explicitly and continuously in design and production with their suppliers, district firms, as privileged suppliers, must respond in kind or lose a key market. It is not, therefore, surprising to see the new disciplines diffuse within districts. But (hereby probably revealing an irresponsible ignorance) I am unaware of broad-gauge studies that trace the higher order effects of these adjustments on the structure of individual district firms and the relations among them and their large customers.

Similar changers emerging from within the districts - endogenously, if you like - are more revealing. Here I am guided by the experience of a two-week stay in the textile district of Prato in 1996. The salient development was the drive to pragmatist formalization of design and production, and the search for adequate institutional structures within which to house the emergent processes, in response to the sheer complexity and pace of product change. Traditionally the spinner, weaver, and finisher were assigned tasks by and communicated largely through the broker who, we saw, marketed the project in which they participated. But when fabrics consist of novel combinations of novel yarns, each under torsion and woven into a structure that precisely balances torque against torque, costly missteps are likely in the absence of richer, more direct communication among the producers themselves. One upshot was the proliferation of explicit handling instructions: booklets that describe how a yarn or fabric is constructed, and how it is to be further worked. Another was the spread of working relations among producers, bypassing the brokers. In some cases the finishers, who transform the look and feel of the fabric by sophisticated chemical and mechanical treatments, were using their position at the end of the production process and their expertise in the preceding steps to become brokers of a new type. They were *both* organizing more direct relations among the other collaborators *and* managing their relations, and those of the "group" around them to the final customer. But there was no sense at the time that the changes were coalescing into a new district structure, let alone a new model district. Indeed the social dislocations that went hand in hand with even piecemeal adjustment - especially the pending displacement of brokers and rise of finishers - was undermining the social order underpinning the district's ability to formulate and decide questions of common concern.

A large body of recent Italian writings on the evolution of districts comes to findings that converge with those of this vignette. The common ground in these debates is the observation of large, "lead" firms in districts:

> Competition from low wage countries and large firm restructuring are forcing the industrial districts to compete almost exclusively in quality-conscious markets. At the same time, changes in demand composition, a need for higher intrinsic production quality, shorter delivery times and a more active product commercialisation push firms to specialize in their core businesses and develop stable relationships with a limited number of suppliers. Groups are becoming common, and at least in some districts there is an increase in external hierarchy. The adequacy of the collective and local governmental institutions so important in the districts' past is

being challenged (see Whitford 2001, p. 27, e-copy of manuscript, and generally for a comprehensive review).

The role of these firms is to formalize relations with the market and increase (the self-) control of subcontractors. One side in the discussion - the *aziendalisti* - argues that these changes lead inevitably to hierarchy, and to the demise of the district community. The *distrettisti* counter the new forms of coordination should be seen as jointly managed or bilateral governance structures not hierarchies (Dei Ottati 1996, p. 58). Hence even

> a district with coordinating lead firms is *still a system* and often is not dependent *only* on those lead firms. It may still need the reproduction of the "old" values and informal mechanisms of coordination, and there may be some collective goods that lead firms and their networks can create only with difficulty (Whitford 2001, p.23).

The view developed here agrees with the statements of the facts but is at odds with both interpretations: it suggests that the "new" district is neither a mere assembly of hierarchical firms and their satellites, nor a system of firms having their true foundation in a local community. Pragmatist governance is in theory both non-hierarchical (joint) and formalizable. My hunch is that a study of districts that took seriously the possibility of an alternative to both the traditional district and the traditional large firm would allow us to see quickly enough if it is worth pursuing the claim that the Third Italy is going pragmatist.

The third body of discrepant evidence, from debates in international trade, growth theory and economic geography, bears directly on the value of diversity as against specialization as a source of innovation, but has little to say about the relation of this dispute to firm organization. At the origin of much of this discussion was a puzzle about the pattern of post-War international trade. The theory of comparative advantage predicts that a country richly endowed with capital relative to its endowment of labour will export capital intensive-goods to and import labour-intensive goods from a country whose relative endowments are the reverse, regardless of the absolute costs of production in either country. But in the post-War period much of the growth in trade was between countries with similar, not complementary factor endowment: the US exported cars and aircraft to Europe, and bought back (different kinds of) cars and aircraft. A plausible explanation for this intra-industry trade was to see it as the result of positive returns to the mass production of different types of highly specialized products: The US churned out big cars, the Europeans made Beetles or their equivalent. The first mover in each case acquired, because of the increasing returns to scale, a cost advantage that prompted further specialization in the same line. As it became clear that knowledge is a key to economic growth, and that knowledge begets knowledge, positive returns and its relation to specialization have become research topics of broad concern (see especially Helpman and Krugman 1985).

One branch of this discussion has taken a surprising turn that bears on the argument here. In economic geography Puga and others are finding that diversity, not specialization promotes innovation. A locale is said to be specialized in this context if, relative to the national distribution of economic activity, it concentrates in an industry or a broad sub-sector of one - in the way Biella specializes in woollen textiles. It is diverse or diversified if it is disproportionately specialized in several different activities simultaneously. Suggestive findings in this research area include these: in their formative and innovative phase, firms locate in cities with diverse economies. Once they have "innovated" - settled on a product line or process - they move to (district-like) locales specialized in the activities they will continue to pursue (Duranton and Puga 1999). Firms adopting new processes, such as programmable automation of metal cutting tools, are, other things being equal, more likely to do so in diverse counties than in counties specialized in metalworking, and so on (Harrison and Kelley 1996).

By themselves these results are hard to interpret. They could be an artefact of aggregation. It is hard to imagine, in the absence of any account of the micro-mechanisms at work, just how the diverse airs of a great metropolis or suburban county foster innovation. It is equally hard to see why proximity to cooperatively competitive neighbours (talking incessantly, it might seem, about their ingenious plans to outdo you) hinders it.

But taken in the context of the foregoing discussion these findings are less perplexing. If firms are using disciplined comparison of difference to test and adjust their assumptions and structures, why not seek out environments where difference is conspicuously provocative? If you are not sure whether your firm is part of the software industry or the financial services industry, why not "concurrently engineer" your company in an environment where you prototype and compare both at the same time? Of course actors that think this way are unlikely to limit themselves (as the economic geography literature implies they might) to locational decisions. If comparative examination of assumptions is what you are after, why not pursue this both within the firm, through introduction of the new disciplines, for instance, and without, in the choice of environment. Firms that acted consistently in this way would in the end contribute to just the kind of transformation of districts that I suggested are already in course: the "young" firms defines itself by innovative expose to diversity in the big city; once mature, it relocates to a district, there to avoid entrapment by unworkable assumptions though the use of the pragmatist disciplines that are, somehow, transforming the district from old to new.

Given all the possibilities for continuity amidst change, and vice versa, in the districts, one reaction to all this might be, why fuss? If "old" districts are so constituted that they can re-constitute themselves as "new" and we understand their original constitution, why the bright-line distinctions and juxtapositions? The real distinction, after all, is between districts, old and new, and the traditional, hierarchical corporation. And no one is claiming that the latter has rehabilitated itself.

Certainly the economic actors find ingenious ways of blurring the lines traditional and craft forms. Think of the German firms that increasingly "apprentice" youngsters to problem-solving teams, rather than master craftsmen, leaving the philosophically inclined to wonder whether socialization into pragmatist institutions produces a "tacit" propensity to doubt assumptions, or something else. Or consider that a numerically controlled machine tool is still a tool, if you are trained to think of instruments as such. Indeed Eric Raymond and others in the Linux world think of programming itself as a craft (Raymond n.d.). Surely the tenaciousness of the vocabulary suggests some continuity in fact?

Without suggesting that craft and pragmatist institutions are incommensurate (the existence of pidgins is a strong argument against such intranslatability), I want to conclude by briefly emphasizing a conceptual difference between them regarding the social embeddedness of economic activity that returns us to themes sounded at the outset.

In the tacit-knowledge view economic activity is embedded in society in two ways. First and most obviously, knowledge of substance and process "sticks" to the social structures of networks, communities, and hierarchies. Because it is unintelligible apart from the experience of those structures, it is tacit. But trust - the disposition to assume that one's potential partners will not take advantage of the vulnerabilities collaboration creates - is socially "sticky" too. Put another way, the disposition to cooperate is coincident with, and therefore limited by, the particular social structures in which embedded. In yet another formulation governance is built into a social order that is distinct from economic exchange itself. The neo-Schumpeterian view of the large firm asserts this in a backhanded way by making the formal institution of property - the exclusive ownership that overcomes holdup problems - the precondition for the community of the large firm. But the thrust is the same.

Pragmatist firms, large and small, dis-embed economic activity in both senses. The new standards, disciplines and pidgins, we saw "unstick" tacit knowledge, allowing local reflection on the assumptions of practice that can cumulate to large-scale changes. How is trust or governance dis-embedded in the pragmatist firm? By the same means. The very same information that permits collaborators to continuously redefine their joint projects across disciplinary and geographic boundaries allows them continuously to assess one another's probity and capacity to meet joint expectations. Governance problems arise in general because of information asymmetries that allow the knowing part to exploit the ignorant one. Pragmatist institutions address these problems largely by symmetricizing information so that the parties are alerted to trouble as it starts (Just how the pragmatist self-correction of project teams or business units aggregates to or otherwise shapes high-level corporate decisions about which assets to own and which strategies to pursue is still an open question. If it is any reassurance, high-priced talent is probably at work on it in a company near you right now).

Supposing lots, suppose this is right. What, if anything, can we say about the structure of economic activity once it is disembedded? Hierarchies are for many obvious reasons optimal in a stable work where projects can be reliably partitioned into narrow tasks. Accidental agglomerations of craft skill are competitive in a world where hierarchy is too inflexible to respond to changing markets, but different or even distant skills are very hard to connect to each other. What is the "optimal" structure in a world where skills are connectable? Or, to use another locution, what is the optimal network in a world where the distinction between strong and weak ties, or local and long-distance ones, tends towards irrelevance? (Watts 1999; Dodds, Sabel and Watts forthcoming).

Those are, given the tendencies sketched above, key questions for the near future. Addressing them will, I think, be or practical importance to economic actors, and a spur to the renewal of reflection of complex self-organization as well (Dobbs, Watts and Sabel forthcoming).

But looking beyond the beyond, addressing these questions won't be the end of the story. For as soon as we begin to discern "optimal" organizational relations or search procedures in the most successful of the disciplined comparisons of difference, we will want to know what social conditions explain why these relations emerge first and/or flourish better in some places rather than others. And the search for the social preconditions of "disembedness" will take us back, by a circuitous but (hopefully) upward spiralling route to some of the oldest and most persistent questions about the foundations - if any - of our modernity.

References

Adler, P. S., Goldoftas, B., et al. (1999), "Flexibility vs. Efficiency? A Case Study of Model Changeovers in the Toyota Production System", *Organization Science*, 10, pp. 43-68.

Barnard, C. I. (1947), *The Functions of the Executive*, Cambridge, Harvard University Press.

Beardsell, M. and Henderson, V. (1999), "Spatial Evolution of the Computer Industry in the USA", *European Economic Review*, 43, pp. 431-456.

Becattini, G. (1990), "The Marshallian Industrial District as a Socio-Economic Notion", in: *Industrial Districts and Inter-Firm Cooperation in Italy*, Pyke, F., Becattini, G. and Sengenberger, W.(eds), Geneva, International Institute for Labour Studies, pp. 37-51.

Brusco, S. (1989), *Piccole imprese e distretti industriali: una raccolta di saggi*, Torino, Rosenberg & Sellier.

Chandler, Alfred D. Jr., Hagstrom, P. et al. (eds) (1998), *The Dynamic Firm: The Role of Technology, Strategy, Organization and Regions*, Oxford, Oxford University Press.

Coase, R. H. (1937), "The Nature of the Firm", *Economica*, 4, pp. 397-405.

Dei Ottati, G. (1996), "The Remarkable Resilience of the Industrial Districts of Tuscany", in: Cossentino, F., Pyke, F. and Sengenberger, W.(eds), *Local and Regional Response to Global Pressure: The Case of Italy and Its Industrial Districts*, Geneva, International Institute for Labour Studies, pp. 37-65.

Dodds, P., Sabel, C. F. and Watts, D. (forthcoming), *Searching for Searchers: Organizational Robustness and Innovation in Highly Complex Solution Spaces.*

Duranton, G. and Puga, D. (1999), *Diversity and Specialisation in Cities: Why, Where and When Does It Matter?*, CEPR (Center for Economic and Policy Research) Discussion Paper.

Duranton, G. and Puga, D. (2001), *Nursery Cities: Urban Diversity, Process Innovation, and theLife-cycle of Products*, CEPR Discussion Paper.

Feldman, M. P. and Audretsch, D. B. (1999), "Innovation in Cities: Science-Based Diversity, Specialization and Localized Competition", *European Economic Review*, 43, pp. 409-429.

Fujita, M., Krugman, P., et al. (1999), "On the Evolution of Hierarchical Urban Systems", *European Economic Review*, 43, pp. 209-251.

Fujita, M. and Thisse, J. F. (2000), "The Formation of Economic Agglomerations: Old Problems and New Perspectives", *Economics of Cities*, Cambridge, Cambridge University Press.

Galison, P. (1997), *Image and Logic: A Material Culture of Microphysics*, Chicago and London, University of Chicago Press.

Harrison, B., Kelley, M. et al. (1996), "Specialization Versus Diversity in Local Economies: The Implications for Innovative Private-Sector Behavior", *Cityscape* 2, pp. 61-93.

Helper, S., MacDuffie, J. P. and Sabel, C. (2000), "Pragmatic Collaborations: Advancing Knowledge while Controlling Opportunism", *Industrial and Corporate Change*, 9, pp. 443-487.

Helpman, E. and Krugman, P. (1985), *Market Structure and Foreign Trade: Increasing Returns, Imperfect Competition, and the World Economy*, Cambridge, Mass., MIT Press.

Huriot, J. M. and Thisse, J. F. (eds) (2000), *Economics of Cities*, Cambridge, Cambridge University Press.

Granovetter, M. (1973), "The Strength of Weak Ties", *American Journal of Sociology* 78 (May), pp. 1360-1380.

Granovetter, M. (2001), "A Theoretical Agenda for Economic Sociology", forthcoming in: *Economic Sociology at the Millenium*, Guillen, M. F., Collins, R., England, P. and Meyer, M. (eds), New York, Russell Sage Foundation.

Kelley, M. and Helper, S. (1997), "Firm Size and Capabilities, Regional Agglomeration, and the Adoption of New Technology", *Economics of Innovation and New Technology*, forthcoming.

MacDuffie, J. P. (1997), "The Road to Root Cause: Shop-Floor Problem-Solving at Three Auto Assembly Plants", *Management Science*, 43, pp. 479-502.

MacDuffie, J. P. and Helper, S. (1999), "Creating Lean Suppliers: Diffusing Lean Production Through the Supply Chain", Remade in America: *Transplanting and Transforming Japanese Production Systems*, Liker, J., Adler, P. and Fruin, M. (eds), New York, Oxford University Press.

Monteverde, K. and Teece, D. J. (1982), "Supplier Switching Costs and Vertical Integration in the Automobile Industry", *Bell Journal of Economics*, 13 (1), pp. 206-213.

O'Connell, L. G. (2001), *Breaking the Bounds of Peripherality: Irish Business in an Era of Multilevel Participative Internationalisation*, Ph.D. thesis, University College Dublin.

Ottaviano, G. I. P. and Puga, D. (1997), *Agglomeration in the Global Economy: A Survey of the "New Economic Geography"*, CEP Discussion Paper.

Piore, M. J. and Sabel, C. F. (1984), *The Second Industrial Divide: Possibilities for Prosperity*, New York, Basic Books.

Puga, D. (2001), *European Regional Policies in Light of Recent Location Theories*, CEPR.

Raymond, R. (n.d.), http://www.tuxedo.org/~esr/writings/.

Sabel, C. F. (1994), "Learning by Monitoring: The Institutions of Economic Development", *Handbook of Economic Sociology*, Smelser, N. and Swedberg, R. (eds), Princeton, NJ, Princeton University Press, pp. 137-165.

Sabel, C. F. and Zeitlin, J. (eds) (1997), *Worlds of Possibility*, Cambridge, Cambridge University Press.

Schumpeter, J. (1926), *Theorie der Wirtschlaftlichen Entwicklung: Eine Untersuchung Über Unternehmergewinn, Kapital, Kredit, Zins und Den Konjunkturzyklus*, München und Leipzig, Duncker und Humblot.

Signorini, L. F. (ed.) (2000), *Lo Sviluppo Locale: Un'indagine della Banca d'Italia sui distretti industriali*, Roma, Meridiana Libri.

Teece, D. (1998), "Design Issues for Innovative Firms: Bureaucracy, Incentives and Industrial Structure", The Dynamic Firm: The Role of Technology, Strategy, Organization and Regions, Oxford, Oxford University Press.

Von Hippel, E. (1988), "Sticky Information and the Locus of Problem Solving: Implications for Innovation", in: *The Dynamic Firm: The Role of Technology, Strategy, Organization and Regions*, Chandler, A. D. Jr., Hagstrom, P. and Solvell, O. (eds), Oxford, Oxford University Press.

Weick, K. E. (1995), *Sensemaking in Organizations*, Thousand Oaks, Sage Publications.

Whitford, J. (2001), The Decline of a Model? Challenge and Response in the Italian Industrial Districts, *Economy and Society*, 30, issue 1, February 2001, pp. 38-65.

Winter, S. G. (2000), "Organizing for Continuous Improvement: Evolutionary Theory Meets the Quality Revolution", in: *The Quality Movement & Organization Theory*, Cole, R. E. and Scott, W. R. (eds), Thousand Oaks, CA, Sage Publications, pp. 49-66.

Watts, D. (1999), *Small Worlds: The Dynamics of Networks between Order and Randomness*, Princeton, Princeton University Press.

Competitive and Synergic Behaviours in the Development of Industrial Clusters: Ecological Modelling and Empirical Evidence

Marco Fortis and Mario A. Maggioni[1]

Abstract. Aim of the paper is, on the one hand, to highlight the economic relevance of industrial districts in Italy in terms of employment, value added and export and, on the other, to apply an original theoretical framework, derived from population ecology modelling, to the analysis of the development of industrial districts in Italy.

In this way it is possible to underline the interplay of agglomeration economies and diseconomies in the growth process of an industrial cluster, to distinguish three main phases of its development process and to stress the complex and different (i.e. synergic, competitive, etc.) interactions which exists between different industries, within the same area, and between different areas within the same industry.

The formal model is able to identify and discuss the existence stability conditions of long run equilibria through a graphical analysis.

The paper continue by describing the structure and evolution of some representative examples of Italian industrial districts (namely Cusio-Valsesia and Lumezzane districts of valves and fittings, Varese district of plastic, rubber and specialised machinery, Castel Goffredo districts of stockings).

A discussion of the main methodological problems connected to the empirical implementation of the original analytical framework, described above, concludes the paper.

---

[1]Sections 1 and 3.1 can be attributed to M. Fortis, Sections 2, 3.2 and 3.3 to M. A. Maggioni. The authors thanks Erika Uberti for outstanding research assistantships and fruitful discussions; Simona Beretta, Alberto Bramanti, Bernhard Böhm, Paul Stoneman, Peter Swann and Massimiliano Riggi for comments and observations on various draft of the paper.

> Ecology is the study of patterns in nature, of how these patterns came to be, how they change in space and time, why some are more fragile than others. Population ecology is concerned with how populations interact with the environment and how these interactions give rise to the larger patterns of communities and ecosystems. *The environment is more than just sun, air, earth and water: it includes other organisms which may help or hinder the survival of a species.* Population ecology is also the study of how these organisms interact (...) in competition and in co-operation.
>
> S. H. Kingsland (1985), *Modelling Nature. Episodes in the History of Population Ecology.*

> No population is alone. Population interact with each other in many ways. These interactions are important. We will see that we cannot understand many population phenomena without considering the interactions. Even if we want to ignore population interactions we cannot. But above all, the *interactions are intrinsically interesting because they produce perhaps the most intricate and fascinating patterns in biology.*
>
> J. Roughgarden (1979), *Theory of Population Genetic and Evolutionary Ecology: an Introduction.*

> Biology is the Mecca of economists.
>
> A. Marshall (1920), *Principles of Economics.*

1. Introduction

As a matter of fact, unlike the other major advanced countries, Italy has a small number of large companies. In 1999, according to Mediobanca's list of industrial and services companies, only four groups posted sales in excess of 20 billion Euro (IFI-FIAT, ENI, Olivetti and ENEL), but only one of them (FIAT) operates primarily in the manufacturing sector, while the others do business above all in the services industry. The latest Fortune's list of the top world companies includes only three Italian companies (Assicurazioni Generali, Fiat and ENI).

Most of the Italian manufacturing system consists of small and medium-sized enterprises, often organised in a number of single-product industrial districts unequalled in any of the other advanced countries.

As highlighted at the OECD Conference held in Bologna last year, Italy has the largest number of workers in small and medium-sized industrial enterprises (enterprises employing less than 250 people) in Europe, with 3.6 million, followed by Germany (3.3 million), France (2.4 million) and Great Britain (2.2 million). It should also be stressed that Italy is the advanced country with the highest percentage ratio of employees of small and medium-sized enterprises to total number of industrial workers, together with Spain and Japan (Italy 73%; Spain

72%; Japan 71%), while the other major countries have a higher rate of workers employed in large companies, with a smaller percentage working in small and medium-sized enterprises (Germany 40%; United Kingdom 48%; France 53%).

Another peculiarity of the Italian economy is that the sectors specialised in the so-called "made in Italy" business (fashion, home furniture, food, traditional mechanics) are the real engine of the Italian manufacturing industry; it has been calculated that the "made in Italy" business employs more than 3 million people. The specialised "made in Italy" sectors consist mainly of small and medium-sized enterprises and of districts. About 200 local manufacturing systems specialised in particular manufacturing sectors employ over 2 million people, or more than 40% of the domestic manufacturing workforce and more than 60% of the typical "made in Italy" sectors.

Moreover, thanks to high-level, quality and innovative productions, the traditional Italian sectors are a big source of value. For instance, it has been calculated that the added value of the Italian apparel, footwear and furniture industry, assuming the same purchase power, is higher than that generated by the German automotive industry or the American pharmaceuticals industry.

In 1999, the "made in Italy" sectors, where most enterprises are small to medium in size (fashion, furniture, household goods, Mediterranean food and traditional mechanics) recorded exports for 235,000 billion lire and a trade surplus of as much as 142,000 billion. This surplus helped to pay the energy and raw materials deficit not only of the "made in Italy" business itself but also that of the whole Italian system, amounting to 63,000 billion, as well as the aggregated deficit of the other manufacturing sectors (means of transport, information technology, electronics, chemicals, paper, rubber, and cement), equal to 53,000 billion, leading ultimately to a total net surplus of 25,000 billion lire in the Italian trade balance. In 2000 the surplus of the "made in Italy" business exceeded 150,000 billion lire and allowed Italy to record a modest surplus in the trade balance in spite of the dramatic worsening of the energy balance.

These figures highlight the major role of the "made in Italy" sectors in the Italian foreign trade. It has also been calculated that about one third of the whole Italian exports and almost 60% of the "made in Italy" sector's come from district system-areas.

The pattern of the Italian industrial districts has drawn the attention of politicians and experts from all over the world.

A few districts in Italy record sales revenues of 4 to 5 billion Euro: Sassuolo in ceramic tiles, Prato and Biella in textiles. Approximately 30 districts post sales revenues ranging between 500 and 1,500 million Euro: Castel Goffredo in women's hosiery, Como in silk products, the areas of Brianza and Alto Livenza (between Pordenone and Treviso) in furniture, Arzignano (Vicenza) and Santa Croce (Pisa) in tanning, Fermo (Ascoli Piceno) in footwear, the Cadore (Belluno) in eyewear and frames, Montebelluna (Treviso) in sports shoes, Verona and the Apuan Alps in marble and granite, Vicenza and Arezzo in jewellery, Bologna in packaging machinery, Milan and Varese in plastics processing machinery, the Murgia region (Bari and Matera provinces) in couches, Pesaro in kitchen furniture,

Rimini, Forlì and Pesaro in wood working machinery, etc. Many other districts, though smaller, are leaders in their sectors, with sales revenues of 50 to 200 million Euro: from Rossano Veneto (Vicenza), world leader in saddles for bicycles, to Bergamo in buttons, from Premana (Lecco) in scissors to Vigevano (shoe-making machinery). Not to mention the typical food districts such as those of Parmigiano Reggiano or Grana Padano cheese, Parma and San Daniele hams, Gorgonzola cheese (Novara), and wines everywhere in Italy.

In their respective sectors, the Italian districts of small and medium-sized businesses hold world trade shares comparable to, or even larger than those of the big automotive, chemical and electronics multinational groups: the Emilian ceramics district and the Castel Goffredo women's hosiery district hold each about 40% of the world exports in their sectors; the Biella and Prato woollen fabrics districts and the Como silk district hold each world exports shares ranging from 15% to 25% in their respective sectors; similarly, world exports shares around 15% are held each by the jewelry districts of Arezzo and Vicenza, the Belluno eyewear district, the Apuane marble district and the Emilia packaging machinery district; shares around 10% are held by the Emilia wood-working machinery district, the Puglia couch district, the Alto Livenza furniture district, the Vicenza tannery district, the Verona marbles district, etc.

2. The Ecological Approach

The introduction has shown the empirical relevance of industrial districts in Italy measured both in terms of employment and value added. This section is devoted to the elaboration of a new theoretical modelling framework able to explain, on the one hand, the location process of firms and, on the other, the development process of industrial clusters.

The section studies the development path of an industrial cluster as the result of two distinct but interrelated processes: an internal dynamics (spurred by firm's location and spin-offs) and an external dynamics (determined by inter-industry and inter-cluster interactions).

2.1. The Effect of Previous Locations on Clustering Dynamics

If one starts the analysis from this macro-economic perspective, the first issues to be addressed concern the existence and the extent of spatial clustering processes, the main questions to be answered being the following: why do firms cluster? And why do clusters have a finite size?

In the population ecology literature, a generic modelling framework describes the growth process of a species in an environment as driven by its reproductive capacity and limited by the available amount of resources and the presence of other interacting species. From a mathematical viewpoint, the core of this situation can be represented by a differential equation (or by a system of differential equations) which describe the changes over time of a variable (birth minus deaths) as function of: the level of the same variable (the size of the population) at each moment in time; a ceiling level (which takes into account the limit imposed by the available amount of resources); and the level of other variables (which represents the interacting species).

Similarly, in the following sections, firms' decisions to locate into a cluster (and consequently the entry rate and the development path of the industrial cluster) are explained in terms (as a function) of the "economic mass" of the cluster, measured through the number of firms already located there.

However, these models do not merely postulate the existence of a stock-flow mechanism between the number of incumbent firms in the cluster and the entry rate, even if one may reasonably assume the number of spin-off firms, which are generated in each period within the cluster, to be a proportion of the number of firms already established there. Neither do they assume that the number of firms is the only relevant variable considered by potential entrants when taking locations decisions. Instead, the number of already located firms reduces potential entrants' search costs, signals the existence of geographic and agglomeration benefits, diffuses information, reduces uncertainty, and causes informational cascades.

What follows is thus a series of models aimed at explaining location decisions and the development path of an industrial cluster when firms' interactions (both effective and perceived) are relevant. For the sake of exposition, we first present a model describing the effects of the location decisions of monoplant firms (i.e. where firm and establishment coincide), belonging to a given industry, within an isolated cluster (i.e. when there is only one potential site for location). In the following sections, the scope of the analysis is enlarged to encompass a number of interactions between two (or more) clusters and/or industries.

2.2. Locational Benefits and Costs and the Development of an Industrial Cluster

Firms decide to settle in a cluster on the basis of the expected profitability of being located there. This profitability depends on net locational benefits - obtained as the difference between gross locational benefits and costs - which, in turn, are based on both observable and unobservable elements.

For simplicity it can be assumed that, in an uncertain world - with limited information regarding local costs and revenues available to the outsiders - profitability expectations for any particular location will be based solely on the number of firms already located there (the number of previous locations being the only observable variable).

Let us assume, as in Arthur (1988 and 1990), that locational gross benefits B_{fq} for firm f locating in cluster q are composed of geographical and agglomeration benefits[2].

Geographical benefits G_{fq} depend on the intrinsic features of the site (such as the quality of local factors of production: capital k_q and labour l_q; the efficiency of the local network of specialised suppliers and business service firms s_q; and the quality of urban and industrial infrastructures u_q). Agglomeration benefits $A_{fq}\left(n_q\right)$ are a concave non-monotonic function of the number of incumbents (i.e. firms already established in cluster q) n_q. Thus:

$$B_{fq} = G_{fq}\left(k_q, l_q, s_q, u_q\right) + A_{fq}(n_q) \tag{1}$$

The assumption of concavity and non monotonicity in A_q implies that, as the number of firms located in cluster q increases, gross benefits firstly increase because of agglomeration economies (due to productive specialisation; scientific, technical and commercial spillovers; reduction in both transport and transaction costs, increases in the quality of the local pool of skilled labour force and in the efficiency of the local credit market); then decrease when congestion more than compensates for agglomeration economies.

Locational costs c_{fq}, symmetrically, include two components: geographical costs g_{fq} (reflecting the cost structure of the cluster in terms of locally prevailing wage w_q and interest rate r_q; average price of business services d_q; and level of land rent and taxation t_q), and agglomeration costs a_q, which are assumed to be a convex non monotonic function of the number of regional incumbents n_q.

---

[2]For analytical convenience we split locational benefits in two classes: geographical and agglomeration benefits. The first class refers to those components which are unaffected by the number of incumbents; while the second refers to those components which depend on the number of incumbents. By adopting this formulation, however, we do not intend to state that agglomeration benefits refer only to spillovers of scientific and technological knowledge and know-how. On the contrary we are convinced that relevant agglomeration benefits derive also from external economies of scale in the use of local resources. The same variable (i.e. labour productivity) has a fixed geographical component, which depends on the quality of local workers, and a variable agglomerative component which depends on the number of firms already located in the cluster.

$$c_{fq} = g_{fq}\left(w_q, r_q, d_q, t_q\right) + a_{fq}(n_q) \tag{2}$$

The assumption of convexity and non monotonicity in a_q implies that, as the number of firms in cluster q increases, locational costs initially decrease until some "optimal" number of users for a given set of urban, industrial and environmental infrastructures and resources is reached. Then they increase due to the competition, between a larger number of firms, for a limited pool of local inputs (i.e. capital, labour, business services, land and public infrastructures) which raises their prices[3].

Net locational benefits can now be calculated as the difference between equations (1) and (2).

$$N_{fq} = B_{fq} - c_{fq} = H_{fq}\left(w_q, r_q, d_q, t_q, k_q, l_q, s_q, u_q\right) + h_{fq}\left(n_q\right) \tag{3a}$$

Assuming that the geographical benefits and costs do not change overtime, if we focus the analysis of the location process on the dynamics of the interactions between the level of available locational benefits, what becomes relevant for describing firms' location decisions is just the net benefit function N_{fq} in the incumbents' space. We can therefore summarise the geographic components H_{fq} with a parameter α_q, which vertically shifts the locational net benefits function, and write the following expression:

$$N_{fq} = B_{fq} - c_{fq} = \alpha_q + h_{fq}\left(n_q\right) \tag{3b}$$

[3]An alternative explanation for the convexity of the locational costs function for firm f runs as follows: the locational costs function is compose by a "fixed" and a "variable" component. The fixed part of the costs (geographic costs) decreases as the number of entrants increase; while the variable part increases (because of competition) as the number of entrants increase. The combination of these two effects produces an U-shaped (convex) cost curve as the interaction between fixed and variable costs of production in standard microeconomics textbooks. A symmetric reasoning may also explain the inverted U-shaped benefits function. This interpretation is surely more realistic than the one used in the paper, however it is not as theoretically efficient as the other one since both components become dependent on the number of incumbents.

It easy to see that the locational net benefits function (3b) is always concave, since N_{fq} is equal to the difference between a concave function $B_{fq}\left(n_q\right)$ and a convex one $c_{fq}\left(n_q\right)$. In other words, each marginal firm, which enters the cluster, increases the average profitability of locating in the cluster only up to a threshold. After that point, any new entrant lowers the average net benefits available to each resident firm and new entrant[4].

This formulation recalls some general results, obtained in the industrial location and urban/regional economics literature (Weber 1929; Isard 1956; Richardson 1978; Papageorgiou 1979; Tauchen and Witte 1983; Miyao and Kanemoto 1987), which show the existence of an optimal dimension of a given spatial agglomeration of firms and/or households because of the concavity of the various benefits functions.

We can therefore state that - if the number of potential entrants is sufficiently large and there are no relevant entry barriers - as net locational benefits initially increase, the growth rate of incumbent firms increases; then it decreases and is finally driven to zero.

If we assume that the number of potential entrants is not constant but it is changing overtime, than the entry pattern will be determined by the rate of birth of potential entrants, since a proportion of such firms will locate in the cluster. If we allow the number of actual entrants to be proportional to the average locational benefits available in the cluster - as a first approximation (or, alternatively, as a macroeconomic hypothesis without explicit micro-foundations[5]) - and we assume that the entry rate of firms into the cluster is proportional to the current level of locational net benefits[6], then the cluster growth will initially be fast, then will slow down and finally will stop[7].

---

[4]However, as it is made graphically evident in Figure 4.2, because of the inverse U shape of the marginal benefits function, there is a range, within the number of incumbents, where marginal net benefits are already decreasing, but still higher than average ones, and average net benefits are still increasing.

[5]Another part of this story refers to the role played by spin-off firms. If we believe that average net locational benefits give incumbent firms some extra profits, workers (or, better, managers) of incumbent firms, which see these extra profits, will decide to start a new firm to appropriate them. Therefore the number of spin-off firms, which are generated by parents firms, is positively related to the level of locational benefits in the cluster, generated by the incumbent firms.

[6]It is evident that, for the moment, we are implicitly assuming firms to behave in a non-strategic manner (i.e. although their entries modify radically the locational benefits available in the cluster, they do not take into account such a modification) and to be totally myopic (i.e. their entries are only dependent on the current level of locational benefits and

If this is the case[8], one would expect the industrial growth of the cluster to follow an S-shaped path with a slow start (when locational benefits are still low), an "explosive" central period of rapid increase (when the average net benefits in the cluster are highest) and a final part when the cluster gradually reaches and then settles to its "equilibrium"[9] size as in Figure 1.

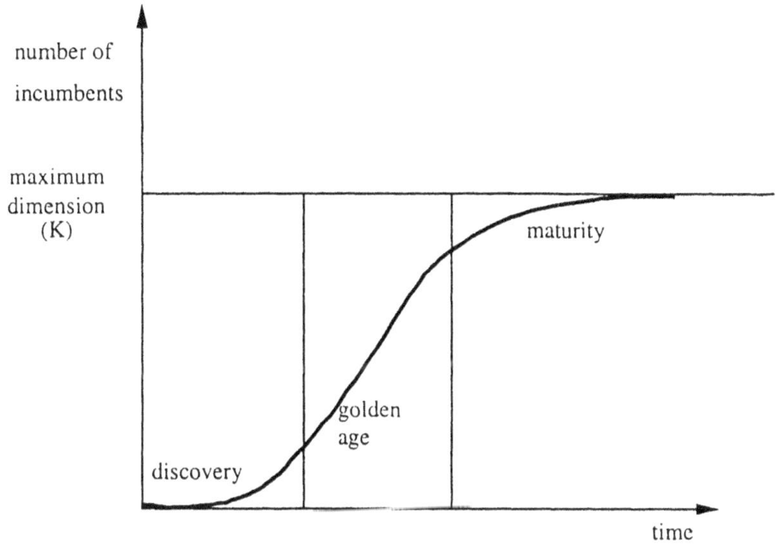

Fig. 1. The development path of an industrial cluster.

---

not on some discounted value of the flow of locational benefits they enjoy from the moment of entry onward). In section 4.3, we will relax some of these hypotheses.

[7]This is not to say that, from a certain moment onward, entries will not take place. On the contrary, once the equilibrium level of the cluster is reached, new entries are still possible, but these happen at the expenses of some incumbents which are driven out of the cluster (or, more probably, out of business). This continuous series of entries and exits may determine a continuous oscillation of the cluster's industrial mass around the equilibrium level.

[8]And we collected some empirical evidence supporting this story (see Maggioni 1993; Maggioni and Porro 1994; Gambarotto and Maggioni 1998; Maggioni 1999).

[9]It should be bore in mind, however, that this notion of equilibrium value can be changed in the long run by radical innovations, exogenous demand shocks and appropriate regional policies.

What is needed now is a formal model of the location process and the development of clusters, which will allow a rigorous testing of all these assumptions.

2.3. A Logistic Model

The simplest growth model for an industrial cluster q - which stresses the relevance of firms spatial interactions - can be expressed in the following format: "the rate of growth of the industrial mass equals the product of the individual firm's contribution[10] to the regional population's growth and the number of firms already in the cluster" (Maggioni 1993, 1999).

If only agglomeration economies and positive spillovers are taken into account (and these are assumed to be constant), then each individual firm's contribution to the level of average locational benefits and, consequently, to the growth of the cluster would be equal to a constant r_q. In this case cluster industrial growth would follow an "explosive" exponential path[11], formally:

$$\frac{dn_q}{dt} = r_q \, n_q(t) \tag{4a}$$

Alternatively one can solve the equation for $n_q(t)$ as a function of the exogenous initial industrial mass of the cluster $n_q(0)$

$$n_q(t) = e^{r_q t} \, n_q(0) \tag{4b}$$

On the other hand, if congestion and competition effects are included, then some modifications to this simple model are required to allow for some "density dependent" factors to progressively depress the level of locational benefits and to slow down the process of industrial growth of the cluster. A simple dynamic model, which takes into account these features, is the logistic equation[12], which can be written as (5a)

---

[10]In terms of changes in the average locational net benefits, due to the interactions of agglomeration economies and diseconomies.

[11]The higher is r_q, the faster is the growth process.

[12]The logistic equation - firstly developed by Verhulst (1845) and Pearl and Reed (1925) for demographic studies, then adopted by the ecological literature since Lotka (1925) - "is the simplest model containing negative density dependence interaction. Further, it is the

$$\frac{dn_q}{dt} = r_q \, n_q(t)\left(1 - \frac{n_q(t)}{K_q}\right) \tag{5a}$$

where r_q is the incipient (or maximum) rate of increase and $K_q = \lim_{t\to\infty} n_q(t)$, is called the cluster "equilibrium" level[13]. Integrating equation (5a) and solving for $n_q(0)$ one obtains (5b):

$$n_q(t) = \frac{K_q n_q(0) e^{r_q t}}{K_q + n_q(0)\left(e^{r_q t} - 1\right)} \tag{5b}$$

Here the individual firm's contribution to regional growth decreases as a linear function of regional population size and is equal to $r_q - \frac{r_q}{K_q} n_q(t)$ (Roughgarden, 1979).

Plotting $n_q(t)$ against time yields an S-shaped curve due to the counteracting roles played by r_q and K_q. When the cluster is small ($n_q(t)$ is near to zero) the term in brackets in equation (5a) is close to one (hence the logistic equation approximately describes an exponential growth path); but as $n_q(t)$ approaches K_q, the term in brackets tends to zero, driving the growth rate to zero and terminating the entry process. Both K_q and r_q play a major role in shaping a logistic growth path: the greater is r_q the steeper is the S shaped curve, the larger K_q the higher the ceiling level of the function (and the equilibrium size of the cluster).

---

first two terms in a power series expansion of a more general growth model where the growth is a function of the actual size of the population" (Dendrinos and Mullally 1985, p. 38).

[13] K in the original ecological jargon is called *carrying capacity*: "a measure of the amount of renewable resources in the environment in units of the number of organisms these resources can support" (Roughgarden 1979, p. 305).

r_q is the incipient rate of growth. In the ecological literature it is often approximated by the difference between the birth and mortality rates of a population. This observation can be translated into the economic framework when net entry (and consequently the intrinsic rate of industrial growth of a cluster) is calculated as the difference between total entries (or start-ups) and exits (or bankruptcies) in the period considered. The same value of r_q can therefore correspond to two very different situations: a steady growing cluster where few new firms enter and no one exits, and a perturbed cluster where a high "birth" rate is almost compensated for by a high "death" rate. Hence r_q is a composite index that describes the cluster growth "potential" and the probability that firms, once entered, survive in the cluster.

K_q defines the regional industrial carrying capacity: the maximum number of profitable firms the cluster can sustain in isolation (i.e. when inter-regional interaction are not considered). K_q will depend upon:

- the finite quantity of geographical benefits (which is related to the limited availability of local "resources" such as: labour, capital, land, and infrastructures);
- the decreasing part of the agglomeration benefits function (which depends on the strategic interactions between firms: competition, congestion and lobbying of incumbents).

K_q is therefore determined by the relationship between the amount of resources (inputs) available in the cluster and the (technical and organizational) efficiency of incumbents in the use of these resources. Therefore in the long run K_q may change as result of the inflow of additional skilled workers, the provision of new advanced public infrastructure, the diffusion of (technical, organisational, etc.) innovations.

For a given cluster q and a given population of M_q outsider firms[14], therefore, we assume that there is an equilibrium level $K_q \leq M_q$ acting as an upper limit to the cluster's growth. In each period t, the number of entries therefore depends both on the actual number of potential entrants $K_q - n_q(t)$ (i.e. the number of outsider firms which can enter the cluster in time t and still make profits) and on the number of firms already located there $n_q(t)$. K_q and $n_q(t)$ in fact

[14]Composed of two main categories: firms already established outside the cluster and potential entrepreneurs inside the cluster looking for the right moment to start their own business.

determine the level of average locational net benefits available to incumbent firms in each period of time.

In this first formulation, the number of located firms directly generates (through agglomeration dynamics) the level of locational benefits; since the entry rate is assumed to be proportional to the level of locational benefits, it also indirectly determines the location of new firms into the cluster.

Such a formulation of the location process is very simple but can be used to empirically estimate key parameters of the location path of different clusters[15]. These estimated parameters could also be used as dependent variables in cross-section analyses in order to assess the influence of different factors on the level of the intrinsic growth rate of a cluster or on its maximum dimension.

However this formulation can be criticised on a number of grounds. The main drawback refers to the "isolation hypothesis": firms' location decisions are modelled as a dichotomous choice, in the sense that there is only one possible site for location and the choice variable is just the timing of the entry. The following sections extend the simple logistic growth path to take into account bilateral interactions.

2.4. Location Process in a Two-Clusters Framework

Several different interactions may arise when a framework with two clusters (1 and 2) is considered. The first concerns the location decisions of a set of industry-specific potential entrant firms confronted with the choice of locating in one out of two clusters. This problem can be formally described through a model of inter-cluster competition with a given number of potential entrant firms[16]. The entry decision is again modelled as dependent on the number of firms located in each cluster, reflecting the effects of agglomeration economies and diseconomies, as described in the previous section.

Formally, the multi-cluster situations[17] can be described through the following system of equations which directly recalls the mono-cluster development process of equation (5a):

---

[15]Where the difference may refers to different industries in the same geographical site, or to different geographical sites in the same industry.

[16]This peculiar approach to the process of multi-clusters development (where the cluster is modelled as an agent and firms as passive industrial mass) reveals its advantages when modelling active policy pursued by different local authorities willing to foster the industrial development of their respective territories.

[17]The same analytical framework can be used to describe the interaction between two industries in the same cluster which compete on a given pool of local limited resources (land, workers, bank credit etc.).

$$\begin{cases} \dfrac{dn_1}{dt} = r_1 n_1(t)\left[1 - \dfrac{n_1(t) + c_{12} n_2(t)}{K_1}\right] \\ \dfrac{dn_2}{dt} = r_2 n_2(t)\left[1 - \dfrac{n_2(t) + c_{21} n_1(t)}{K_2}\right] \end{cases} \tag{7a}$$

where r_1 and K_1 are respectively cluster 1 incipient growth rate and equilibrium levels and c_{12}, is the "competition coefficient" which measures the extent to which cluster 2 competes as an alternative location to cluster 1.

However - for expositional purposes - it is more convenient to replace this formulation (which underlines the similarities between the two-clusters case and the model of development of one isolated cluster depicted in 4.5a) with an alternative formalisation which is able to model different multiregional interactions through simple differences in the sign of parameters. We thus replace (7a) by (7b), where time references are dropped for the sake of clarity:

$$\begin{cases} \dfrac{dn_1}{dt} = \left(a_1 - a_{11} n_1 - a_{12} n_2\right) n_1 \\ \dfrac{dn_2}{dt} = \left(a_2 - a_{22} n_2 - a_{21} n_1\right) n_2 \end{cases} \tag{7b}$$

In system (7b), n_1 and n_2 are the "economic masses" (number of incumbents) of each cluster, a_1 and a_2 are the intrinsic rates of increase of each cluster in isolation (r_1, r_2 in system 4.7a), a_{11} and a_{22} - the parameters on the quadratic terms which reflect the concavity of the net agglomeration benefits function - i.e. that reflect the inhibiting effects that a firm's entry has on the growth rate of the same cluster[18] (intra-cluster competition parameters); a_{12} and a_{21} show the inhibiting effects that one firm - locating in a cluster - has on the growth of the other cluster[19] (inter-clusters competition parameters).

---

[18] $\dfrac{a_1}{a_{11}}$ and $\dfrac{a_2}{a_{22}}$ in this formulations are the clusters' maximum size or isolated equilibrium values (K_1 and K_2).

[19] Different interpretations for these parameters are possible: the simplest relates the values of inter-regional competition parameters to the fact that - given a certain number of potential entrants - each firm that locates in cluster 1 does not locate in cluster 2. Then the inter-clusters coefficients should be equal, i.e. $a_{12} = a_{21}$. Another interpretation explicitly

According to the relative value of these parameters, three main outcomes[20] are possible, as shown by phase diagrams in Figure 2.

Phase diagrams are useful tools to describe systems of differential equations. At any moment in time the state of the system is fully described by the number of firms located in each cluster (n_1 and n_2). Within the phase plane one can further draw for each cluster the locus of points (i.e. combination of n_1 and n_2), which is called the isocline, where the cluster's industrial mass does not change (i.e. where $\frac{dn_1}{dt} = 0$ and $\frac{dn_2}{dt} = 0$). The multiple equilibria of the system are thus identified by the intersections of one isocline with either the axes or the other isocline[21]. The arrows in the phase planes indicate the direction in which the system, at that point, will move; the colours and shapes of the dots show whether an equilibrium is either stable or unstable or a saddle point.

---

relates to the possible differences in the logistic growth path of the two clusters in isolation (which reflect differences in the net benefits function). In this case inter-clusters competition coefficients can be different.

[20]Selected from the analysis of 13 possible consistent cases. For a more complete analysis, see Capelo (1989, pp. 165-169) and Beltrami (1987, pp. 72-73).

[21]The intersection of an isocline with an axis is an equilibrium point since one cluster in deserted and the other does not change its industrial mass, the intersection of two isoclines is an equilibrium point since both clusters do not register any changes in the number of located firms.

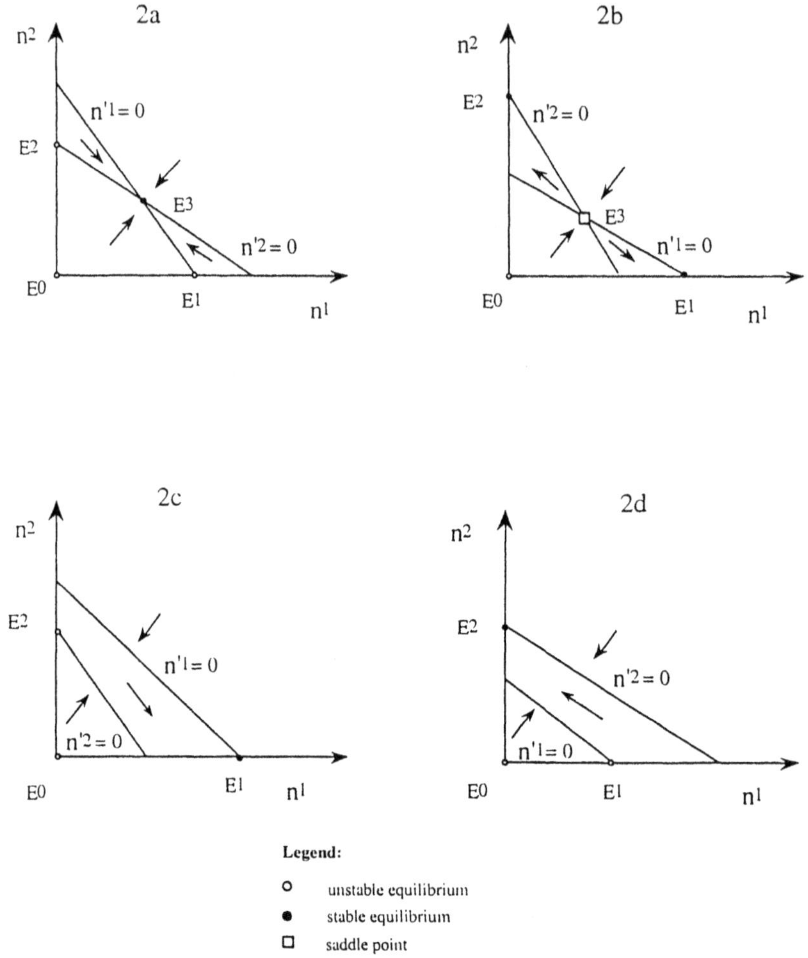

Fig. 2. Inter-regional competition.

Fig. 2a shows the persistent coexistence of two clusters (both recording a positive number of locations) as a stable equilibrium when:

$$\frac{a_1}{a_{12}} > \frac{a_2}{a_{22}} \quad \text{and} \quad \frac{a_1}{a_{11}} < \frac{a_2}{a_{21}} \tag{8a}$$

Fig. 2b shows the coexistence of two clusters as an unstable equilibrium, which is extremely vulnerable to local perturbations and initial biases[22] when:

$$\frac{a_1}{a_{12}} < \frac{a_2}{a_{22}} \quad \text{and} \quad \frac{a_1}{a_{11}} > \frac{a_2}{a_{21}} \tag{8b}$$

Fig. 2c and 2d show the total exclusion of one cluster from the location process that occurs only in the other cluster (irrespective of the initial state) which reaches its carrying capacity leaving the opponent deserted [23]. In particular, only cluster 1 is entered when:

$$\frac{a_1}{a_{12}} > \frac{a_2}{a_{22}} \quad \text{and} \quad \frac{a_1}{a_{11}} > \frac{a_2}{a_{21}} \tag{8c}$$

while only cluster 2 is entered when:

$$\frac{a_1}{a_{12}} < \frac{a_2}{a_{22}} \quad \text{and} \quad \frac{a_1}{a_{11}} < \frac{a_2}{a_{21}} \tag{8d}$$

Since the industrial growth of the cluster in isolation follows a logistic path, then the isoclines are linear and the stability of equilibrium depends on the equilibrium point being above the line[24] joining the maximum dimensions of each cluster in isolation: $\left(\frac{a_1}{a_{11}}, 0\right)$ and $\left(0, \frac{a_2}{a_{22}}\right)$[25].

This presentation of the possible outcomes of firms location processes in a two-clusters framework would seem to suggest that inter-cluster competition leads -

[22]More precisely the equilibrium defining the coexistence of two clusters is a saddle point; any initial state off the stable manifold, or any perturbation that move the system far from it, leads to the "deprivation" of one out of two clusters.

[23]This case includes two opposite situations of only cluster 1 surviving and only cluster 2 surviving.

[24]Whose equation is $n_1 = \left(1 - \frac{n_2}{a_2/a_{22}}\right)\left(\frac{a_1}{a_{11}}\right)$.

[25]Alternatively, if one expresses the clusters size in "equivalent numbers" (i.e. as fractions of their maximum dimension), the equilibrium is stable if the total equivalent number at the joint equilibrium is greater than unity (i.e. greater than the equivalent number at the isolated cluster's equilibrium).

almost necessarily - to a situation of locational monopoly, in which only one cluster is chosen as the preferred site by industry-specific firms while the other remains deserted. Although the economic history of industrial clusters displays few cases supporting this result, the coexistence of multiple non-empty clusters is a more common phenomenon. This must therefore be caused by factors still missing from the above analysis.

One way to explore the problem assumes the intrinsic growth rates of two clusters to be equal (i.e. $a_1 = a_2$). Then the persistent coexistence of both clusters is achieved when:

$$a_{11} > a_{21} \qquad \text{and} \qquad a_{22} > a_{12} \tag{8e}$$

These two inequalities state that, to obtain persistent coexistence of both clusters, an increase in the numbers of incumbents in either cluster must inhibit its own growth more than it inhibits the growth of the other (e.g. that the degree of intra-cluster competition and congestion must be larger then the degree of inter-regional competition). This result is likely to be achieved if the clusters are, at least partially, "attracting" different types of firms or, in other words, if firms have significantly heterogeneous locational preferences.

This observation has been acknowledged within the population ecology literature for a long time under the name of the "Gause principle"[26]. This principle states - in the original biological terms - that in a situation where "two species are resource-limited, (...) only one species can survive on a single resource" (Levin 1970, p. 413). Translated into an economic framework, the Gause "principle of competitive exclusion" states that - if firms are completely homogeneous in their locational preferences - then the most likely outcome is the unanimous choice of one single cluster (the one ensuring at the start the highest level of geographical benefits). The existence of a plurality of non-empty clusters inhabited by firms belonging to the same industrial sector can therefore be explained in terms of the existence of different types of firms that have different locational needs and preferences[27].

The degree of competition between two clusters is therefore proportional to the degree of locational homogeneity of the firms' population. The same concept can be expressed in a rather different way: the degree of inter-regional competition is

---

[26]Being formally stated by Gause (1934), even though an earlier and slightly different formulation is ascribed to Volterra (1926).

[27]It is very interesting to compare this result to very similar ones obtained independently and following a totally different approach by Arthur (1988 and 1990).

inversely proportional to each cluster's specificity[28]. According to this observation, each cluster should foster the development of specific regional characteristics in order to develop a "niche" strategy able to limit the degree of inter-regional competition. This observation, which has a general value, has a particular interest for any public authority willing to start a new industrial cluster in an area where there are already other established clusters. The identification and the attraction of a specific subset of firms is the crucial element, which determines the success, or the failure, of the regional development project.

2.5. Co-operative Interactions

The analytical representation, used in equation (7b) to describe the location decisions of a set of potential entrants between two alternative locations, can be easily encompassed to describe other significant interactions which may arise during the development process of clusters and/or industrial clusters. These different relationships can be modelled simply by changing the signs (or values) of the inter-regional coefficients a_{12} and a_{21} in the system of equations (7b) as shown in table 1.

Table 1. Different types of bilateral interactions.

Types of interactions	Sign and values of interaction coefficients a_{12} and a_{21}
Competition	- -
Mutualism	+ +
Commensalism	+ 0
Predation	+ -
Neutralism	0 0
Amensalism	0 -

[28]Where regional specificity means that each cluster is the ideal location for a different type of firm.

The regional economics literature records several contributions on predatory relationships (see for example; Dendrinos and Mullally 1985; Maino 1989; Nijkamp and Reggiani 1992 and 1998). This section therefore concentrates on co-operative relations that - despite their empirical relevance - have been neglected by the main stream of population ecology and, consequently, by the early pioneers of "ecological" regional economics[29]. These relationships describe a set of co-operative interactions where the development of a certain set of firms, such as an industrial cluster in a given nation (or one industry in a given cluster) is not obtained at the expense of the development of another industry (or cluster) but the two processes enjoy synergistic effects.

Both ecological and economic systems record empirical evidence of mutualistic relationships between two populations over a long period of time. There are plenty of historical cases and empirical evidence of these phenomena: from the development of the textile districts in the mining clusters of Britain and Belgium in the 19th century[30], to the synergistic development of an industrial and a residential bordering area[31], from the development of a subcontractors' belt around a main industrial core, to the long timed establishment of a cluster of integrated producers and craftsmen in one specific quarter within a city[32].

---

[29]There are two different reasons why mutualistic and amensalistic relations have been ignored. The first reason is an historical one: both Lotka and Volterra - generally acknowledged as the forerunners of population ecology - limited their analyses to predation and competition. The second reason is a theoretical one: in its purest formulation, a mutualistic model does not have a stable equilibrium and one or both populations undergo "an unbounded exponential growth in an orgy of mutual benefaction" (May 1976, p. 95). A growth model without stable equilibrium seemed of no interest to regional and applied economists. However, more sophisticated models of mutualistic interaction allow for saturation in the magnitudes of reciprocal benefits and obtain - as general result - a stable equilibrium with both populations (usually clusters, but they can equally be industries) developing larger than in isolation. It must be said, furthermore, that this equilibrium is less stable - in the sense that perturbations are more slowly damped - than in correspondent systems without mutualistic relations.

[30]Textiles firms, in these locations utilised the female components of the miners' families as source of cheap labour force.

[31]The first demanding workers and offering wages, the second demanding goods and offering labour. This interaction goes well when there are not relevant negative externalities (such as pollution, noise etc.) spilling from the industrial cluster to the residential one. In such a case we may well describe the situation as amensalistic.

[32]For an example of such clustering dynamics, see Storper and Walker (1989, p. 81) which report an amazing map showing the agglomeration in the London watch and clock making quarter of Clerkenwell, in 1861.

In particular, we think that the model analytically described in expression (9) could well describe the behaviour of two technologically related industries within the same industrial cluster, each receiving positive benefits from the development of the other one[33]. We think that the analysis of inter-industry relations is as important as (if not more than) the analysis of inter-regional competition since one of the main findings of the literature on industrial clusters, and of the original empirical evidence presented in section 6.3, is that the physical proximity of firms belonging to different but interrelated industries $i = (1,2)$ act as an engine sustaining the cluster's long run growth[34].

The above remarks can thus be described more formally as follows:

$$\begin{cases} \dfrac{dn_1}{dt} = (a_1 - a_{11}n_1 + a_{12}n_2)n_1 \\ \dfrac{dn_2}{dt} = (a_2 - a_{22}n_2 + a_{21}n_1)n_2 \end{cases} \tag{9}$$

In this model each industry in isolation would follow a logistic growth pattern, but the presence of one industry in the same cluster has a positive influence on the rate of growth of the other one. The system's dynamics - which is described through a phase diagram in Figure 3 - always displays three trivial unstable equilibria (Figure 3a): the origin (corresponding to the underdevelopment of both industries) and each industry's isolated maximum dimension. However, under certain assumptions concerning the slope of isoclines (Figure 3b), a fourth stable equilibrium (E^3) emerges which enables both industries to reach a higher level in the cluster. The coexistence of two industries in the same cluster allows both industries to grow larger than would have been possible in the isolated case, but intra-industry competition effects prevent the system from experiencing explosive growth.

[33]However it is also theoretically possible that inter-industries relations, within a cluster, become conflicting because of competition on non industry-specific inputs such as real estates, financial services, fiscal and accounting consultants etc. The prominence of co-operative versus competitive interaction must therefore be empirically assessed.

[34]Swann defines such a process as "convergence between technologies" and shows that "single technology clusters, while they might grow faster in the formative stages, did not have the lasting power of diversified clusters" (Swann 1998, p. 64).

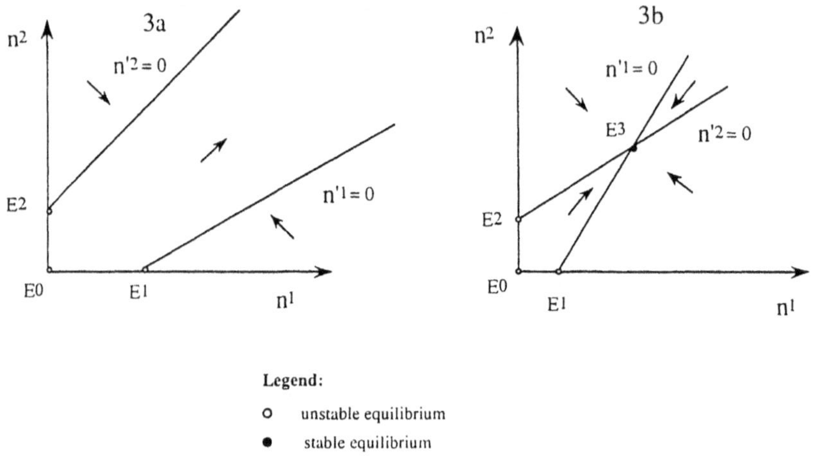

Fig. 3. Inter-industry facultative mutualism.

The conditions for an equilibrium solution in which both industries survive can be re-expressed as a strict positivity requirement on the co-ordinates of E^3 in the (n_1, n_2) space, formally:

$$\frac{(a_2 a_{11} + a_1 a_{21})}{(a_{11} a_{22} - a_{12} a_{21})} > 0 \tag{10a}$$

$$\frac{(a_2 a_{12} + a_1 a_{22})}{(a_{11} a_{22} - a_{12} a_{21})} > 0 \tag{10b}$$

Given that all coefficients are positive, the two conditions reduce as follows:

$$a_{11} a_{22} > a_{12} a_{21} \tag{10c}$$

A stable equilibrium in a mutualistic interaction is therefore achieved when the product of the two intra-industry competition coefficients exceeds the product of the inter-industries cooperation coefficients. The same analytical framework may describe the mutualistic growth of two neighbouring clusters, specialised into complementary productions, where the driving forces of the joint development are vertical and horizontal productive linkages established between clusters.

Other stimulating results are obtained when the symmetry assumption required by facultative[35] mutualism is relaxed. An interesting case concerns a situation in which the first cluster in isolation will grow according to a logistic path and its growth is further fostered by the presence of industrial activity in the neighbouring one, whilst the second cluster is strictly dependent on the existence of the industrial activity in the first cluster for its very survival. A typical example of this kind is the relationship existing between the "core" of an industrial area (where the large companies are located) and the surrounding area, "periphery", crowded by several small firms (i.e. sub contractors and suppliers) - specialised in accessory productions - which are vitally linked to the industrial success of the core[36]. Similar relationships (which do not imply any spatial differentiation) may well arise between two different industries in the same cluster. A typical example (thoroughly discussed in Swann, 1993) is constituted by the relationships between the manufacturing core of an industrial cluster and the network of specialised service sectors. Since these specific business services must be "consumed" in the very place of their production and cannot be exported to other areas, the existence of a service sector within the cluster is crucially dependent on the existence of a prosperous manufacturing core.

These interactions can be formally described as follows:

$$\begin{cases} \dfrac{dn_1}{dt} = \left(a_1 - a_{11}n_1 + a_{12}n_2\right)n_1 \\ \dfrac{dn_2}{dt} = \left(-a_2 + a_{21}n_1\right)n_2 \end{cases} \qquad (11)$$

Where n_1 is the economic mass (number of located firms) of the manufacturing core cluster and r_q is the economic mass of the specialised service sector.

With the aid of a phase diagram (see Figure 4a), it is possible to see that the system has three equilibria. The first is trivial, unstable and corresponds to the origin; the second, stable, corresponding to the isolated maximum dimension of the core manufacturing industry; the third, which is a saddle point, corresponds to

[35]Where the term "facultative" refers to the fact that each industry, in isolation, develops and reaches its equilibrium level. The presence of an industry is therefore no necessary condition for the development of the other one.

[36]The Motor vehicle industry (which is clustered around Detroit in the USA, Coventry in the UK, Besançon in France, and Torino in Italy) displays a core-periphery spatial configurations in almost every countries.

a balanced co-operative relationship[37], in which manufacturing and service sectors coexist with mutual benefits. When the system lies in this unstable mutualistic equilibrium, small positive perturbations lead the system to an unlimited growth while small negative perturbations may cause the "extinction" of the service sector. This is the only possible outcome when $\frac{a_2}{a_{21}} < \frac{a_1}{a_{11}}$ (see Figure 4b).

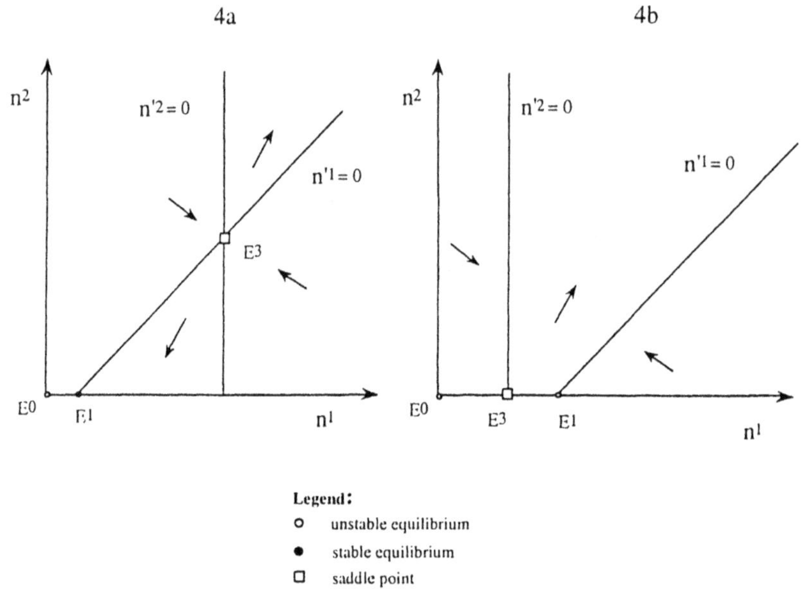

Fig. 4. Core-periphery dynamics (inter-industry commensalism).

2.6. Predatory Interactions

The relationships between two industries in the same cluster may also display a predatory pattern with one industry growing at the expense of the other.

Predatory interactions have been widely studied in the ecological and in the economic literature (one does not need to recall the names of Alfred Lotka, Vito Volterra and Richard B. Goodwin) and they may well describe the sort of interaction existing between a rising and a declining industry in a given area.

---

[37]Which exists if and only if $\frac{a_2}{a_{21}} > \frac{a_1}{a_{11}}$ (see Figure 4.5a).

The story is as follows: an established industry in a certain area produces, as a sort of spin-off, the birth of an interrelated industry (i.e. the footwear industry requires specialised machinery for cutting and sewing leather). If the derived industry has a higher value added per worker (or, more generally, if firms in this sectors are more productive and profitable) then it may well happen that the development of the second industry determines endogenously the crisis of the first one since it may cause the rise of local wages, rent and interest rate to such a level which is not sustainable for the original industry.

Analytically one may model a prey-predator system as a mix of systems (7b) and (9) since its essence lies in the combination of a competitive with a synergetic relationship.

$$\begin{cases} \dfrac{dn_1}{dt} = \left(a_1 - a_{11}n_1 - a_{12}n_2\right)n_1 \\ \dfrac{dn_2}{dt} = \left(a_2 - a_{22}n_2 + a_{21}n_1\right)n_2 \end{cases} \tag{12}$$

Where n_1 is the prey (or the older and currently declining industry) and n_2 is the predator (or the younger and rising one).

Predatory interactions typically display oscillatory dynamics. A phase diagram (see Figure 5) illustrates two possible configurations: case (a) where the coexistence of both population is a stable equilibrium surrounded by a converging spiral, case (b) where the coexistence of both population is an unstable equilibrium surrounded by a diverging spiral.

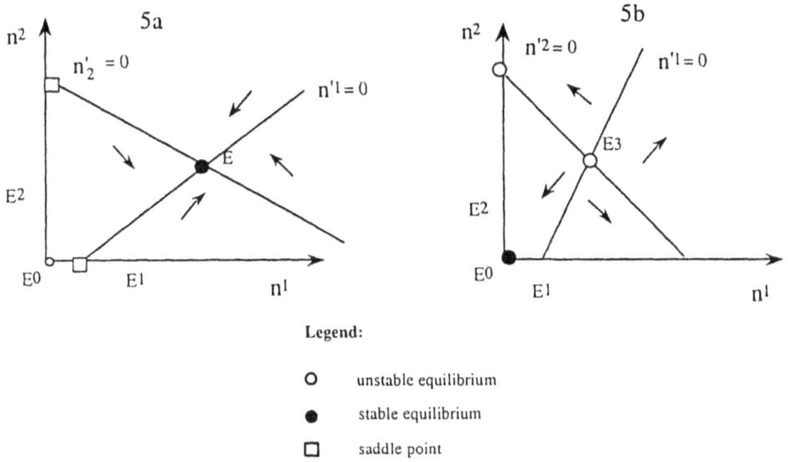

Fig. 5. Intra-cluster structural dynamics (inter-industry predation).

In the two situations, if the system starts far from the equilibrium, both predator and pray number oscillate. However, while in the first case the oscillations have a decreasing amplitude and, in the long run, the system reaches its equilibrium; in the second case the oscillations continuously increase their amplitude and the system dynamics may well determine the extinction of one or both populations.

Besides the one above mentioned, various type of interactions coexist within the development dynamics of industrial clusters. The qualitative identification of such interactions and the quantitative measurement of their effect on the growth rate of industries are thus left to the empirical analysis of section 3.3. The effects of the variability of macro-economic conditions on the behaviour of the systems have been extensively discussed in Gambarotto and Maggioni (1998).

2.7. Multi-Cluster Interactions: Some Methodological Remarks

Previous sections have shown some results that can be obtained by using a series of ecologically derived models to analyse the growth of an industrial cluster (or industry) in isolation and of a system of two interacting clusters (or industries).

It would seem therefore natural to extend these results to the case in which more than two clusters are involved. Unfortunately, this extension is not straightforward for empirical as well as theoretical reasons. The firsts refer to the geometrical increase of the relevant interaction parameters, while the second concerns the dynamic behaviour of the system which becomes qualitatively complex.

Two different approaches can be used to overcome this problem. The first relates to the possibility of constructing a large computational model, assigning plausible parameters' values and then simulating the dynamic of the system; the second consists of abandoning the description of individual cluster behaviour - within a multi-cluster complex - and focusing entirely on some overall aspects of the resulting structure, such as the relation between the increase in system complexity and its overall stability[38].

Even though it is difficult to draw some general conclusions, here are some principles and guideposts for multi-cluster analysis which could be derived from the original population ecology literature (see Levins 1968; Levin 1970; May 1974; Roughgarden 1979) which may give some indications on possible behaviour of such a model.

Equilibria may exist only within limited ranges of the interaction and environmental parameters (such as cluster's incipient growth rates, maximum

[38]In this section "increased complexity" is, rather loosely, defined in terms of increased number of cluster, more interactions and therefore more parameters. Stability is described as a lower level of fluctuations in the economic mass of the clusters, and as the ability of the system to recover from perturbations.

dimension, and macro-economic variability). Outside this restricted range, equations may undergo sustained oscillations and describe a collapsing system in which one or all clusters are completely deserted by firms.

Even when the value of parameters ensures an equilibrium solution, this equilibrium may be unstable to large perturbations in the "economic mass" of the cluster. This feature involves the concept of "resilience[39]" of the system and highlights the difference between global and local stability conditions.

Fixed parameter values have been assumed throughout the above analysis; on the contrary, in the real world, both interaction and environmental parameters may fluctuate through time, causing further perturbations to the system.

Studies based on a generalisation for *n*-agents of equation (7b), in which the coefficients are assigned randomly, show that as the number of agents involved increases, the probability to find a "feasible" equilibrium (i.e. where all clusters register a positive number of located firms) decreases. Biological models also show that dynamic stability typically decreases with the increase in the number of populations, or in the number and strength of inter-populations interactions[40].

The previous statement should not be interpreted as a general law stating that, in the real world, complex systems are less stable than simpler one. If a long run perspective is assumed - and we allow the market to act as a selection device - then it is likely that, in a relatively stable environment, a complex and fragile system will persist forever; whilst, in an highly unpredictable and exogenously determined environment the same system will collapse.

The Gause principle of competitive exclusion holds for multi-cluster systems as well. In its original ecological formulation it reads as follows: "no stable equilibrium can be attained in an ecological community in which some ε components are limited by less than ε limiting factors" (Levin 1970, p. 419-20)[41]. In economic terms this means that only one cluster can be sustained through a single-industry development which is based on one single type of firms. In other words a system of ε clusters is stable only if there are at least ε different types of firms.

---

[39]Which can be defined as a measure of the magnitude of the populations' perturbations that the system will tolerate before collapsing into some qualitatively different dynamic regime.

[40]"Thus, as a mathematical generality, increasing complexity makes for dynamic fragility rather than robustness" (May 1974, p. 160).

[41]Some similarities (not simply formal) can be found between Gause principle and the requirement on the number of objectives not to exceed the number of instruments in the theory of economic policy, and the relation between number of factors and number of commodities in international trade theory.

3. The Empirical Evidence

This section is twofold. On the one hand it is devoted to the description of the structure and evolution of four Italian industrial districts; on the other hand it contains an explorative descriptions of the implementation of the econometric estimation of the relationships described in section 2 and of the main problems (both methodological and related to the sample size) encountered in such an empirical analysis. Finally some first empirical results, derived from both econometric and calibration techniques.

3.1. Structure and Evolution of Three Industrial Districts: A First Descriptive Analysis

With about 300 firms and over 2,000 billion lire of annual sales revenues, the women's hosiery district of the Mantova area, with its centre at Castel Goffredo, is one of the successful industrial districts of the "*made in Italy*" business. The district, the absolute leader in its sector in Europe, contains all the leading companies operating in the business in area of a few square kilometres: the largest Italian groups, among which is Golden Lady (owner of the Omsa and Sisì trade marks), Csp International (now listed at the Milano Stock Exchange), Pompea, Levante, Calzificio Real, a number of small and medium-sized firms, as well as Filodoro, a company controlled by multinational Sara Lee, that settled in the district to enjoy the location economies and benefits available in a highly-skilled area.

The historical origins of this industrial district go back to the first women's hosiery factory (Noemi) built at Castel Goffredo in the early '30s, when the area's prevailing activity was still farming. The young were the first to leave the rural areas and start working in the industry, thus contributing to the development of the first textile settlement, which, however, had a rather short life. When Noemi closed down in 1958, the workers, employed until then in the various manufacturing stages of the hosiery industry, became entrepreneurs and set up small handicraft shops that would later slowly develop into the first firms of the district; this way, through a typical budding process, they gave rise to a network of small enterprises, each specialized in a single manufacturing process. Later, early in the '70s, the panty hose appeared in the domestic marketplace and was such a success that the local manufacturers were initially hardly able to meet the demand. The number of factories rose, also helped by the local allied activities such as the manufacture of machinery for the textile and stocking-weaving industry in particular (Lonati, based in Brescia, the first manufacturer of tubular hosiery machines in the world, was the biggest example). The Castel Goffredo pattern became more and more synonymous with quality, attracting big foreign dealers and encouraging the local operators to look outside the domestic boundaries to discover a strong export vocation that is still a typical feature of the district. One of the success factors of Castel Goffredo is the close relationship between the local manufacturers and large companies such as Du Pont and Nylstar (a fifty-fifty joint-venture between French Rhodia and Italian Snia) for the supply of nylon, the raw material for women's hosiery. With Lycra, for example, hosiery has become high-tech: the advantages of comfort and longer life were immediately appreciated by women, whose purchases led the sector to new production records. Paradoxically, however, the growing product quality is causing problems to some companies in the Mantova district: the increasingly strong panty hoses have been causing a sales decline, with some companies losing as much as 10-20% in the last few years. In addition, east European manufacturers are entering the business with a production of lower quality, which however is shrinking the Italian market shares in countries like Poland, Romania and Russia. In order to oppose this sales drop, the Castel Goffredo companies have changed their strategy, diversifying into the underwear industry, where they make seamless garments using a micro-fibre very similar to that used for hosiery. It is obviously a very expensive switch that requires capital investments in machinery, research work in materials and communications, which, however, has been successful for the district's larger companies, for some of which this new line of products already accounts for 40% of the total sales revenue.

Even better known than Castel Goffredo are the two industrial districts where most of the Italian manufacture of brass and bronze valves and fittings is concentrated: one is located in north-east Piemonte, around Orta Lake, astride the provinces of Novara and Vercelli (Valsesia) with some branching into Verbania; the other one, with slightly lower sales revenues and workforce, located around Lumezzane, in the province of Brescia.

The origin of the Cusio-Valsesia district of valves and fittings manufacture lies in the long tradition of brass and bronze processing, which dates back to the first bell casting activities at Valduggia as early as in the 15th century. The valves and fittings industry gradually developed around the villages in the low Cusio region, Valsesia and the low-central Novara region in the last century, especially starting from the '50s. In the last ten years, in particular, the district consolidated its leadership in the international market, with a number of leading companies pulling the whole district, giving a major contribution to its worldwide reputation thanks to medium-high quality levels, a strong process and product innovation rate and, as far as sanitary faucets are concerned, a high design level.

According to a balance sheet survey and questionnaires collected by the Fondazione Comunità e Innovazione and the Research Centre in Economic Analysis, International Economics and Economic Development (CRANEC) of Università Cattolica (Milano), in 1994 the Cusio-Valsesia district of valves and fittings, without considering the allied industries, posted sales for about 1,850 billion lire of finished products, which, according to estimates based on the projections of a smaller sample of businesses, rose to 2,200 in 1995, were more or less unchanged in 1996, rose further to 2,400-2,450 billion a year over the 1997-99 period and reached about 2,600 billion lire in 2000. Brass valves for sanitary fixtures accounted for 43% of the sales revenues and chromium-plated sanitary faucets accounted for the remaining 57%.

With approximately 1,700 billion lire's worth of sales abroad, the Cusio-Valsesia district accounts for approximately one third of the Italian exports of the sector and has a share of about 5% in the world exports and of 9-10% of the EU's exports of item 84.81, representing the most important valves and fittings manufacturing area in the world. The shares are even larger if brass and bronze valves and fittings only are considered, which is the sector the district is specialized in. The exports are about 60% of the total sales, with some companies reaching peaks of 80% or more and, in a few cases, almost the total sales. The main exportation markets are the EU countries (Germany, Great Britain, France and Spain in this order), the United States and the Middle East, with growing exports to Eastern Europe, the Far East and South America in the last few years.

According to ISTAT's intermediate industry census, in 1996 the Novara, Vercelli and Verbania provinces recorded 431 companies in the valves and fittings sector, employing 8,101 workers. More difficult is the assessment of the impact of the allied industries because many activities (e.g. metal treatment and polishing) are shared with another industrial district in the area, namely that of cookware and household goods, concentrated in the Omegna area. It is a fact, however, that, again in 1996, the moulding, casting, metal treatment and polishing businesses in the three mentioned provinces employed a workforce of 6,897 distributed over 1,139 local units, a lot of them very small in size, and most of them are more tied to the valves and fitting business than to that of household goods. It can therefore be estimated that the Cusio-Valsesia district of valves and fittings involves not less than 12,000 jobs in direct activities and allied industries, excluding services such as transport, advertising, etc.

The growth of the valves and fitting sector in the Novara, Vercelli and Verbania provinces has been very fast: according to the ISTAT censuses, the jobs were 3,228 in 1971, 5,390 in 1981, 6,684 in 1996 and, as already said, 8,101 in 1996. Based on a projection of the figures of the top 44 companies, the direct jobs are likely to have risen to about 8,800 in 1999-2000. Equally significant was the employment growth in the casting, moulding, galvanic treatment and polishing of metals in the three provinces: the jobs rose from 3,146 in 1971 to 6,897 in 1996 and there is no doubt that most of the increase can be ascribed to the growth of the valves and fitting business in the area.

Through the figures of the ISTAT census of 1996 we can also assess the district from the viewpoint of the size of the businesses. Considering only the valves and fittings sector proper (i.e. without the allied industries), we find: 3 big companies (with more than 250 workers), 34 medium-sized enterprises (50 to 249 workers), 132 small enterprises (10 to 49 workers) and 262 micro-enterprises (less than 10 workers). Again according to the 1996 ISTAT census, the main manufacturing centre in the district is San Maurizio d'Opaglio (see Figure 6), with about 2,000 workers employed in the valves and fittings sector and allied industries, followed by Gozzano (about 1,500 workers plus allied industries), Pogno (1,000 workers), Valduggia (900 workers) and Pella (500). Other important municipalities in the district are Invorio, Borgomanero, Briga Novarese, Fontaneto d'Agogna, Borgosesia, Quarona and Grignasco.

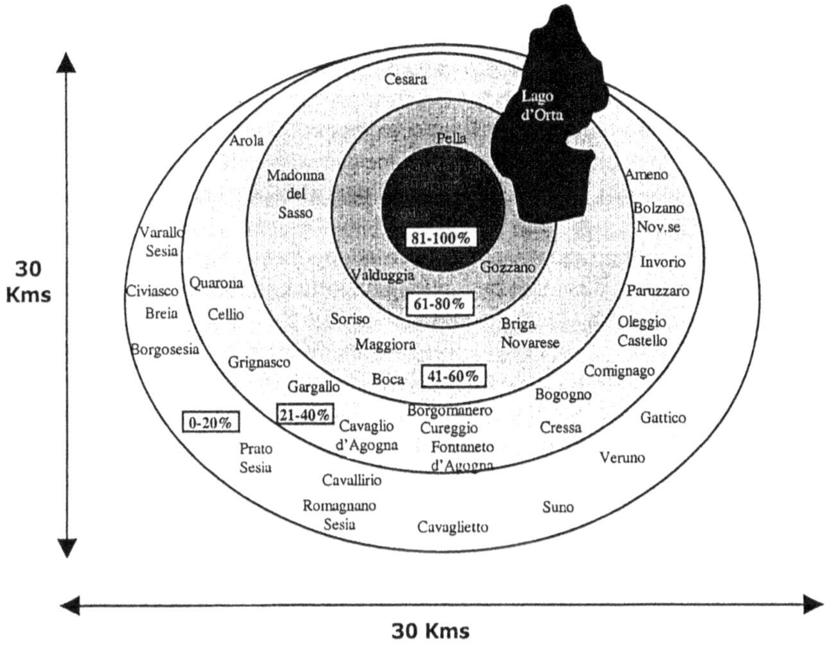

Fig. 6. Epicenter of the Cusio-Valsesia district of taps and valves based on the % impact of the local units specialized in the main business and allied industries on the total local units of each municipality: year 1996.

In addition, if the workforce figures of each municipality in 1991 and 1996 are compared, a growth appears in the relative weight of the most external municipalities as a result of new production settlements outside the ring of the central core of the district's municipalities. As a result of this trend, though the number of workers has been rising in the 5 central municipalities (Pogno, S. Maurizio d'Opaglio, Pella, Valduggia and Gozzano), the relative impact of the workers on the total workforce of the district decreased from 58% in 1991 to 56% in 1996, while that of the peripheral municipalities rose from 42% to 44%.

The historical factors of success of this district include some elements common to the two distinct businesses of sanitary faucets and valves. First of all, the previously mentioned tradition of brass and bronze processing in the area. This tradition has been developing over the years and has produced a very high specific know-how in the manufacture of both valves and faucets, training of highly qualified technicians and workers and setting up a strong network of relationships inside the district for the supply of components, semi-finished parts and services. Another strong point of the district as a whole is its proximity to the manufacturing centres of the raw material, the brass bar, and the availability of a scraps recycling system. In addition, the leading role of the bigger manufacturers was a major factor in building the world reputation of the district and the quality of the output, which meet the highest international standards. In the sector of sanitary faucets, design is an important element for success, on which the district concentrated great efforts to be able to compete with the big foreign manufacturers. Compared to the valves sector, that of sanitary faucets has seen the proliferation of small businesses, sometimes only assemblers, whose development was possible thanks to the strong presence of component manufacturers and providers of services (galvanizing, polishing).

Let's review now the Lumezzane and Brescia district in Lombardia, which is a bit smaller than that of Piemonte but equally important in the international market. Located north of Brescia, along the Trompia and the Sabbia valleys, this industrial district is mainly concentrated in Lumezzane, the pivot of the district, Sarezzo and Villa Carcina, with branches both to the north through Marcheno and Gardone Val Trompia and to the south as far as Concesio and Brescia itself.

As highlighted for the Piemonte district, also here tradition was a decisive factor in the birth and development of the industrial district. Iron working flourished here thanks to the abundance of minerals and of water already in ancient times. It was arms manufacture that made these valleys well known as early as in Roman times. Later, in the course of the centuries, the manufacture of farming tools, cutlery and crockery developed and, more recently, that of brass and bronze valves and fittings. In addition, the old tradition of iron and metal working in the Brescia province allowed this district to specialize in the upstream steps of the manufacturing process, i.e. in casting and moulding of the raw material (brass bar), with the valves sector prevailing over that sanitary faucets.

In order to complete the analysis of the Italian valves and fittings industry, the Fondazione Comunità e Innovazione and Università Cattolica (Milano) studied this second industrial district with the same methodology as used for the other one. An initial quantitative assessment of the district was obtained through an extensive survey made in 1995 by means of questionnaires sent to the enterprises and an analysis of the balance sheets of the bigger companies operating in the sector in the Lumezzane and Brescia area: the total sales revenues in 1995 amounted to 1,670 billion lire and the exports to over 850 billion. In addition, the intermediate ISTAT census of 1996 recorded for item 29.13, "manufacture of valves and fittings", 185 local units and 4,531 workers in the Brescia province. For the following years, instead, the figures were updated on the basis of a sample of 35 companies, 19 of which manufacturing valves, 8 faucets and 8 intermediate components. From the data of the sample it was possible to estimate the total figures of the district both for 1994, i.e. one year before that of the first survey, which was 1995, and for the following years until 1999. Starting from 1,350 billion of sales revenues in 1994, the Lumezzane district recorded a fast growth (+23%) and reached 1,670 billion in 1995; the 1996 figure was slightly lower (1,650 billion), while increases to 1,890 billion (+14%) and 1,950 billion (+3%) were recorded in 1997 and 1998, respectively. The latter level was maintained in 1999.

The same trend was recorded in exports, which exceeded the 1,000 billion threshold in the last few years. Like the Piedmontese district, that of Lumezzane is a strong exporter: on the whole the ratio of exports to total sales revenues is about 50%, but a large number of enterprises reach much higher percentages, with peaks of 98%. Considering the sectors, instead, the impact of exports on sales revenues is higher for sanitary faucets (60% on the average) than for valves (50-55%) and intermediate components (about 35%). The last-named sector, in particular, is represented by the enterprises that operate in auxiliary processing such as metal polishing, colouring or chromium-plating, or manufacture intermediate components rather than finished products. For all these firms, usually small in size, the outlet is the local market, which explains the smaller percentage of sales abroad.

As mentioned before, unlike the other district, that of Lumezzane is more specialised in valve manufacture. The enterprises operating in this sector are many more than those manufacturing sanitary faucets, and account for 60% of the sales revenues of the whole district. Some of the bigger valve manufacturers, in addition, take care also of the initial step of the manufacturing process, i.e. casting and moulding of the brass bar; these are capital intensive operations which require the installation of casting or moulding works, so the companies having such plants tend to use them also for other companies in the sector, to which they sell raw product for subsequent processing.

Finally, concerning the allied industries, it is very difficult to estimate the number of workers employed in allied industries because of the overlapping of the valve and fitting sector with other mechanical industries present in the area, such as that of cookware manufacture.

At this point, after the women's hosiery district of Castel Goffredo and the two valve and fitting districts, we are presenting briefly the case of the Varese province, with special reference to the plastics material processing industry and related machinery. The chemical, plastics and rubber industry is undoubtedly a major sector in this area of Lombardia; it is a broad and diversified business that ranges from chemical commodities to pharmaceutical products, from semi-finished to finished products. Apart from the drugs industry, that has been developing especially thanks to multinationals that selected the Varese province for the existing technical and production background and its favourable location between Milano and Switzerland, it is the manufacture of rubber and above all of plastics articles that saw the biggest growth, making it one of the Italian areas with the highest concentration of plastics conversion companies and a real industrial district. To give an idea of the extension of this sector, in 1996 (intermediate ISTAT census figures) the enterprises operating in the "manufacture of rubber and plastic articles" in the Varese province were 831, with a total workforce of about 11,900 and a research conducted by Università Cattolica of Milano in 1998 estimated that the 200 bigger companies in the district posted sales revenues of about 3,400 billion lire.

Historically, the conversion of plastic materials and rubber entered the industrial system of the Varese province immediately after World War II, but the real growth was recorded in the last two decades, with the breakthrough of plastics in the manufacture of intermediate and finished product. A major factor allowing the birth and consolidation of this industry, besides the diversification of the expertise available in the region and the presence of a few big companies operating in the sector, was the strong concentration in the province (the highest in Italy) of manufacturers of calenders, i.e. the equipment that - starting from the raw material, here plastics - makes sheets of the desired thickness for subsequent moulding. The manufacture of plastics conversion equipment is in line with the long mechanical tradition of the Varese province. The mechanical industry, that was initially a support for the textile industry, especially for maintenance of looms of British or German origin, has been expanding to become one of the main businesses in the province. It is emphasised that the mechanical industry of this province has been closely tied to the history of the means of locomotion: from the early Caproni, Marchetti and Macchi airplanes to the Isotta Fraschini cars, the Cagiva motorbikes and the Agusta helicopters. Now the most widely spread and qualified mechanical production is tied to the conversion of plastic, a sector, again according to the ISTAT census of 1996, counting 452 local enterprises with a workforce of over 7,300. With the mechanical industry it is more difficult to quantify the production value because the mentioned Università Cattolica research surveyed only a very small portion of the district's companies, but it estimated overall sales revenue of about 650 billion lire for the 40 leading equipment manufacturers in 1998.

Therefore, on the whole, the Varese district of plastics and related mechanical industry consists of about 1000 local units with over 19,000 workers.

3.2. The Data

The data used in the following section concern the level of employment and the number of local units in different "clusters" from 1951 to 1991 (for some districts data for 1996 are also included). Data are taken from the "Censimento delle attività produttive", the Italian businesses' census which is run each 10 years[42]. The term cluster here identifies the intersection of a given industry with a given territorial area (Provincia). It is therefore clear that in the empirical analysis we are counting employees and local units which do not strictly belong to the cluster (which is a subset of the provincial territory) but data at a finer geographical disaggregation (Comune) are available only from 1991.

Data thus suffer from two main problems: the first, as explained above, concerns the imprecise identification of the cluster, the second regards the small numerosity of the available sample (6 observations at most). A similar analysis has been conducted on US high-tech cluster in Maggioni (1999). There, by using County

[42] 1996 is the first attempt to build an "intermediate census" to shorten the time lag from 10 to 5 years.

Business Patterns, it has been possible to get more precise identification, yearly observations and longer time series (about 40 observations).

The empirical exercise has therefore been forced to use a combination of econometric analysis and calibration techniques. Some results were intrinsically fragile and statistically not very significant. Further research on this issue is still needed. We therefore decided to present only a subset, which yields the more robust and significant results.

3.3. Econometric Analysis and Calibration Techniques

The analytical formulation of the abovementioned interactions when the number of interacting entities is equal to 2 is as follows:

$$\begin{cases} \dfrac{dn_r}{dt} = \left(a_r - a_{rr}n_r - a_{rs}n_s\right)n_r \\ \dfrac{dn_s}{dt} = \left(a_s - a_{ss}n_s - a_{sr}n_r\right)n_s \end{cases} \tag{13}$$

where n_r and n_s are the "economic masses" (number of plants or level of employments) in cluster r and s and all parameters are as in equation (7b). In particular a_r and a_s are the intrinsic rates of increase of each cluster in isolation; a_{rr} and a_{ss} are the intra-cluster competition parameters which reflect the inhibiting effects that a firm's entry has on the growth rate of the same cluster (because of congestion effects); a_{rs} and a_{sr} (inter-cluster competition parameters) show the inhibiting effects that one firm, locating in a cluster, has on the growth of the other region (inter-cluster competition parameters).

In theory, one then would like also to take into account the inter-industry interactions which develop within the same cluster. If this is the case, then a two-regions, two sectors can be modelled as follows

$$\begin{cases} \dfrac{dn_r^i}{dt} = \left(a_r^i - a_{rr}^i n_r^i - a_{rs}^i n_s^i + a_r^j n_r^j\right)n_r^i \\ \dfrac{dn_s^i}{dt} = \left(a_s^i - a_{ss}^i n_s^i - a_{sr}^i n_r^i + a_s^j n_s^j\right)n_s^i \\ \dfrac{dn_r^j}{dt} = \left(a_r^j - a_{rr}^j n_r^j - a_{rs}^j n_s^j + a_r^i n_r^i\right)n_r^j \\ \dfrac{dn_s^j}{dt} = \left(a_s^j - a_{ss}^j n_s^j - a_{sr}^j n_r^j + a_s^i n_s^i\right)n_s^j \end{cases} \tag{14}$$

where the subscripts r, s refer to areas, and the superscripts i, j refer to industries.

It appears evident that, even though this formulation is linear, and expresses the agglomeration diseconomies without exogenously determined parameters, its applicability is limited by the large number of parameters which have to be estimated.

We thus proceeded by regressing the growth of the economic mass of a given cluster (i.e. a couplet industry-provincia) on the more relevant interaction coefficients (these being either inter-regional or inter-industry) as derived from theoretical *a-priori*.

We therefore decided to test two sort of interactions:

- the interactions between two districts in the same industry (looking for signs of territorial competition) namely: Cusio-Valsesia versus Lumezzane, for the taps and fitting industry and Vigevano versus Macerata and Ascoli Piceno for the footwear industry;
- the interactions between two industries in the same cluster (looking for signs of competitive or synergetic inter-industry relation). In particular we focus on the relation between the sector producing a final good (these being plastic and rubber products or footwear) and the sector producing machinery specialised for that particular type of good. The two chosen area are: Varese for the plastic and rubber/specialised machinery sector and Vigevano for the footwear/specialised machinery sectors.

From the regression we got estimates for the coefficient of the system equation (13). By simple calculations it is then possible calculate the coefficients of the isocline for each population, i.e. the combination of the economic masses of the two populations which ensure that the population under examination do not varies over time.

Analytically one must find the solution of the following equations:

$$\frac{dn_1}{dt} = 0 \text{ and } \frac{dn_2}{dt} = 0$$

which have a trivial solution for n_1 and n_2 equal to zero and a non trivial soiution which identify the following two lines:

$$n_2 = -\frac{a_1}{a_{12}} - \frac{a_{11}}{a_{12}} n_1$$

$$n_2 = -\frac{a_2}{a_{22}} - \frac{a_{21}}{a_{22}} n_1$$

and plot them in the phase plan. The intersection of two isoclines is an equilibrium point for the system, since in this point both populations do not change overtime.

If one then plots in the same space the actual measure of the two populations in the cluster one may suggest the future evolution of the cluster buy predicting the direction of variation of the two carrying capacities. Figure 7, 8, 9 10 and 11 present the main results.

When the econometric estimates were either fragile or not particularly significant or made impossible by the small size of the sample, we used an alternative technique which implies the determination of the coefficients for each equation in system (12) by substituting the variables' value for different triplets of years and moving the triplets down the observations in order to "test" the stability and robustness of these coefficients to changes in the years-base.

Figure 7 and 8 represent the synergetic relationships existing between the plastic and rubber industry and the specialised machinery sector in Varese (where the measured variable is the number of plants) obtained through both calibration and econometric estimates.

Both techniques show the existence of synergetic relationships: the development of one sector fosters the development of the other one. However the two pictures differ in the dynamical structure. The estimates diagram (Figure 7) shows a saddle point equilibrium (approximately 350 machinery plants and 650 plastic and rubber plants) and suggests that while the presence of the machinery industry is vital for the existence of the plastic and rubber producers (which buy most of their equipment within the cluster) the relevant market for the machinery industry (which historically developed after the plastic and rubber one) is the world; therefore the connections with local producers of plastic and rubber products are convenient but no more necessary.

The calibration diagram (Figure 8) suggests a pure symmetric synergetic interactions in which the carrying capacity of both population grows indefinitely.

Figures 9, 10 and 11 show the other side of intra-cluster relationships. Figure 9 shows a competitive interaction between footwear industry and the specialised machinery in Vigevano, when the economic mass is measured by the number of plants. Figure 10 and 11 show predator-prey relationships between these two sectors. The lung run forecasting for this sectors in both cases shows the development of the machinery sector at the expenses of the footwear industry.

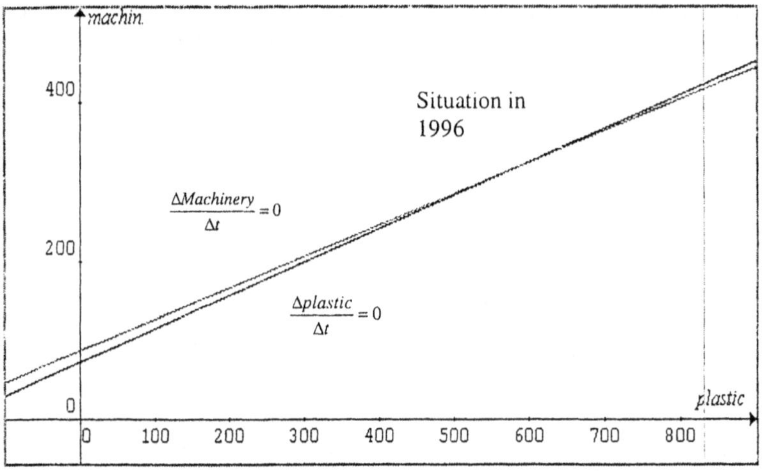

Fig. 7. Long run evolution of Plastic and Machinery in Varese, number of plants (estimates).

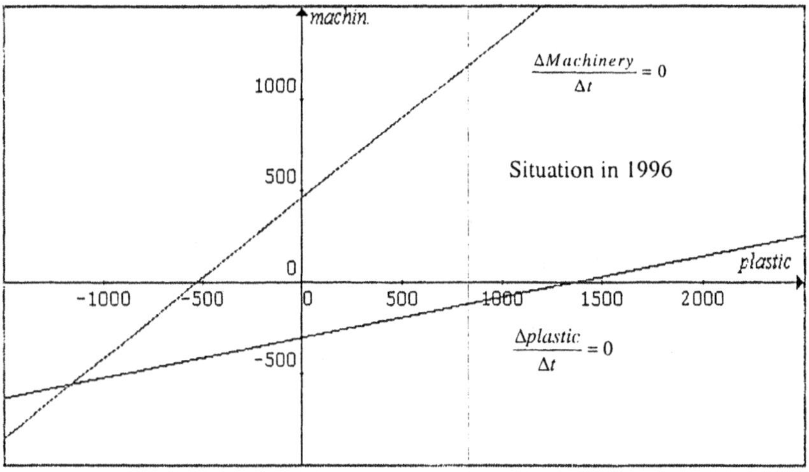

Fig. 8. Long run evolution of Plastic and Machinery in Varese, number of plants (calibration).

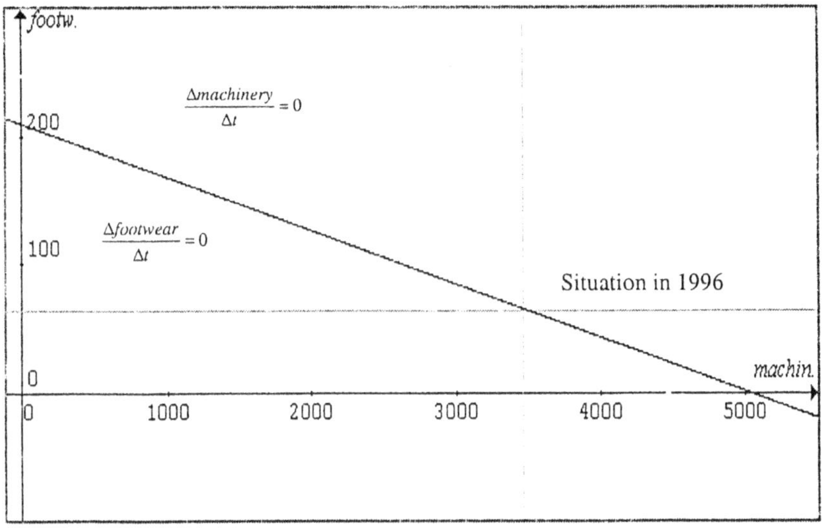

Fig. 9. Long run evolution of Footwear and Machinery in Vigevano, number of plants (estimates).

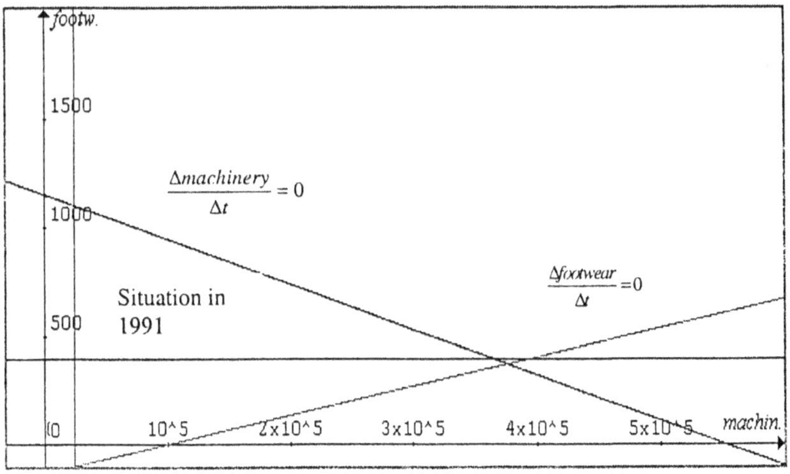

Fig. 10. Long run evolution of Footwear and Machinery in Vigevano, employees (calibration).

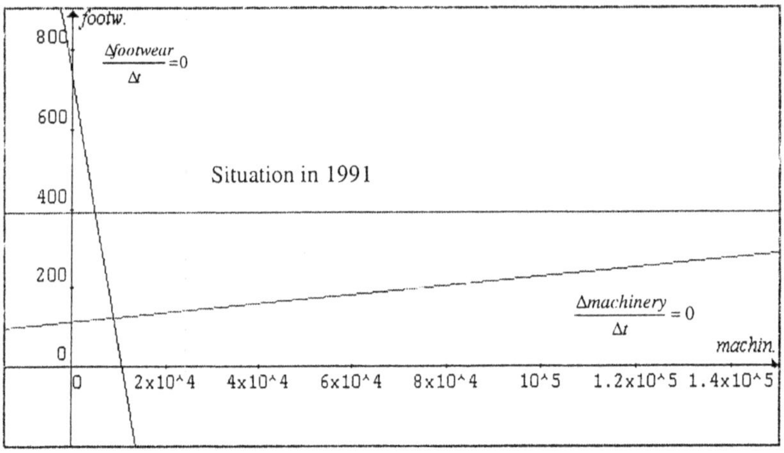

Fig. 11. Long run evolution of Footwear and Machinery in Vigevano, employees (calibration).

4. Conclusion

The paper has shown the relevance of industrial cluster of SMEs in the economic structure of Italy. The analysis of such a phenomenon, and of its most peculiar organisational model, the industrial district, have often be based either on isolated case studies, which were very difficult to compare, or on cross-sectional econometric analyses, which obtain some general results at the expenses of much of the individual cluster specificities.

The use of population ecological models may solve some of these problems being explicitly devoted to the analysis of inter-industry and inter-cluster relationships. The estimation of such models entails several methodological difficulties which are made worst by the limited availability of data. However, if this paper succeeds in getting scholars' attention on these techniques and make them part of the toolbox of industrial economists and the regional scientists, we'll have reached our main target.

References

Arthur, W. B. (1988), "Urban System and Historical Path Dependency", in: *Cities and Their Vital Systems*, Ausubel, J. H. and Herman, R. (eds),Washington D.C., National Academy Press, pp. 85-97.

Arthur, W. B. (1990), "Silicon Valley Locational Clusters: When Do Increasing Returns Imply Monopoly?", *Mathematical Social Sciences*, 19, pp. 235-251.

Beltrami, E. (1987), *Mathematics for Dynamic Modelling*, Boston, Academic Press.

Capelo, A. C. (1989), *Modelli matematici in biologia*, Padova, Decibel.

Dendrinos, D. S. and Mullally, H. (1985), *Urban Evolution. Studies in the Mathematical Ecology of Cities*, Oxford, Oxford University Press.

Fortis, M. (1998), *Il "made in Italy" – Quando stile e creatività non sono solo moda*, Bologna, Il Mulino.

Fortis, M., Clerici, A. and Nodari, A. (1999), "Il distretto della rubinetteria-valvolame del Piemonte Nord Orientale: Cusio e Valsesia", in: *Aree distrettuali prealpine – Meccanica, tessile, gomma e plastica*, Fortis, M. (ed.), Milano, Franco Angeli.

Fortis, M. and Nodari, A. (1999), "Il distretto bresciano della rubinetteria-valvolame: Lumezzane", Fortis, M. (ed.), *Aree distrettuali prealpine – Meccanica, tessile, gomma e plastica*, Milano, Franco Angeli.

Fortis, M., Bassetti, G. and Nodari, A. (1999), "Il distretto mantovano delle calze femminili: Castel Goffredo", in: *Aree distrettuali prealpine – Meccanica, tessile, gomma e plastica*, Fortis, M. (ed.), Milano, Franco Angeli.

Fortis, M. and Nodari, A. (2001), *The Role Played by the Industrial Districts in the Italian Valve, Tap and Fitting Sectors*, paper presented at CEIR (Comitato Europeo Industrie Rubinetterie) General Committee, Venezia, 11-12 June.

Gambarotto, F. and Maggioni, M. A. (1998), Regional Development Strategies in Changing Environments: an Ecological Approach, *Regional Studies*, 32, pp. 49-61.

Gause, G. F. (1934), *The Struggle for Existence*, Baltimore, Williams & Wilkins.

Isard, W. (1956), *Location and Space-economy*, Cambridge (Mass.), MIT Press.

Kingsland, S. H. (1985), *Modelling Nature. Episodes in the History of Population Ecology*, Chicago, The University of Chicago Press.

Levin, S. A. (1970), "Community Equilibria and Stability, and an Extension of the Competitive Exclusion Principle", *American Naturalist*, 108, pp. 207-228.

Levins, R. (1968), *Evolution in Changing Environments: Some Theoretical Explorations*, Princeton, Princeton University Press.

Lotka, A. J. (1925), *Elements of Physical Biology*, Baltimore, Williams & Wilkins.

Maggioni, M. A. (1993), *Ecological Dynamics and Critical Mass Processes in the Location of High Tech Firms*, paper presented at the 40th RSAI (Regional Sciences Association International) Conference, North American Section, Houston, 11-14 November.

Maggioni, M. A. (1999), *Clustering Dynamics and the Location of High-Tech Firms*, Ph.D Thesis, Coventry (UK), University of Warwick, July, forthcoming by Springer Verlag, Heidelberg.

Maggioni, M. A. (1994), "Modelli ecologici per l'analisi della dinamica industriale regionale", in: *Modelli d'analisi e d'intervento per un nuovo regionalismo*, Pasquini, F., Pompili, T. and Secondini, P. (eds), Milano, Franco Angeli, pp. 79-105.

Maggioni, M. A. and Gambarotto, F. (1997), "Sviluppo locale, dinamiche globali e ruolo dell'operatore pubblico: un modello ecologico", in: *La dinamica dei sistemi produttivi territoriali: teorie, tecniche, politiche*, Bramanti, A. and Maggioni, M. A. (eds), Milano, Franco Angeli, pp. 280-306.

Maggioni, M. A. and Porro, G. (1994), Dinamiche di crescita regionale: il ruolo delle aspettative in modelli di tipo ecologico, *Quaderno della ricerca di base "Modelli di sviluppo e regional competition"*, Milano, Università Bocconi, 6.

Maino, R. (1989), "Aree interne ed interdipendenze settoriali. Uno schema interpretativo fondato su modelli dinamici del tipo preda-predatore", in: *Aree interne, tutela del territorio e valorizzazione delle risorse,* Becchi Collidà, A., Ciciotti, E. and Mela, A. (eds), Milano, Franco Angeli, pp. 81-110.

Marshall, A. (1920), *Principles of Economics*, London, McMillan.

May, R. M. (1974), *Stability and Complexity in Model Ecosystems*, Princeton, Princeton University Press.

May, R. M. (1976), *Theoretical Ecology, Principles and Applications*, Philadelphia, Saunders.

Metcalfe, J. S. (1994), "Competition, Fisher's Principle and Increasing Returns in the Selection Process", *Journal of Evolutionary Economics*, 4 (4), pp. 327-346.

Miyao, T. and Kanemoto, Y. (1987), *Urban Dynamics and Urban Externalities*, London, Harwood Academic Publishers.

Nijkamp, P. and Reggiani, A. (1992), *Interaction, Evolution and Chaos in Space*, Berlin, Springer-Verlag.

Nijkamp, P. and Reggiani, A. (1998), *The Economics of Complex Spatial Systems*, Amsterdam, Elsevier.

Papageorgiou, Y. Y. (1979), "Agglomerations", *Regional Science and Urban Economics*, 9, pp. 41-49.

Pearl, R. and Reed, L. J. (1925), *Skew-Growth Curves*, Proceeding of the National Academy of Natural Sciences of the USA, 11, pp. 16-22.

Quadrio Curzio, A. and Fortis, M. (eds) (2000), *Il made in Italy oltre il 2000 - innovazione e comunità locali*, Bologna, Il Mulino - collana promossa dalla Fondazione Comunità e Innovazione.

Quadrio Curzio, A., Fortis, M. and Maggioni, M. A. (eds) (1996), *I distretti economici delle prealpi e delle alpi centrali*, Sondrio, Credito Valtellinese - Collana socio-economica.

Richardson, H. W. (1978), *Regional and Urban Economics*, Harmondsworth, Penguin Books.

Roughgarden, J. (1979), *Theory of Population Genetics and Evolutionary Ecology: an Introduction*, New York-London, Macmillan.

Storper, M. and Walker, R. (1984), "The Spatial Division of Labour: Labour and the Location of Industries", in: *Sunbelt/Snowbelt: Urban Development and Regional Restructuring*, Sawers, L. and Tabb, W. K. (eds), Oxford, Oxford University Press, pp. 19-47.

Storper, M. and Walker, R. (1989), *The Capitalist Imperative. Territory, Technology and Industrial Growth*, New York, Basil Blackwell.

Swann, G. M. P. (1993), *Can High Technology Services Prosper if High Technology Manufacturing Doesn't?*, Centre for Business Strategy Working Paper Series, London, London Business School, 143.

Swann, G. M. P. (1998), "Towards a Model of Clustering in High-Technology Industries", in: *The Dynamics of Industrial Clustering*, Swann, G. M. P., Prevezer, M. and Stout, D. (eds), Oxford, Oxford University Press, pp. 52-76.

Tauchen, H. and Witte, A. D. (1983), An Equilibrium Model of Office Location and Contact Patterns, *Environment and Planning A*, 15, pp. 1311-1326.

Verhulst, P. F. (1845) *Recherches mathématique sur la loi d'accroissement de la population*, Nouveaux Mémoires de l'Académie Royale des Sciences et Belles-Lettres de Bruxelles, 18 (2), pp. 3-38.

Volterra, V. (1926), *Variazioni e fluttuazioni del numero di individui in specie di animali conviventi*, Roma, Memorie Accademia dei Lincei, 2, pp. 31-113.

Weber, A. (1929), *Theory of the Location of Industry*, Chicago, Chicago University Press.

On the Ubiquitous Nature of Agglomeration Economies and Their Diverse Determinants: Some Notes

Giulio Bottazzi, Giovanni Dosi and Giorgio Fagiolo[1]

Abstract. This highly preliminary work attempts to study the multiple drivers of agglomeration phenomena in contemporary economies and proposes a tentative taxonomy where the conditions of knowledge accumulation, often specific-to-specific locations and specific sectors, play a paramount role. We discuss the achievements and limitations of current theorizing on spatial location of economic activities, and we propose a simple model, which is estimated on Italian data, highlighting the rich intersectoral diversity of agglomeration forces, together with, in a few cases, also lack of them.

1. Introduction

The embeddedness of economic processes into underlying spatial dimensions ought to be sufficiently straightforward not to require much further elaborations. Indeed, *spatial dimensions* include both literally geographic aspects - related to the physical locations of agents - and more metaphorical metrics - regarding e.g. technological and institutional "distances"; mechanisms of inclusion/exclusion between networks, organizations and, of course, nations; degrees of information and knowledge sharing, etc. Having said that, it is equally easy to acknowledge that the economic discipline is far from offering anything resembling robust accounts of spatial localization of economic activities, or, even less so, their underlying generating dynamics. Needless to say, there is no possible claim of any systematic answer here to such tangled questions. Much more modestly, we shall add a few further question marks, together with some hints on hopefully novel interpretative conjectures (Indeed, in what follows, we shall somewhat indulge on our naiveté as newcomers to the field!). If *space* - however defined - *matters*, it is also because particular "places" in it, persistently affect (i) identities, capabilities

[1]Support to the research by the Italian Ministery of University and Research (MURST, Project No. 2-13-2-E4099GD) is gratefully acknowledged. Roberto Monducci from the Italian Statistical Office (ISTAT) has been, as usual, exceptionally supportive. Marco Lippi, Carolina Castaldi, Deborah Tappi and several participants to the conference "Complexity and Industrial Clusters: Dynamics and Models in Theory and Practice", Milan, June 2001, organized by the Fondazione Comunità e Innovazione, have offered precious comments.

and behaviours of individual agents; (ii) interaction patters; and, ultimately, (iii) individual *and collective* performances. In turn, this means that sheer geography, together with institutional and technological specificities, ought to be studied in their long-term effects upon economic structures and relative efficiencies. Even when going along with many our fellow economists in (the admittedly despicable practice of) blackboxing institutional diversities across nations and regions, one should nonetheless be able to track intersectoral differences in the agglomeration drivers exerted by technological factors. This is the first point that we shall address below (Section 2). Second, an assessment - albeit quite telegraphic - of the state-of-the-art of diverse strands of "economic geography" might help in flagging out achievements, standing shortcomings and challenges ahead (Section 3). Theories on agglomeration (*and dispersion*) forces urgently demand stronger links with empirical predictions. This is what we begin to explore in Section 4, while in Section 5 we put our techniques to work on three - quite diverse - industrial sectors on Italian data (i.e. primary metals, transport equipment and furniture). Broader conjectures will be put forth in the conclusions (Section 6) of a report - we want to emphasize - which is very much preliminary and "work-in-progress".

2. Empirical Evidence on Agglomeration Phenomena: On Some Facts and Puzzles

A survey of the enormous empirical literature on agglomeration economics in general and industrial districts in particular is well beyond the scope of these notes[2]. Here let we just mention four sets of empirical regularities.

2.1. Discrete Types of Agglomeration Structures

The first robust piece of evidence concerns the variety of agglomeration phenomena yielding equally diverse "types" of local structures, including the following broad classes.

1. *Horizontally Diversified Agglomerations.* They comprise a good deal of the "made in Italy" districts, presenting remarkable ever-changing product varieties generally produced by a multiplicity of small and medium firms (e.g. clothing, textiles, jewelleries, tiles, etc.).
2. *Agglomerations of Vertically Disintegrated Activities.* Again, largely overlapping with the former, they include a quite few "made in Italy" districts whereby activities previously vertically integrated within individual firms undergo a sort of "Smithian" process of division of labour *cum* branching out of different firms. In some analogy with the old "pin

[2]On the Italian evidence about industrial districts, see, among the others, Signorini (2000), Brusco and Paba (1997) and Onida et al. (1992).

story" of Adam Smith, division of labour and spatial agglomeration rest upon: (a) economies of specialization; (b) input-output links; and (c) user-producers exchanges of knowledge (see shoe making and textile/clothing among others).

3. *Hierarchical Spatially Localized Relations.* They generally involve an "oligopolistic core" and subcontracting networks (although not necessarily mechanisms of technological dominance and rent-extraction by such a "core"). In Italy, transport equipment, white goods, etc. are good cases to the point.

4. *Agglomeration Phenomena Based on Knowledge Complementarities - at least partly fuelled by "exogenous science"*- the "Silicon Valley" in the U.S. being the most famous example. Incidentally, note that in Italy this type of agglomeration is almost non-existent.

5. *Agglomerations as Sheer Outcomes of Path-Dependence* - for example due to spatial inertia in the birth and death of firms - without however any particular advantage of agglomeration itself[3].

Different types of agglomeration clearly hint at possibly different drivers of agglomeration itself and their different sectorial specificities.

2.2. Intersectorial Differences in the Importance of Agglomeration Economies

A second, related, set of empirical regularities concerns the intersectorial differences in the revealed importance of spatial agglomerations. Figure 1 summarizes the statistical evidence on the contribution of districts production to exports in Italy at quite high degrees of sectorial disaggregation[4], highlighting a characteristic skewed distribution. Complementary evidence on the contribution of individual districts to the total Italian exports of the sectors in which they are specialized (Figure 2) confirms the idea of a significant divide between a group of "districts activities" and the rest of industrial production for which agglomeration economies appear to be much less relevant[5].

---

[3]This is indeed the thesis of Klepper (2000) concerning the role of the Detroit area in automobiles.

[4]While the contribution to exports rather than production is admittedly less than perfect, it allows - in Italy - those higher levels of disaggregation often corresponding to "industrial districts".

[5]Table A1 (see Appendix A) provides the complete list of 4-digit sectors accounted in Figure 1 with the respective contributions by districts to the total Italian exports.

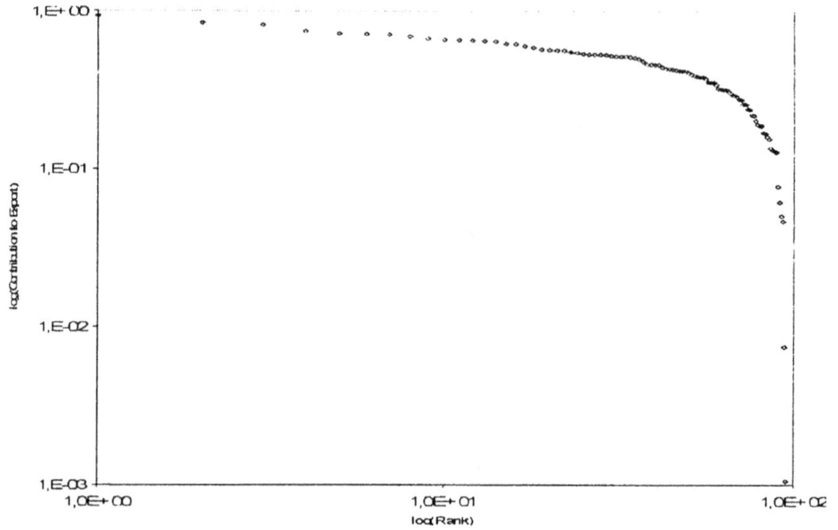

Fig. 1. Log of Districts' contribution to Total Italian exports by product category vs. Log of Rank (Ateco 91 Classification). Year: 1996.
Source: ISTAT, Censimento Intermedio dell'Industria e dei Servizi. See also App. A, Table A1.

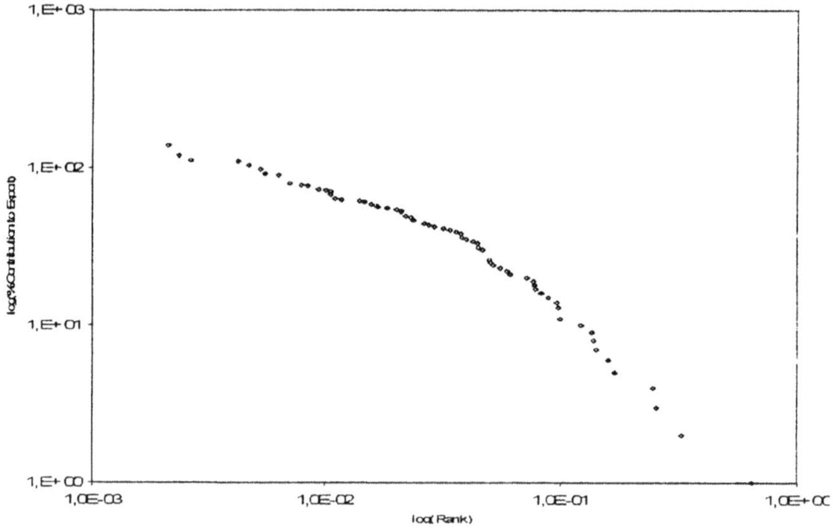

Fig. 2. Log of Percentage Contribution to exports of specialized districts by ATECO micro-sector vs. Log of Rank (Total Italian Exports by the micro-sector =100). (Ateco 91 Classification).
Source: ISTAT, Censimento Intermedio dell'Industria e dei Servizi.

2.3. The Importance of Agglomeration Maps into Diverse Sectorial Patterns of Innovation

Third, the foregoing intersectorial differences in agglomeration economies interestingly map into taxonomic differences in the sectorial patterns of innovation, as proxied by Pavitt's categorization (Pavitt 1984)[6]. In particular, as shown in Figure 3, agglomeration economies appear particularly relevant in "scale-intensive sectors" - hinting at forms of *hierarchical agglomeration* discussed above - and in "supplier-dominated sectors" - which tend to include most of the so-called "made in Italy" activities[7]. Conversely, they appear the least relevant in "science-based" sectors.

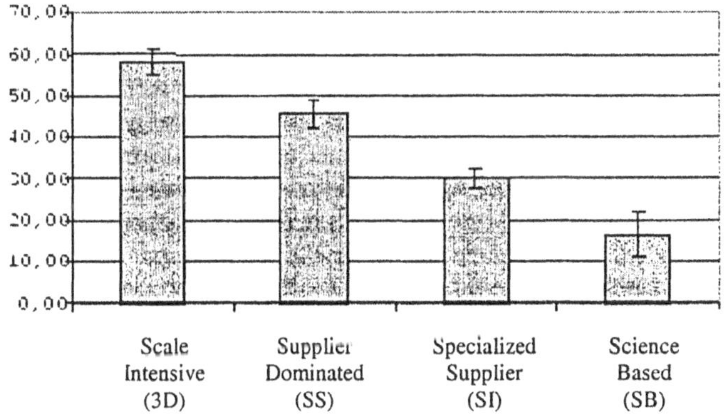

Fig. 3. Mean of Percentage Contribution to exports according to Pavitt's Taxonomy.

2.4. Some Form of "Life-cycles" in Agglomeration Phenomena?

Widespread evidence supports some sort of *metastability* in agglomeration phenomena, in the sense that in quite a few circumstances they persist on time-scales of order of magnitude greater than those of the processes supporting them, even though they tend to vanish on much longer time-scales. So, for example, Brusco and Paba (1997) report on several industrial districts in Italy (especially but not only in the South) which existed after World War II but disappeared

---

[6]The Italian ATECO 91 classification of sectors into Pavitt's taxonomy is available from the authors upon request.

[7]"Supplier-dominated" is an unfortunate and somewhat misleading name for a set of industries which might well be characterized by a lot of product differentiation (e.g. related to fashion) and organizational innovations, but at the same time acquire most of their technological innovations via intermediate and capital inputs produced elsewhere. Textiles, clothing, furniture, toys, etc. are good examples of such a set of industries.

thereafter. At the international level, sectors like steel, automobiles, tires and many others appeared quite spatially concentrated near the time of their birth, but became geographically much more footloose. In fact, it could well be that geographical stickiness falls whenever: (a) specific technological paradigms become fully established; and (b) international oligopolistic firms emerge, incorporating the core knowledge associated with production and innovation[8]. There are two general conjectures which stem from the foregoing evidence, namely:

1. The drivers of agglomeration economies are often nested in the nature of sector-specific patterns of knowledge accumulation.
2. Relatedly, cross-sectional differences in agglomeration forces ought to be at least partly explained on the grounds of underlying differences in such processes of technological and organizational learning. They affect among others the relative importance of phenomena such as localized knowledge spillovers; inter- vs. intra-organizational learning; knowledge complementarities fuelled by localized labour-mobility; innovative explorations undertaken through spin-offs and, more generally, the birth of new firms.

To what extent are such knowledge-related drivers accounted for in the current literature? This is what we shall briefly discuss in the next section.

3. Space, Geography and Agglomeration: Some Telegraphic Comments on the State-of-the-Art

In a nutshell, one might identify four main questions that scholars concerned about the "spatial dimension" of economic interactions have been all trying to address, albeit from different perspectives, for more than a century, namely: (i) What is the role of mere "chance" in the observed spatial concentration of economic activities? (ii) Relatedly, why and when could one observe persistent spatial patterns that cannot be explained by resorting to pre-existing heterogeneity in agents and locations (i.e. by some kind of "comparative advantage theory" alone)? (iii) Could one neatly identify agglomeration (centripetal) and dispersion (centrifugal) forces lying at the heart of the processes generating sustained spatial concentration (and possibly its destabilization)? And, (iv) how and when emerging spatial structures of production and innovation tend to become self-sustained over time? (And, conversely, what make them wither away?). As well known, pioneering works such as Von Thunen (1826) and Marshall (1890), tried to investigate the main economic forces driving geographical differentiation and agglomeration. For instance, Von Thunen's simple analysis of land use - by stressing the importance of space constraints in decentralized economies - addressed the relationships between micro and macro geographical outcomes. Even more importantly, Marshall's discussion of his famous "localization

[8]Some related remarks by one of us may already be found in Dosi (1982).

externalities" (or "external economies") triad[9] became a cornerstone in the theory of economic agglomeration. From then on, however, diverse trajectories of exploration emerged. A first large class of models hinges upon the basic idea that many different spatial agglomeration patterns (from concentration of economic activities in few locations to hierarchical structures) can be explained as the solution of a static, well defined, trade-offs between identifiable agglomeration and dispersion forces. This intuition, rooted once again in Von Thunen's work, has become the core of the analyses provided by "central-place" theory developed by Christaller (1933) and Lösch (1940) - and more generally by "regional science" models of Isard (1956) and by Henderson's (1974) treatments of urban systems. More recently, it has inspired static models with non-market externalities such as Papageorgiou and Smith (1983) and Fujita (1988, 1989). For instance, central place theory stresses the importance of economies of scale in the process of agglomeration of any economic activity and transportation costs in a community of farmers, while Henderson (1974), represents the economy as a system of cities and formalizes as an "inverted U" the relation between individual gains (due to some form of Marshallian localization externalities) and individual losses (due to some form of crowding effect) caused by concentrating activities in a single location. However, in order to explain hierarchical features of central places (an their supposed optimality) central place theory resorts to quite counterintuitive ideas as "nested hexagonal patterns" (!!), while urban theory relies on the existence of exogenous institutions such as city corporations. More elegantly, Papageorgiou and Smith (1983) and Fujita (1988, 1989) envisage agglomeration patterns as the outcome of the trade-off between some form of locally positive informational spillovers arising among agents endowed by heterogeneous information (e.g. social or technological spillovers) and congestion effects arising in spatial systems when concentration is too high (use of limited land, commuting costs, wage rate and land rent, etc.). A second class of models that has become prominent in the last few years, now known under the perhaps misleading heading of "New Geographical Economics"[10], acknowledges instead some form of increasing returns to scale (or indivisibilities) at the level of individual agents as both the incentive triggering agglomeration and the force able to sustain concentration (once the latter has emerged). The basic challenge of this stream of research - derived mainly from the theory rather than from any empirical puzzle - was to provide a satisfactory treatment of increasing returns to scale and monopolistic competition in a static equilibrium framework *cum* fully rational agents. By bridging monopolistic competition models à la Dixit and Stiglitz (1977) and Samuelson "iceberg-like" trade costs, such models have been able to

[9]That is: (i) backward/forward linkages associated to the trade-off between market-size and market-access; (ii) informational spillovers and (iii) advantages of thick markets for specialized local providers of inputs.

[10]See among others Krugman (1991a, 1991b, 1993), Krugman and Venables (1995a, 1995b, 1996), Venables (1996, 1998), Ottaviano and Thisse (1998), Fujita, Krugman and Venables (1999), Puga and Venables (1996). See also Ottaviano and Puga (1998), Fujita and Thisse (1996).

account for agglomeration patterns and inter-locational specialization by positing a self-reinforcement process - stemming from some form of market externality - which finds its counterpart in dispersion forces caused by competitive pressures implied by either agglomeration itself or immobility of some factors (e.g. labour). Although the wide proliferation of models sharing this common framework does not make easy to provide a taxonomy of both assumptions and results, the basic argument[11] hinges upon a circular causation between: (i) the decision of a firm to concentrate in a given area; and (ii) the positive net increase in profits enjoyed by firms deciding to follow it thereafter. Indeed, localized market-size effects in presence of imperfect competition usually offset the decrease in profits due to fiercer competition (unlike what would have happened under perfect competition)[12]. In a two-region, two-industries economy, this can account for the emergence of core-periphery patterns with agglomerations sustainable as stable equilibria and all (horizontally-differentiated) industrial goods produced in one region. However, multiple equilibria arise and small asymmetries in initial parameters can be amplified to give rise to very different spatial patterns. In particular, crucial to equilibrium selection are the roles played by the degree of immobility of production factors, transportation costs, degree of differentiation of the horizontally differentiated industrial good and the importance of indivisibilities in production. The most important contribution made by "New Geographical Economics" models is perhaps having provided closed-form solutions to the problem of describing in a common framework monopolistic competition, transportation costs and increasing returns at the microeconomic level. In fact, this same formal machinery (of variants thereof[13]) has been recently applied to diverse open issues in spatial economics such as industrial clustering in open economies (and its relationships with economic growth)[14], the emergence of the evolution of urban systems and cities formation[15], industrial specialization in an array of imperfectly competitive sectors[16]. However, any hope for analytical tractability and clearly interpretable closed-form results strongly clashes with many attempts to generalize the basic framework (see the case of many locations discussed in Fujita and Thisse 1996; see also the discussion in Martin 1999). Even

---

[11]See in particular Krugman (1991a, 1991b).

[12]An alternative but complementary argument (see Venables 1996) stresses cost-linkages instead of demand-linkages as the engine of agglomeration. Agglomeration in a given region with the associated fiercer competition does not only lower revenues, but also decreases costs due to the existence of input-output links between firms. Hence, the effect on net profit can once again be positive.

[13]The basic model sketched above has been extended to allow for endogenous wage determination in a general equilibrium setting (Krugman and Venables, 1995b; Puga and Venables, 1996); and for initial comparative advantages (Krugman 1993).

[14]See for instance Fujita, Krugman and Venables (1999), Chap. 4, and Englmann and Waltz (1995).

[15]See Fujita, Krugman and Mori (1999), and Fujita, Krugman and Venables (1999), Chap. 3.

[16]See Krugman and Venables (1996) and Venables (1998).

though these models have been dealing only with one element of the Marshall's proposed triad[17], thus neglecting issues as informational spillovers and the advantages of thick markets for specialized local providers of inputs, they need to rely on highly disputable oversimplifications in order to retain elegance and analytical tractability, such as the special Dixit-Stiglitz framework itself and Samuelson's scheme of transportation costs (i.e. no description of a separated transportation sector), to name a few[18]. In our view, the commitment to mainstream formalism of an "equilibrium *cum* fully rational agents" framework (and to analytical solutions) has strongly limited the explicative power of "New Geographical Economics"[19]. In fact, notwithstanding verbal acknowledgment of the very dynamic nature of any spatial agglomeration processes (Fujita et al. 1999), the treatment of dynamics in their models is admittedly unsatisfactory. So, instead of following their own mainstream prescriptions and endogeneizing dynamics as the outcome of intertemporal maximization problems by economic agents, Fujita et al. (1999) discuss issues of stability and selection among multiple equilibria in an essentially static framework. Finally, the list of empirically testable implications that such models are able to provide is intrinsically quite poor and always of an indirect nature (Martin 1999, p. 70). Even more dramatically, "New Geographical Economics" formalizations do not lend themselves to generate testable implications about industry distributions across locations, so that there is very little hope that any predictions about e.g. rank-size relationships could be ever taken to the data employing such models. Despite "New Geographical Economics" has often been claimed to be the most prominent attempt to provide a unified framework for spatial economics (as well as the sole robust answer to the resurgence of interest in agglomeration issues recently prompted by real-world concerns such as European market integration), there actually exist at least three other broad streams of research that in the last decades have been trying, in partly complementary ways, to open up the black box of spatial issues in economics with quite eclectic theoretical spectacles. First, many scholars have been providing rich and qualitative analyses, mostly (but not entirely) empirically focused, of urban/regional development and industrial agglomeration phenomena. In particular, by thoroughly analysing the role of externalities and technological spillovers, as well as the importance of social, cultural and institutional forces in shaping the rise and the decline of industrial districts, these authors have been providing a huge amount of illuminating insights

---

[17]That is the backward/forward linkages associated to the trade-off between market-size and market-access.

[18]Notice also that these models do not allow for any treatment of uncertainty. On the contrary, static models based on spatial competition (i.e. the so-called "shipping" and "shopping" models) introduced some uncertainty in consumers' perceptions of payoffs from locational choices. See e.g. Anderson, de Palma and Thisse (1992).

[19]A similar point about "New Geographical Economics" has been also made by Martin (1999). However, we do not agree with his analysis of Polya-Urns type models (Arthur 1994), as the latter class of formalizations clearly avoid any commitment to concepts like full rationality and general equilibrium. See also below.

about specific examples together with some attempts to offer interpretative frameworks able to grasp the larger picture[20]. Second, a long stream of literature on multinational investment - from the pioneering works by Vernon (1966), all the way to the recent contributions by Cantwell and colleagues[21] - is rich of insights on the interaction between technologies, corporate strategies and locational features. It is indeed surprising that geography-centred investigations have largely neglected such complementary contributions. Third, building on the seminal work by Brian W. Arthur and Paul David, a more theoretically grounded literature has been attempting to analyse the nature of economies/diseconomies of agglomeration in a truly dynamic framework in which persistent spatio-temporal patterns are conceived as emerging out of direct interactions among very stylised, boundedly-rational, heterogeneous economic agents. By acknowledging the history (or "path") dependent nature of the observed uneven spatial distribution of economic activities, the basic argument stresses the importance of dynamic increasing returns implied by some form of agglomeration economies/diseconomies (Arthur 1994, Chs. 4 and 6) and/or local network externalities (David et al. 1998; Cowan and Cowan 1998). Even more importantly, by recasting the analysis of agglomeration phenomena in a truly dynamical setting, one is able to appreciate the subtleties of the trade-offs between purely random factors and more systematic, historical forces (or, put it differently, the issue of necessity vs. chance) underlying the emergence of spatially ordered structures. Without entering into the mathematical details[22], the basic argument envisages a discrete-time economy with a finite set of regions (say R) and an enumerable population of firms deciding where to locate. To keep things as simple as possible, assume that firms i = 1, 2, enter sequentially the decision stage[23]. Firm i entering at time t has an idiosyncratic perception of the gain associated to the choice of locating in region r = 1,.., R, equal to the sum of an intrinsic, time-independent, attractiveness term q_i^r (e.g. the *ex-ante* geographical benefit) and some function g of the number of firms $y_i^r(t)$ that have so far decided to locate there (i.e. a measure of agglomeration economies - if $g' > 0$ - or, respectively, diseconomies - if $g' < 0$). Assume that firms choose to locate in the region associated to the best perceived gain. If the intrinsic attractiveness terms are randomly drawn from some given distribution F, then one can easily work out the function mapping current regional shares into the probability that each region will be chosen next. This extremely simple framework (and extensions thereof) can

[20]We refer here to a huge body of literature covering both "economic geography" studies (see Lee and Willis 1997 for a survey; and the references in Martin 1999), and in particular, the Italian studies on industrial districts, see Antonelli (1990, 1994), Brusco (1989), Sforzi (1989), Becattini (1990), Brusco and Paba (1997), Signorini (2000), Tattara (2001).

[21]See e.g. Cantwell (1989) and Cantwell and Iammarino (1998).

[22]See however Arthur (1994), Dosi, Ermoliev and Kaniovski (1994) and Dosi and Kaniovski (1994) for more detailed discussions.

[23]The assumption of sequential one-time decisions is not actually crucial. See e.g. David et al. (1998) for an example in which firms are allowed to revise their current choice from time to time.

indeed provide a rather wide array of predictions, in particular concerning the ability of economies/diseconomies of agglomeration to shape the long-run concentration patterns. Together, it is also able to account for early, small, mainly non-predictable, events as they interact with more systematic forces in conveying observable structures. Despite their highly stylised nature (firms are indeed conceived as very naïve entities; microeconomic foundations are rather poor; etc.), this class of models begins to open-up the black box of spatial phenomena in economics by focusing on inherently dynamical decentralized systems populated by simple interacting agents. To summarize: multiple strands of theoretical and historical literature do highlight the renewed richness of the investigation of spatial phenomena in economics. However, it is also fair to say that major persistent shortcomings of the theory concern, at the very least: (i) thorough and relatively general accounts of the interaction patterns between forms of knowledge accumulation and types of agglomeration phenomena, and (ii) the ability of yielding robust empirical predictions concerning agglomeration patterns conditional on underlying technological and organizational characteristics of diverse industrial activities. Let us now turn to the latter issue and suggest a formal machinery able to detect the different revealed strength of agglomeration in different sectors.

4. A Stochastic Dynamical Model of Plants Location

As mentioned in the foregoing section, a class of models aimed at empirical predictions on the grounds of explicit dynamics of firm location is the one presented in Arthur (1994, Chs. 4 and 6). There are however some drawbacks in such a methodology. First, the prediction of this type of model generally concerns the asymptotic state of the system, i.e. its state when an infinite number of firms has chosen its location. The comparison of different asymptotic outcomes does represent a valuable theoretical method to compare the effect and the relative strength of the different determinants of the location dynamics (agglomeration economies, scale economies, etc.). Nevertheless, it is obvious that its predictive power concerning actual empirical distribution should be taken with some caution since one is often facing empirical processes involving a relatively small number of firms (so that "infinity" might well be a misleading approximation). Moreover, the mathematical tools based on Polya-like processes are particularly well-suited to the description of sequential entry settings where each decision is primarily influenced by the choices of earlier entrants: due to its infinite memory, this machinery does in fact provide an elegant formulation of "history-dependant" processes. However, it is less suited to all circumstances where individual decisions are much less irreversible and/or the stochastic component of the process does not tend to zero (as in Polya dynamics), due to persistent entry *and mortality*. Here, in order to account for these drawbacks, we shall explore a distinct (Markovian) framework wherein, first, we will consider a finite number of firms and locations and second, we will describe industry dynamics as belonging to an invariant dynamical process. That is, we shall try to capture the idea that the actual distribution of plants on the territory keeps changing with the passing of time, even if generated by a stationary probability distribution. The probabilistic character of our model, when it comes to firms location choices, can be thought as taking in account the plausible existence of different "unobservable" constraints that shape the locational incentives of different plants (or firms). Notice that the ongoing displacement of plants (firms) can be thought of as the actual change of location of a given production activity, as well as the death of a plant/firm in a given location and the birth of a similar one in a different place. In order to simplify the treatment, we consider the number of firms constant along the system evolution. (Indeed, one can think to the number of firms as an "average" over the period of observation[24]). Suppose to have N firms distributed over M distinct locations. Consider the occupation number vector

$$\mathbf{n} = (n_1, \dots, n_M) \qquad n_i > 0 \quad \sum_{i=1}^{M} n_i = N \tag{1}$$

which provides the number of firms belonging to each location. If, in a "heroic" simplification, one assumes that all firms are identical, the vector $\mathbf{n}$ completely specifies the state of the system. The dynamical evolution of such a system can be described as a finite Markov chain. Let $P(\mathbf{n}' \mid \mathbf{n})$ be generic element of the transition matrix, i.e. the probability that if the state of the system is $\mathbf{n}$ at time t, its state at time $t + 1$ would be $\mathbf{n}'$. This probability does not depend explicitly on time and its specification completely defines our model. In order to capture the effect produced on the distribution of plants by the presence of agglomeration economies, the probability that a firm moves to a given location should depend on the number of firms already located there. Moreover, it is straightforward to introduce some degree of heterogeneity among locations: this can be done by allowing for intrinsic "geographical benefits", in general heterogeneous across locations (in analogy with Arthur 1994). Finally, for the sake of simplicity, we assume that just one firm changes location at a given time[25]. At each time step, a firm is chosen at random with equal probability $1/N$ and exits the industry. Then a randomly chosen, new firm enters a location with a probability proportional to the sum of the number of firms already there and a "geographical benefit" term. The transition element thus becomes:

[24]Notice that if one is on the contrary interested in the actual time evolution of the plant/firm distributions, the exact specification of the entry/exit dynamics becomes mandatory.

[25]Notice that this assumption has no effect whatsoever on the form of the invariant distribution of the process.

$$P(\mathbf{n}+\Delta_k-\Delta_i|\mathbf{n})=\begin{cases}\dfrac{n_i}{N}\dfrac{a_k+n_k}{A+N-1} & k\neq i\\ \dfrac{n_i}{N}\dfrac{a_k+n_k-1}{A+N-1} & k=i\end{cases} \tag{2}$$

where a = (a_1,..., a_M) is the array of intrinsic "benefits" for the M locations, $A = \sum_{k=1}^{M} a_k$ and Δ_i = (0,..., 1,..., 0) is the unitary vector with i-th component equal to 1. Some considerations are in order. First, notice that the intrinsic benefit a_k is proportional to the probability of choosing location k when the latter is empty[26]. Second, we have chosen the simplest linear relationship between the probability of choosing a location and the number of firms already there[27]. In fact, a more realistic account of location-specific returns to agglomeration should allow for the dependence of the a terms upon the number of plants/firms located there and/or for threshold effects. However, as we shall see, even the foregoing simpler approximation does not fare too badly with the data[28]. For positive a's, each location has a positive probability of receiving the entering firm: hence, any possible state of the system is reachable, in a suitable number of steps, with finite probability starting from any other state and the Markov chain defined by (2) is irreducible. If $p(\mathbf{n}, t)$ is the probability to find the system in state $\mathbf{n}$ at time t, its evolution reads:

$$p(\mathbf{n},t+1)=\sum_{n'}P(\mathbf{n}|\mathbf{n}')p(\mathbf{n}',t). \tag{3}$$

The invariant distribution $\pi(\mathbf{n}, t)$ is thus obtained by imposing the detailed balance condition:

$$P(\mathbf{n}|\mathbf{n}')\pi(\mathbf{n}')=P(\mathbf{n}'|\mathbf{n})\pi(\mathbf{n}) \tag{4}$$

The explicit expression for the invariant distribution, known as the M-dimensional Polya distribution, can be easily obtained and reads:

---

[26]This model is known as Ehrefest-Brillouin model and has been introduced in Garibaldi and Penco (2000) as a generalization of the famous Ehrefest model of statistical physics. A similar simplified version has been introduced in Kirman (1993).

[27]This model overlaps with the Arthur's one with linear returns function, see Arthur (1994).

[28]Moreover, negative values for the a's can be in principle considered, in order to describe the presence of agglomeration diseconomies characterizing the distribution of firms over the different locations. Nonetheless, the purely random exit dynamics constitutes, as such, a limit to the actual concentration of plants/firms in a given site, since more populated sites are also more likely to yield dying ones.

$$\pi(\mathbf{n}, a) = \frac{\Gamma(N+1)\Gamma(A)}{\Gamma(A+N)} \prod_{i=1}^{M} \frac{\Gamma(a_i + n_i)}{\Gamma(a_i)\Gamma(n_i + 1)} \tag{5}$$

The values of the parameters a_k determine the nature of the distribution: for lower values of the a parameters, the effect of agglomeration becomes more relevant. In the limit $a_k \rightarrow +\infty$ and $a_i/a_k \rightarrow 1$ for any i and k, "agglomeration economies" disappear and the expression in (5) reduces to a multinomial distribution, while for $a_k = 1, \forall\ k$ it becomes what is known in statistical physics as the Bose-Einstein distribution.

5. Some Empirical Evidence

The complete multivariate distribution in (5) does provide a complete probabilistic description of our model. However, in order to obtain a quantity which can be more easily compared with empirical data, it is better to consider the marginal probability distribution of the occupancy number of a given site. The problem is to compute the probability that a site with "intrinsic value" a would end up, after the placement of N firms, with exactly h among them. By summing over all the residual degrees of freedom, one gets (see Bottazzi 2001):

$$p(h; a, A, N) = \binom{N}{h} \frac{\Gamma(a+h)}{\Gamma(a)} \frac{\Gamma(A-a+N-h)}{\Gamma(A-a)} \frac{\Gamma(A)}{\Gamma(A+N)} \tag{6}$$

In the following we shall use this expression, whose parameter will be set by a fitting procedure, in order to compare the prediction of our model with the empirical observations. We shall use a database from the Italian Census of Manufacturers for the year 1996 containing business units (BUs) belonging to $M = 784$ "local system of labour mobility" (LSLM, for a definition see Sforzi 2000) and to $L = 25$ different sectors for a total of 591,110 local units (plants). Here, we present some experiments over three sectors - primary metals, transport equipment and furniture - which one should expect to display different degrees of agglomeration economies and different drivers of the latter (see also the taxonomic discussions of Section 2). Let $n_{i,l}$ be the number of BUs in LSLM i operating in sector l. Moreover we denote with $n_{.,l}$ the total number of BUs operating in sector l and with $n_{i,.}$ the total number of BUs belonging to i-th LSLM. As already mentioned, instead of considering average quantities measuring the "strength of agglomeration" of a given sector (as done, for instance, in Sforzi, 1991) we shall analyse the complete "occupancy distribution" of the BUs in the various LSLM, i.e. we compute the observed frequency $f_{obs.}(h;l)$ with which a LSLM hosting exactly h BUs active in sector l appears in our data:

$$f_{obs.}(h;l) = \frac{1}{M}\sum_{l=1}^{M} \delta_{n_{i,l},h} \tag{7}$$

where δ is the Kronecker (index) function, and we compare this expression with the theoretical prediction of (6), once having of course specified the parameters a characterizing the theoretical distribution. As a first benchmark, one could consider all the LSLM as equal (i.e. with the same "intrinsic appeal") and obtain a theoretical expression directly from (6) putting $a_i = \beta$ and $A = M\beta$. This model would depend on a single parameter β which measures the "strength" of the agglomeration effect - with a low β meaning high agglomeration economies. However, tests of this model yield quite bad agreement with data, and the theoretical description constantly underestimates the observed distribution tails. The reason for this becomes apparent if one plots, for a given sector l, the number of BUs $n_{i,l}$ of a LSLM against the total number of BUs in all the other sectors, except the one under consideration (that is $n_{i,.} - n_{i,l}$). Under the previous assumption of *a priori* equiprobability, no dependence should appear between the two variables, since BUs belonging to different sectors should choose their locations independently. On the contrary, a strong positive correlation appears which contradicts the purported identity between the various LSLM's: we plot the result of this analysis in Fig. 4, for the three chosen sectors. The parameters fitted from a log-linear regression are reported in Table 1.

Table 1. The "agglomeration" parameter β and the slope α of the linear regression by sector and by LSLM.

	Primary Metals	Transport Equipment	Furniture
β	10.00	0.32	0.50
α	0.91 ± 0.01	0.70 ± 0.04	0.70 ± 0.01

As an alternative, let us assign to each location an "intrinsic attractiveness" which is proportional to the number of BUs which are located there and belong to all sectors but the one under analysis:

$$a_{i,l} = \beta_l \frac{n_{i,.} - n_{i,l}}{N - n_{.,l}} \tag{8}$$

where l is the sector under analysis and the β_l coefficient captures, as above, the intensity of agglomeration economies. This procedure meant to capture also "geographical" effects that make a location intrinsically preferable compared to others, in terms of better industrial infrastructures, sheer overall size, etc.. It is likely that these advantages are location-specific and apply to all sectors under consideration. After controlling for "horizontal" locational effects, one can derive a measure of relative advantages between locations. Following this idea, the predicted "weighted" frequencies become:

$$f_l^{wg}(h) = \frac{1}{M}\sum_{i=1}^{M} p(h; a_{i,l}, A_l, n_{.,l}) \tag{9}$$

where p is the distribution in (6) and $A_l = \sum_{i=1}^{M} a_{i,l}$. As can be seen in Figs. 5 through 7, the accordance of the theoretical prediction with data is quite high. The values of the β 's used for the theoretical curves are reported in Table 1. Out of the three sectors chosen, the first, primary metals, does show an almost total lack of agglomeration effects, while the other two, transport equipment and furniture, seems to display rather high agglomeration economies. Notice, however, that the nature of such an agglomeration is actually very different for the latter two sectors: transport equipment displays a scattered locational patterns made-up of relatively few firms (possibly hinting at the "hierarchical agglomerations" mentioned in Section 2), while furniture highlights more "district-like" patterns.

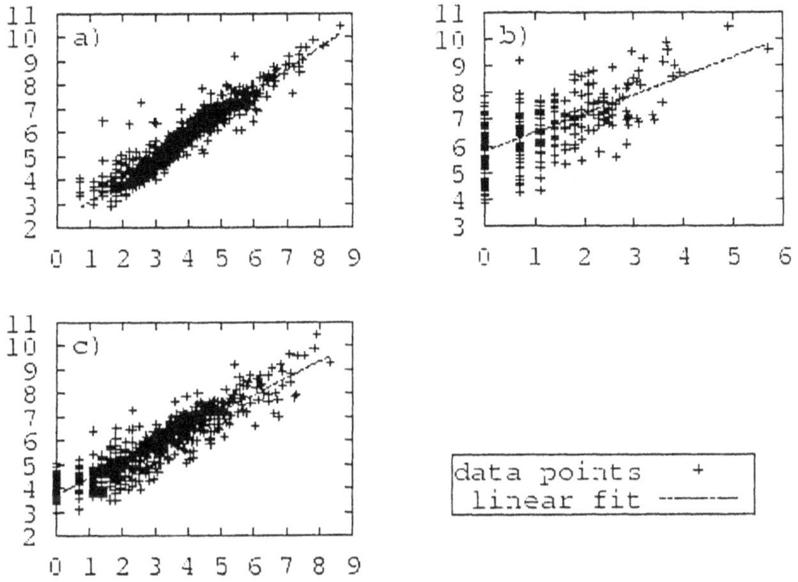

Fig. 1. Total number of plants in a LSLM (Local System of Labour Mobility) vs. the number of plants in that location pertaining to a specific sector. **a)** primary metal, **b)** transport equipment, **c)** furniture. All the variables are in log scale.
Source: ISTAT, Censimento Intermedio dell'Industria e dei Servizi, 1996.

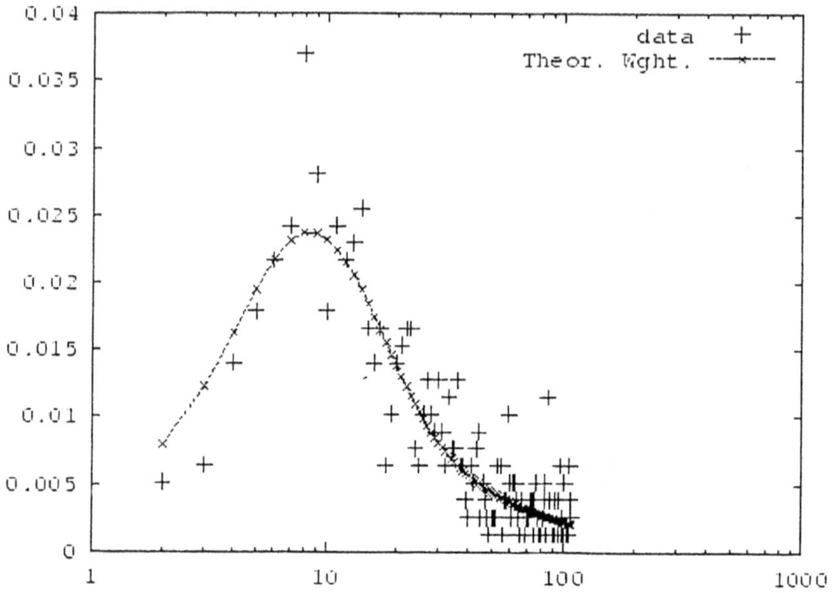

Fig. 5. Frequencies and number of firms by LSLM (Local System of Labour Mobility). The primary metal sector.
Source: ISTAT, Censimento Intermedio dell'Industria e dei Servizi, 1996.

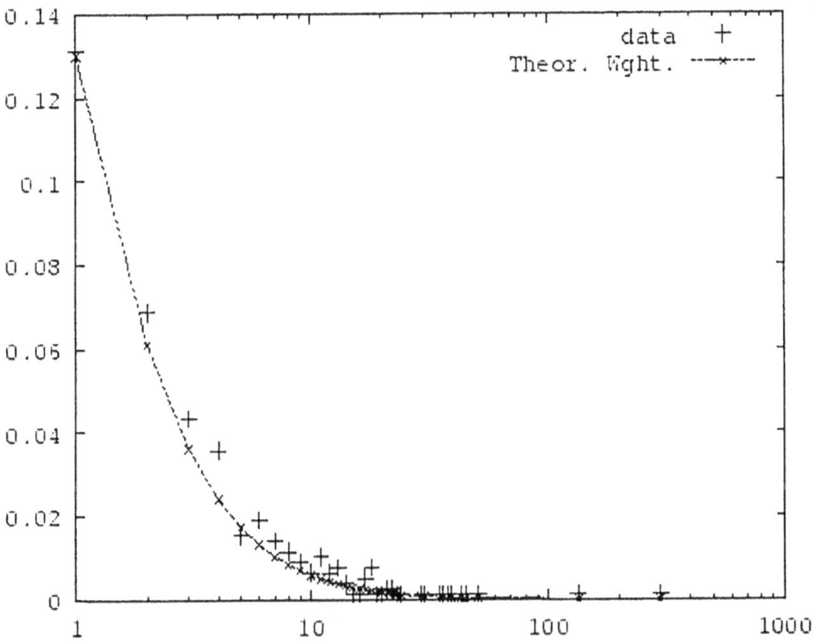

Fig. 6: Frequencies and number of firms by LSLM (Local System of Labour Mobility). The transport equipment sector.
Source: ISTAT, Censimento Intermedio dell'Industria e dei Servizi, 1996.

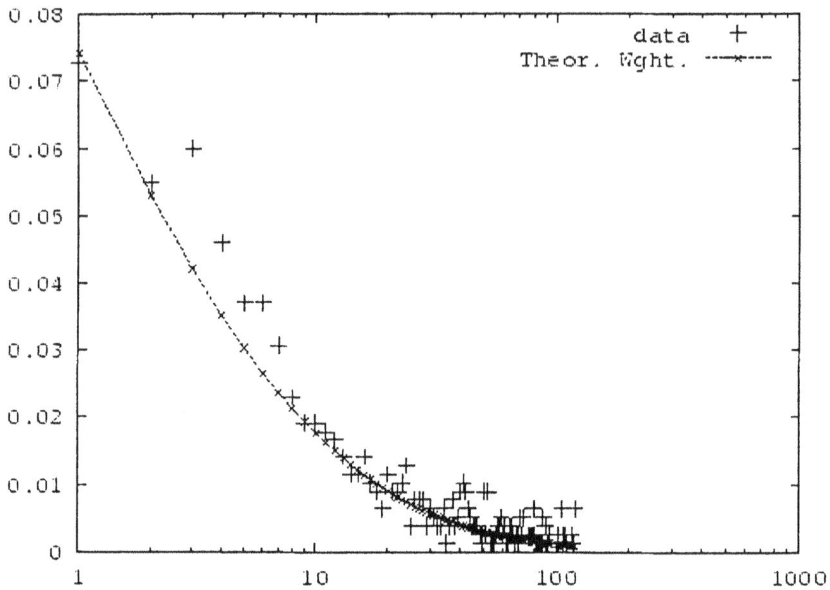

Fig. 7. Frequencies and number of firms by LSLM (Local System of Labour Mobility). The furniture sector.
Source: ISTAT, Censimento Intermedio dell'Industria e dei Servizi, 1996.

6. Conclusions

We mentioned in the Introduction the highly preliminary nature of these notes. In this spirit, we have tried to, first, flag out some taxonomies of agglomeration structures and agglomeration drivers; second, suggest some links between the observed patterns and the underlying dynamics of knowledge accumulation; third, identify some related achievements and limitations of current theorizing; and, finally, fourth, develop a formal machinery able to statistically characterize the revealed intensities of agglomeration forces in different sectors. Indeed one could consider such a contribution as a small step toward bridging "spatial" analyses on the one hand, and investigations of knowledge-driven patterns of industrial change, on the other.

Appendix A. Background Evidence on the Distribution of Exports by Sector and by District.

Table A1. Rank of Italian Districts contribution to export by product category (Ateco 91 Classification). Year: 1996.
Source: ISTAT, Censimento Intermedio dell'Industria e dei Servizi.

Rank	Product category	Contr.
1	Sports goods	0,935
2	Tanning and dressing of leather	0,842
3	Ceramic tiles and flags	0,820
4	Repair of watches, clocks and jewellery	0,746
5	Musical instruments	0,720
6	Other agricultural and forestry machinery	0,715
7	Knitted and crocheted fabrics	0,711
8	Footwear	0,690
9	Nonwovens and articles made from nonwovens, except apparel	0,669
10	Other furniture	0,658
11	Other textiles	0,655
12	Tanks, reservoirs and containers of metal	0,654
13	Veneer sheets; plywood, laminboard, particle board, fibre board	0,648
14	Knitted and crocheted underwear	0,643
15	Other textiles products	0,620
16	Nonwovens and articles made from nonwovens, except apparel	0,618
17	Weapons and ammunition	0,598
18	Cutlery	0,586
19	Man. Of other basic iron, steel and ferro-alloys	0,570
20	Other products of wood	0,566
21	Meat and meat-based products	0,562
22	Textiles manufacturing (clothes excl.)	0,561
23	Steel tubes	0,549
24	Nonwovens and articles made from nonwovens, except apparel	0,544
25	Sawmilling and planing of wood, impregnation of wood	0,535
26	Refractory ceramic products	0,531
27	Other manufacturing products	0,530
28	Other fabricated metal products	0,530
29	Lighting equipment and electric lamps	0,527
30	Household and sanitary goods and of toilet requisites	0,520
31	Other transport equipment	0,517
32	Machine - tools	0,515
33	Accumulators, primary cells and primary batteries	0,515
34	Dressing and dyeing of fur; articles of fur	0,511
35	Other special purpose machinery.	0,505
36	Cement	0,497
37	Optical instruments and photographic equipment	0,480
38	Cutting, shaping and finishing of stone	0,460
39	Electric motors, generators and transformers	0,460
40	Wooden containers	0,458
41	Games and toys	0,457
42	Other plastic products	0,442

Table A1. (continued)

43	Pesticides and other agro-chemical products	0,432
45	Other articles of concrete, plaster and cement	0,428
46	Other non-metallic mineral products	0,423
47	Production and distribution of electricity	0,418
48	Electricity distribution and control apparatus	0,418
49	Other general purpose machinery	0,417
50	Insulated wire and cable	0,406
51	Bricks, tiles and construction products. in baked clay	0,398
52	Other electrical equipment	0,391
53	Beverages	0,385
54	Non-electric domestic appliances	0,383
55	Manufacture and processing of other glass including technical glassware	0,382
56	Bicycles	0,373
57	Pharmaceutical preparations	0,353
58	Processing and preserving of fish and fish products	0,353
59	Paints, varnishes and similar coatings, printing ink and mastics	0,352
60	Watches and clocks	0,341
61	Paper and paperboard	0,322
62	Parts and accessories for motor vehicles and their engines	0,320
63	Precious metals production	0,317
64	Other food products	0,317
65	Bodies (coachwork) for motor vehicles; trailers and semi-trailers	0,312
66	Soap and detergents, cleaning and polishing preparations	0,302
67	Builders' carpentry and joinery	0,294
68	Man-made fibres	0,292
69	Printing	0,287
70	Basic iron and steel and of ferro-alloys	0,276
71	Instr. and appliances for measur., checking, testing, navigating, exc.industrial process.	0,275
72	Milk and cheese products	0,258
73	Luggage, handbags and the like, saddlery and harness	0,257
74	Other rubber products	0,240
75	Processing and preserving of fruit and vegetables	0,238
76	Other organic basic chemicals	0,220
77	Printing of newspapers	0,218
78	Television and radio receivers, sound or video recording or reproducing apparatus	0,202
79	Prepared feeds for farm animals	0,191
80	Other chemical products	0,188
81	Tobacco industry	0,188
82	Motor vehicles	0,170
83	Railway and tramway locomotives and rolling stock	0,168
84	Steam generators (central heating excl.)	0,160
85	Crude oils and fats	0,155
86	Starch and related products	0,136
87	Renting of office machinery and equipment including computers	0,132
88	Basic pharmaceutical products	0,130

Table A1. (continued)

89	Aircraft and spacecraft	0,128
90	Nuclear fuels treatment	0,078
91	Building and repairing of ships	0,062
92	Electronic valves and tubes and other electronic components	0,050
93	Television and radio transmitters and apparatus for line telephony and line telegr.	0,047
94	Refined petroleum products	0,008
95	Coke and coal related products	0,001

References

Anderson, S. P., dePalme, A. and Thisse, J. F. (1992), *Discrete Choice Theory of Product Differentiation*, Cambridge, MIT Press.

Antonelli, C. (1990), "Induced Adoption and Externalities in the Regional Diffusion of Information Technology", *Regional Studies*, 24, pp. 31-40.

Antonelli, C. (1994), "Technology Districts, Localized Spillovers, and Productivity Growth: The Italian Evidence on Technological Externalities in Core Regions", *International Review of Applied Economics*, 12, pp. 18-30.

Arthur, W. B. (1994), *Increasing Returns and path-dependency in Economics*, Ann Arbor, University of Michigan Press.

Barca, F. (ed.) (1997), *Storia del Capitalismo Italiano dal dopoguerra ad oggi*, Roma, Donzelli.

Becattini, G. (1990), "The Marshallian Industrial District as a Socio-economic Notion", in: *Industrial Districts and Inter-Firm Cooperation*, Pyke, F., Becattini, G. and Sengenberger, W. (eds), Geneva, International Institute for Labour Studies.

Bottazzi, G. (2001), "A Stochastic Model of Firms Location", Laboratory of Economics and Management (LEM) Working Paper, S. Anna School of Advanced Studies, Forthcoming.

Brugnoli, A. and Fachin, S. (2000), "Concentrazione e Diffusione: la Geografia Industriale Italiana, 1981-1991", *Economia e Politica Industriale*, 27, pp. 103-128.

Brusco, S. (1989), *Piccole Imprese e Distretti Industriali*, Torino, Rosenberg and Sellier.

Brusco, S. and Paba, S. (1997), "Per una storia dei distretti industriali italiani dal secondo dopoguerra agli anni novanta", in: *Storia del Capitalismo Italiano dal dopoguerra ad oggi*, Barca, F. (ed.), 1997.

Cantwell, J. A. (1989), *Technological Innovations and Multinational Corporations*, Oxford, Basic Books.

Cantwell, J. A. and Iammarino, S. (1998), "MNCs, Technological Innovation and Regional Systems in the EU: Some Evidence in the Italian Case", *International Journal of the Economics of Business*, 5, pp. 383-408.

Christaller, W. (1933), *Central Places in Southern Germany*, Jena, Fischer.

Cowan, R. and Cowan, W. (1998), "On Clustering in the Location of R&D: Statics and Dynamics", *Economics of Innovation and New Technology*, 6, pp. 201-229.

David, P. A., Foray, D. and Dalle, J. M. (1998), "Marshallian Externalities and the Emergence of Spatial Stability and Technological Enclaves", *Economics of Innovation and New Technology*, 6, pp. 147-182.

Dixit, A. K. and Stiglitz, J. E. (1977), "Monopolistic Competition and Optimum Product Diversity", *American Economic Review*, 67, pp. 297-308.

Dosi, G. (1982), "Technological Paradigms and Technological Trajectories", *Research Policy*, 11, pp. 147-162.

Dosi, G. and Kaniovski, Y. M. (1994), "On "badly behaved" dynamics", *Journal of Evolutionary Economics*, 4, pp. 93-123.

Dosi, G., Ermoliev, Y. M. and Kaniovski, Y. M. (1994), "Generalized urn Schemes and Technological Dynamics", *Journal of Mathematical Economics*, 23, pp. 1-19.

Ekelund, R. B. and Hebert, R. F. (1993), "Cycles in Development of Spatial Economics", in: *Does Economic Space Matter?*, Ohta, H. and Thisse, J. F. (eds), New York, St. Martin's Publ.

Englmann, F. C. and Waltz, U. (1995), "Industrial Centers and Regional Growth in the Presence of Local Inputs", *Journal of Regional Science*, 35, pp. 3-27.

Fabiani, S. and Pellegrini, G. (1997), *Education Infrastructure, Geography and Growth: an Empirical Analysis of the Development of Italian Provinces*, Roma, Banca d'Italia, Temi di Discussione, n. 323.

Feller, W. (1968), *An Introduction to Probability Theory and Its Applications*, New York, Wiley & Sons.

Fujita, M. (1988), "A Monopolistic Competition Model of Spatial Agglomeration", *Regional Science and Urban Economics*, 18, pp. 87-124.

Fujita, M. (1989), *Urban Economic Theory. Land Use and City Size*, Cambridge, MA, Cambridge University Press.

Fujita, M. and Thisse, J. F. (1996), "Economics of Agglomeration", *Journal of the Japanese and International Economies*, 10, pp. 339-378.

Fujita, M., Krugman, P. and Mori, T. (1999), "On the Evolution of Hierarchical Urban Systems", *European Economic Review*, 10, pp. 339-378.

Fujita, M., Krugman, P. and Venables, A. I. (1999), *The Spatial Economy: Cities, Regions, and International Trade*, Cambridge, Massachussets, The MIT Press.

Gabszewicz, J. J. and Thisse, J. F. (1986), "Spatial Competition and the Location of Firms", in: *Location Theory, Fundamentals of Pure and Applied Economics*, Gabszewicz, J., Thisse, J. F., Fujita, M. and Schweizer, U. (eds), London, Harwood Academic Publishers.

Garibaldi, U. and Penco, M. A. (2000), "Ehrenfest urn Model Generalized: an Exact Approach for Market Participation Models", *Statistica Applicata*, 12, pp. 249-272.

Guiso, L. and Schivardi, F. (2000), *Information Spillovers and Factor Adjustment*, Banca d'Italia, Temi di Discussione, n. 368.

Henderson, J. V. (1974), "The Sizes and Types of Cities", *American Economic Review*, 64, pp. 640-656.

Isard, W. (1956), *Location and Space-Economy*, Cambridge, Massachussets, The MIT Press.

Kirman, A. P. (1993), "Ants, Rationality and Recruitment", *The Quarterly Journal of Economics*, 108, pp. 137-156.

Klepper, S. (2000), *Entry by Spinoff*, Pittsburgh, Carnegie-Mellon University, Mimeo.

Krugman, P. (1991a), "Increasing Returns and Economic Geography", *Journal of Political Economy*, 99, pp. 483-499.

Krugman, P. (1991b), *Geography and Trade*, Cambridge, Massachussets, The MIT Press.

Krugman, P. (1993), "First Nature, Second Nature, and Metropolitan Location", *Journal of Regional Science*, 33, pp. 129-144.

Krugman, P. (1996), *The Self-organizing Economy*, Cambridge, Blackwell.

Krugman, P. and Venables, A. J. (1995a), "Globalization and the Inequality of Nations", *Quarterly Journal of Economics*, 110, pp. 857-880.

Krugman, P. and Venables, A. J. (1995b), *The Seamless World: A Spatial Model of International Specialization*, CEPR, London, Discussion Paper, No. 1230.

Krugman, P. and Venables, A. J. (1996), "Integration, Specialization and Adjustment", *European Economic Review*, 40, pp. 959-968.

Lee, R. and Willis, J. (1997), *Geographies of Economies*, London, Arnold.

Lösch, A. (1940), *The Economics of Location*, Jena, Fischer.

Marshall, A. (1890), *Principles of Economics*, London, Macmillan.

Martin, R. (1999), "The New "Geographical Turn" in Economics: Some Critical Reflections", *Cambridge Journal of Economics*, 23, pp. 65-91.

Onida, F., Viesti, G. and Falzoni, A. M. (eds) (1992), *I distretti industriali: crisi o evoluzione ?*, Milano, Egea.

Ottaviano, G. I. P. and Puga, D. (1998), "Agglomeration in the Global Economy: A Survey of the "New Economic Geography"", *World Economy*, 21, pp. 707-731.

Ottaviano, G. I. P. and Thisse, J. F. (1998), *Agglomeration and Trade Revisited*, Louvain, CEPR and CORE (Center for Operations Research and Econometrics) Discussion Paper, 1903.

Papageorgiou, Y. Y. and Smith, T. R. (1983), "Agglomeration and Local Instability of Spatially Uniform Steady-States", *Econometrica*, 51, pp. 1109-1119.

Pavitt, K. (1984), "Sectoral Patterns of Technical Change: Towards a Taxonomy and a Theory", *Research Policy*, 13, pp. 343-373.

Piore, M. J. and Sabel, C. (1984), *The Second Industrial Devide*, New York, Basic Books.

Puga, D. and Venables, A. J. (1996), "The Spread of Industry: Spatial Agglomeration in Economic Development", *Journal of the Japanese and International Economies*, 10, pp. 440-464.

Sforzi, F. (1989), "The Geography of Industrial Districts in Italy", in: *Small Firms and Industrial Districts in Italy*, Goodman, E. and Bamford, J. (eds), London, Routledge.

Sforzi, F. (1991), "The Marshallian Industrial Districts in the Italian Economy", in: *Industrial District and Inter-firm Cooperation in Italy*, Pyke, F., Becattini, G. and Sengenberger, W. (eds), Geneva, International Institute for Labour Studies.

Sforzi, F. (2000), "Local Development in the Experience of Italian Industrial Districts", in: *Geographies of Diversi ties. An Italian Perspective*, Various Authors, Roma, CNR-IGU (International Geographical Union).

Signorini, L. F. (ed.) (2000), *Lo Sviluppo Locale*, Roma, Meridiana/Donzelli.

Tattara, B. (ed.) (2001), *Il Piccolo che Nasce dal Grande*, Milano, Franco Angeli.

Venables, A. J. (1996), "Equilibrium Locations of Vertically Linked Industries", *International Economic Review*, 37, pp. 341-359.

Venables, A. J. (1998), "Geography and Specialization: Industrial Belts on a Circular Plain", in: *Regional Integration*, Baldwin, R., Cohen, D., Sapir, A. and Venables, A. J. (eds), Cambridge, Massachussets, Cambridge University Press.

Vernon, R. (1966), "International Investment and International Trade in the Product Cycle", *Quarterly Journal of Economics*, 80, 190-207.

Von Thünen, J. H. (1826), *Det Isolierte Staat in Beziehung auf Landwirtschaft und Nationalokonomie*, Hamburg, Perthes.

III. Success Cases Around the World

The Entrepreneurial Spark: Individual Agents and the Formation of Innovative Clusters

Maryann P. Feldman and Johanna Francis[1]

Abstract. This paper outlines the development of a high-tech industrial cluster in the U.S. Capitol region through the efforts of entrepreneurs who adapted to both constructive crises and new opportunities. We examine the initial spark of entrepreneurship and how it influences the formation of high technology clusters. The perspective taken is that entrepreneurs are a critical element in the formation of clusters and their actions are important in the analysis of clusters as complex adaptive systems. In addition, we consider how this perspective and the role of entrepreneurship may influence economic development policy.

1. Introduction

Once established, industrial clusters benefit from virtuous, self-reinforcing processes that support and sustain them (Anderson et al. 1988; Arthur 1994). Yet, we have a limited understanding of how these clusters emerge, take hold and transform a regional economy (Kargon, Leslie and Schoenberger 1992). Much of the economic development discourse is informed by attempts to replicate the characteristics associated with a fully functioning regional system in what may be considered a mechanistic economic development machine - line up inputs such as venture capital, social networks and support services and economic development will follow. However, these factors appear to lag rather than lead cluster formation (Feldman forthcoming). More importantly, this perspective ignores the importance of entrepreneurs as economic change agents, able to create or attract the necessary resources and institutions to support their ventures, as well as the rich historical and regional context in which they operate. Although the literature has largely ignored the individual change agent involved in the development of regional economies (see Appold 2000 for a review), this is at odds with our understanding of the importance of such agents (Kay 2000), the co-evolution of technology and

[1]Prepared for the conference on *Complexity and Industrial Clusters – Dynamics and Models in Theory and Practice* organized by the Fondazione Comunità e Innovazione under the aegis of the Accademia Nazionale dei Lincei, held in Milan, Italy on June 19 and 20, 2001. We are indebted to the individuals who agreed to interviews and generously shared their time and expertise in identifying salient issues as well as providing and validating information.

institutions (Nelson 1993) and way in which entrepreneurs actively interact with their local environments (Saxenian 1994).

This paper examines the impact of the initial spark of entrepreneurship and how it influences the formation of high technology clusters. The perspective taken is that entrepreneurs are a critical element in the formation of clusters and their actions are important to the analysis of clusters as complex adaptive systems. In addition, we consider how this perspective and the role of entrepreneurship may influence development policy. Historical studies document the genesis and development of successful clusters, however this line of inquiry is not well integrated into conceptualisations of regional change. Moreover, emulation does not necessarily lead to success. Silicon Valley's early successes in the electronic industry, spawned replicates across the country, many of which never took hold (Leslie and Kargon 1994). Although critical components of the Silicon Valley success are readily identified, it is not merely a matter of replicating them in new locations. It is not only the components themselves that matter, but the way in which they are combined to give rise to a self-reinforcing innovative system (Kargon, Leslie and Schoenberger 1992). Thus, our perspective, in contrast, focuses on the role of change agents and relies on concepts from the developing literature on complex systems. In effect, we propose that entrepreneurs are the key that unlocks untapped potential in the local environment and sparks economic growth.

We also focus on individual decisions to form a company and examine how these decisions are influenced by regional context. The specific case considered is the development of the U.S. Capitol region[2]. By any number of measures, this region, which was previously dominated by government employment, lacked the attributes conventional wisdom associates with an entrepreneurial environment. Yet, the Capitol Region has established technological leadership based on entrepreneurial activity in biotechnology and the Internet[3] - two new industries that have seeded and established themselves in the past twenty years. We focus on the evolution of this region, specifically on the development of an industrial cluster as a series of three phases. In the initial phase, early entrepreneurial efforts emphasized serving existing government agencies and most entrepreneurs were from the private sector. The second phase is dominated by increased entrepreneurial activity as an adaptation to a series of changes in the external environment. These efforts were a response to creative crisis due to federal downsizing as well as to new opportunities precipitated by changes in the legal framework - the local effect of changes in the policy environment. During this stage, entrepreneurs defined resources to promote and protect their interests. In

[2]The U.S. Capitol Region is considered here as the Consolidated Metropolitan Statistical Area (CMSA) which includes Washington D.C., Northern Virginia and the Maryland Suburbs including Baltimore City and its environs. Two counties in West Virginia were added in 1990.

[3]Within the region, there is evidence of geographic differentiation. Biotechnology is primarily concentrated in the Maryland suburbs in Gaithersburg and along the I-270 Corridor. The Internet companies are concentrated in the Northern Virginia suburbs.

this way, the independent actions of entrepreneurs are catalytic components of a self-organizing system. The final stage is the establishment of a critical mass of resources, some developed within the region and other resources, such as venture capital, moving to the area. In this phase, the full virtuous cycle of self-reinforcing activity is in play. The critical mass of start-up activity has spawned the necessary infrastructure to sustain it, which has in turn attracted more activity to the region.

By considering the early entrepreneurial efforts through which biotechnology and ICT took hold in the region, the approach taken is appreciative history-friendly theorizing (Malerba et al. 1999; Teubal and Andersen 2000). The emphasis is on the role of individual entrepreneurs as agents of change who make decisions to start companies, shape local environments and institutions and develop the resources and relationships that further their interests. It is argued in this paper that viewing entrepreneurs as agents of change is critical to understanding not only the entrepreneurship event but also the creation of a positive local environment. For external validity, we compare and contrast our findings with the genesis of Silicon Valley (Leslie and Kargon 1997; Saxenian 1994).

The next section of the paper offers an interpretive history of the development of entrepreneurship in the U.S. Capitol Region. We then turn to a consideration of the characteristics of complexity highlighted in the literature using examples from the U.S. Capitol Region and Silicon Valley. The intention is to provide prescriptive information for those regions trying to spark entrepreneurship and an economic transition with an emphasis on the uniqueness of each region and the ability of creative individuals to capitalize on that unique environment.

2. Entrepreneurship Comes to Washington: An Interpretative History

The region around the U.S. Capitol has recently emerged as a hotbed of entrepreneurial activity, wealth creation and economic growth. This recognition reflects a transformation of the region from an economy dominated by public sector employment that was neither innovative nor supportive of private sector activity (Feldman forthcoming). In an attempt to understand this transformation, we have studied the genesis of entrepreneurial efforts. Our analysis is an interpretive summary based on interviews with entrepreneurs in biotechnology and ICT. The interviews gathered information on: where entrepreneurs were employed prior to starting their companies, what the motivation was for starting their companies, what resources they used in developing their companies and the technologies and the subsequent spin-off activity they generated.

There are some methodological issues to mention. First, it is important to note that this is a retrospective study. We are limited to identifying firms in existence now or that were at one time prominent enough to leave a record. While we are able to trace these firms back to their founding, we have no knowledge of similar firms that were started but may have failed, been acquired or merged into other firms prior to our study[4]. This approach does allow us to consider the roots of successful entrepreneurship and the ways in which entrepreneurial activity took hold, but it cannot address the failure of enterprises that died without leaving a record. While each of these companies has its own unique and compelling founding story, our objective is to discern trends and patterns.

The Capitol region has the third largest concentration of biotech companies in the United States (PricewaterhouseCoopers 1998). Leading companies in the region include Human Genome Sciences (HGS) and Celera Genomics Corporation, two key actors in the international effort to map the human genome. In addition, another local company, MedImmune, is currently the world's eighth largest dedicated biotech company with six FDA approved products on the market. In total, there are approximately 300 Small and Medium Sized (SME) biotech firms in the region as of 2001. The establishment of the industry was chosen to be the mid-1970s, as Stanley Cohen and Herbert Boyer invented their genetic engineering techniques in 1973. These techniques gave rise to the modern commercial biotech industry.

The earliest entrepreneurs in the Capitol region started firms during the time when many prominent national firms such as Amgen and Genentech formed. These early entrepreneurs were previously employed at large supplier firms to the National Institutes of Health (NIH). The presence of the NIH, as the U.S. agency whose mandate is to oversee health and medical research, is a defining characteristic of the Capitol region. It employs a large number of researchers at the agencies' home campus in Bethesda, Maryland. The NIH has been a spawning ground for new company start-ups, particularly in the last ten to fifteen years. Other government institutes and agencies, such as the Walter Reed Army Institute for Research (WRAIR) and the U.S. Food and Drug Administration have also been a significant source of biotechnology entrepreneurs. Although the region's universities have generated new companies, their involvement is more recent and was not apparent at the earliest stages. While the initial entrepreneurs came from large corporation and government institutions, new start-up firms became particularly fruitful in generating second, third and fourth generation start-ups.

The Information and Communications Technologies (ICT) industry also has a strong presence in the Capitol Region with a concentration in Northern Virginia. According to some sources, the region may be regarded as the birthplace of the Internet[5]. Prominent companies in the region include MCI, AOL, NexTel, and PSINet. Over 400 SMEs may be considered as ICT firms located in the area[6].

---

[4]Such a study would involve access to a source of historical data on firms such as tax or employment records to discern when the firms came into existence and when they ceased to exist. This approach would be limited because the smallest and most typical form of start-up, the sole proprietorship, might not be captured.

[5]For example, the Virginia Economic Development Partnership uses this slogan (http://yesvirginia.org/wva-be.html).

[6]There is no accepted definition of the ICT industry and estimates of the number of entities in the region vary widely and appear to be influenced by media-hype. This is the authors' conservative estimate.

Companies in the region supply half the total worldwide Internet backbone (PricewaterhouseCoopers 1998).

The modern computer networking technologies that are the backbone of the Internet and ICT emerged in the early 1970s from the U.S. Department of Defense Advanced Research Projects Agency (variously called ARPA and DARPA) (see Kahn and Cerf 1999 for more detail). Individuals leaving the Department of Defense (DOD) and the allied services formed the earliest start-ups. In addition, individuals from private industry, both within the region and from outside, figure prominently. Area universities are notably absent from this list.

While biotech and ICT are very different industries, the objective here is to discern patterns in the origins of the companies and to explore the temporal development of entrepreneurial activity. Several notable patterns emerge. First, entrepreneurs hail from a variety of different prior organizations. Government agencies served an important incubator function in both industries; however, they were not the sole source of entrepreneurial talent. There is evidence of great diversity in entrepreneur's backgrounds. Second, the earliest start-ups were service firms, not originally involved in the types of R&D activities that generate new industries. Firms such as Bethesda Research Labs and AMS were not launched as product development firms although they evolved in that direction over time. Thus, the industry had rather humble beginnings - not the type of start-up that would attract much attention from investors, the media or local economic development officials. Third, entrepreneurship picks up momentum. Over time, generations of new firms spun-off from the earliest start-ups and entrepreneurs who cashed in from one new venture created other new companies.

Two basic features in this interpretive history stand out: although the Capitol region did not have the generally regarded "prerequisites" for high technology development, a confluence of events created an opportunity for entrepreneurial individuals to create start-ups. Second is that the organizations and entrepreneurial ventures co-evolved. Both the biotech and ICT industries had humble beginnings in the region and initially there was a low level of activity. However, as entrepreneurship caught hold, the industry as we observe it today began to emerge and the necessary components to sustain it grew up in the region.

The genesis of entrepreneurship was very different in the Capitol region than Silicon Valley. Studies of the evolution of Silicon Valley such as Kargon, Leslie and Schoenberger (1992), emphasize the early input of Frederick Terman, who is viewed as the founder of the Silicon Valley phenomena. Terman had the goal of transforming Stanford, circa 1946, into the MIT of the west. As the dean of engineering at Stanford, he had the desire to foster scientific innovation as part of the academic process but also to tie entrepreneurial effort and academic accomplishment together by taking promising young students to visit high technology start-ups run by Stanford graduates. These young students had a first hand glimpse at research in application and were able to talk to the Stanford graduates who had patented their academic research and moved it into an entrepreneurial enterprise. Terman noted that many early successful young entrepreneurs had little formal education, and suggested that someone with formal engineering education and some business knowledge might do even better

(Packard 1965). Further, in 1954, Terman created the honours cooperative program, giving local companies the opportunity to send their engineers back to school to gain advanced degrees. He also supported the creation of the Stanford Industrial Park (Kargon, Leslie and Schoenberger 1992).

In contrast to the Capitol region, the connection between academe and industry in Silicon Valley was firmly entrenched early on and entrepreneurship was encouraged and supported. Looking back on the genesis of Silicon Valley, one administrator commended that it had been no accident, "A clearly visible impact of Joint Services Electronics Program activity at university centres has been the evolution of industrial activity, usually not far from those centres, entrepreneured, manned and carried forward by students (and faculty) who have gone into those industries" (Shostak 1987). Although the right conditions are often discussed as the secret to the success of Silicon Valley, the early contributions of a pioneer such as Frederick Terman cannot be ignored. He did not create the conditions for development that we recognize in mature clusters, but rather he knew how to extract the maximum advantage from the existing regional resources and institutional context within which he operated and taught others to do the same (Kargon, Leslie and Schoenberger 1992). Through his initial foundation and promotion of Stanford as a major engineering school, he put into play the process that would develop the "golden triangle" of research, industry and federal government into a vibrant and self-reinforcing cluster. Although this golden triangle was widely emulated, not many of those attempts were successful. Some research universities have a greater predilection for theory and are not interested in industry application, *per se*. Leslie and Kargon, (1994) document an attempt by Bell Labs to develop close industry ties with Princeton, an attempt that largely failed. "Princeton was the wrong sort of university partner" (Leslie and Kargon 1994). Although the New Jersey corridor did not turn into another Silicon Valley as envisioned in the 1960s, it did evolve into a cluster of research centres and one of the densest concentrations of electronics talent in the U.S. Similar patterns are observed in the Capitol region. Initially, one of the major research universities, Johns Hopkins, was not interested in industry-university collaboration and had a climate suited to theoretical discovery but not inclined to practical application (Desrochers and Feldman 2000).

3. Emergence of a Cluster

Lacking the strong cooperative environment and culture of university-industry technology transfer that characterizes the history of Silicon Valley, the Capitol region developed as a creative response to outside factors. In this section, we discuss these factors and how they coalesce to form a complex adaptive system of innovation.

Between 1970 and 1990, the Capitol region was affected by a series of exogenous shocks to its employment structure. Some of these shocks were: the downsizing of the federal government, the initiation of federal outsourcing for services, especially computer and telecommunication services that could be adapted to the commercial sector, changes in intellectual property that allowed access to high opportunity technologies and the favourable treatment of small firms with regard to securing government contracts or financing. Each shock provided a further impetus for firm formation (see Feldman forthcoming).

Beginning during the Carter administration in the 1970s, there was a pronounced downsizing in federal employment that continued through the Reagan presidency. As a result, federal employment became less secure, compensation levels were frozen and future prospects deteriorated. Many of the affected individuals were victims of location inertia - they had strong personal ties to the region. However, simultaneously an opportunity for entrepreneurship emerged. The Civil Service Reform Act of 1978 contained an initiative to outsource the production of goods and services to the private sector. Thus, individuals in the prime of their careers found entrepreneurship a viable employment option. The threshold for entrepreneurial risk-taking had in effect been lowered by government downsizing and outsourcing. When the federal "cushion" became less comfortable, the incentives to leave were higher and there were ready markets to tap. Federal procurement spending in the metropolitan Washington area grew by 114.3% from 1983 to 1997, creating enormous opportunities for private sector firms while nationally, federal procurement spending increased by 3.1% during this same period (Haynes, Fuller and Qiangsheng 1997, p. 149). The Reagan administration defence was materially different from other defence initiatives as it focused on the technical and software attributes of armaments systems such as electronics, design and systems management rather than armaments production[7]. While this initiative stimulated economic growth throughout the United States, the National Capitol Region was one of the major beneficiaries (Stough, Campbell and Haynes 1998)[8].

The changes in employment structure and incentives were simultaneous with new opportunities for the commercial exploitation of intellectual property rights that accrued from publicly funded research. In 1980, as a response to declining American competitiveness, a new era in the transfer of publicly funded intellectual property to industrial firms began with the passage of the Stevenson-Wydler Technology Innovation Act, and the Bayh-Dole University and Small Business Patent Act. Thus, the large number of federal and university labs in the Capitol Region were allowed to license their innovations to private firms. Most companies appear to have started with personal funds rather than venture capital, a finding consistent with the literature (Bhide 1999; Blanchflower, Oswald, and Stutzer 2000). The Small Business Innovation Development Act of 1982 established the Small Business Innovation Research (SBIR) Program. Under this Act, all federal agencies with an annual R&D budget greater than $100 million are required to set aside a percentage of R&D funds for small business. The Act thus greatly increased the funding available to technologically oriented small business. Lerner

---

[7]This affected ICT but also biotechnology. Consider the firm Martek that is a spin-off from the defence contractor Martin Marietta and was funded by DOD.

[8]For examples see Feldman (2001).

(1996) estimates that the SBIR program has provided over $6 billion to small high-technology firms between 1983 and 1995.

In conclusion, entrepreneurship in the region was a response to exogenous factors: underemployed skilled labour brought about by changes in federal employment policy coupled with new opportunities for the private sector to contract with the federal government and commercialise new technologies. The two industries considered here responded to different pressures. The advent of entrepreneurship was reactive and adaptive. While both sectors benefited from great opportunity for commercial products, biotechnology was more influenced by CRADAs and opportunities for licensing and joint product development while ICT benefited more from outsourcing opportunities. In both cases, locational inertia kept the entrepreneurs in the area. Over time, the region has developed the supporting conditions the literature associates with entrepreneurial environments. The next section considers the ways in which these factors developed.

4. Regional Entrepreneurship as a Complex System

Although we know a great deal about mature clusters (such as Silicon Valley, CA and Research Triangle Park, NC), we understand very little about how such clusters develop and why they occur in certain areas and not others (Leslie and Kargon 1994). By focusing on the entrepreneur as the spark or agent of change, we provide a rationale for the emergence of clusters and an understanding of the dynamic that directs their development. The entrepreneur can be viewed as an agent who takes existing resources and re-defines, re-combines and re-deploys them to create new products, services, and organizations (Russell 1996). Through the process of creating new companies, entrepreneurs spark regional industrial transformation, a transformation that exhibits path dependence, adaptivity, and self-organization.

Interestingly, these observations about the role of the entrepreneur and innovation in industrial transformation were already present in Schumpeter (1942). Schumpeter placed entrepreneurship and its connection with dynamic uncertainty in the centre of economic reasoning, by tracing all disrupting economic change to innovations and identifying the innovator with the entrepreneur (Blaug 1985). Entrepreneurs, in Schumpeter's view, can be considered the "spark-plugs" or "flash-points" of creative destruction (Schumpeter 1942). When an environment provides a mix of resources and opportunities, entrepreneurial individuals will move to capitalize on those opportunities. Their input creates the initial changes in the regional environment that begin to build on each other. Bygrave and Hofer (1991, p. 19) suggest, "The essence of the entrepreneurial process is a fundamental discontinuity in the industry involved".

The entrepreneur is also active within a regional context. Leslie and Kargon (1994) propose that technology clusters are better thought of as ecologies of mutually dependent institutions. The entrepreneur then, operates in and activates his eco-system to his purpose. It is in the creative feedback response of the entrepreneur to his environment, that the character of the cluster is determined. If the incentives and the environment are right, these initial changes cascade into a fully functioning cluster able to adapt to adverse shocks. Looking at clusters such as Silicon Valley, the Texas conurbation, Research Triangle Park, and the Capitol region, we see that entrepreneurship responded to each unique environment, creating clusters each with their own signature characteristics (Leslie and Kargon 1994; Feldman forthcoming).

Before 1980, there was very little ICT or biotech activity in the Capitol region. Although some of the necessary components for those industries were in place - resources for example - there was not sufficient incentive for entrepreneurial activity. Unlike Stanford's open appreciation and interaction with industry in the Santa Clara valley, the major research universities in the Capitol region were more conservative and theoretically oriented (Desrochers and Feldman 2000). Without entrepreneurial incentives or an open culture of technology transfer, there was no way to ignite the development of biotech or ICT clusters. If the entrepreneurial event were a linear system - where you line up the inputs and produce a given and certain output - then the environment and initial conditions would not matter. Nevertheless, entrepreneurship, when considered as part of a larger complex adaptive system of innovation, is context sensitive and history or region dependent.

Feedback effects from the interactions of the agents and their environment characterize the way in which the cluster evolves. These interactions, in the case of the Capitol region, produced companies that specialized in vaccines and genomics, as well as in Internet technology. The conditions that created the high-technology Silicon Valley cluster generated relative specialization in other areas such as software applications, hard drives and computer chip technology (Kargon, Leslie and Schoenberger 1992). The fact that the development process matters for the character of the mature industry suggests a non-linear dynamic applies and that history determines the direction of the industry trajectory. From the initial spark of entrepreneurial activity, to the adaptation of the entrepreneur to environmental factors, to the emergence and self-organization of the industry around the initial start-ups, we see a complex interdependent mix of components that develops in a non-determinate way. Looking at activity circa 1980 or even 1985 cannot give us much information about what the mature cluster will look like. For example, when we consider the Capitol region in 1970, there was no indication biotech or ICT industries were germinating. Instead, we see very little entrepreneurial activity, virtually no venture capital, a large well-paid government sector and a relatively hostile academic environment.

There were some positive pre-existing conditions in the region such as a long tradition of science, fed by the NIH, Department of Defense, and other government institutions as well as universities, a highly educated work-force and a large number of medical institutions. But in order to activate those resources, the set of exogenous events described above were needed to change incentives for entrepreneurial activity and create an opportunity for individuals who were enterprising and adaptive. The first entrepreneurs, who re-combined the regions high-skilled labour and innovative activity in new ways to create the first biotechnology and ICT companies, seized this opportunity. These early entrepreneurs did not have an environment rich in social and venture capital, nor a network of organizations to support them, nor regional development offices to facilitate their ventures, so they adapted and pieced together their companies from the building blocks that did exist. Entrepreneurship, in this sense was a break point, creating a bifurcation or catastrophe point that set in motion factors that would transform the regional economy.

Entrepreneurs do not know initially which direction their company will go, how to set up its structure or even what products or services will ultimately be offered and successful (Zaltman et al. 1973). Rather, the process of entrepreneurship is a classic trial and error or learning-by-doing process. In this sense, the learning process and the adaptation to new events and to the existing environment are important determinants in the final development. The evolution of the new firms depends itself on the start-up process, the regional (and national/international where relevant) demand as well as the available resources in the region. The cluster and the characteristics of the cluster therefore emerge over time from the individual activities of the entrepreneurs and the organizations and institutions that co-evolve to support them. For example, in the Capitol region, several large military institutions work on vaccines. This relative expertise and the ability to patent innovations generated by the military labs, created a niche for start-ups in the region. Additionally, the Capitol region is home to one of the largest concentrations of medical expertise, hospitals, clinics and medical networks, facilitating the intensive FDA required testing of new biotech drugs.

Further, the cluster self-organizes around the entrepreneurial activities. Entrepreneurial activity is creative and innovative and therefore, the specific needs of the entrepreneur cannot be predicted *a priori*, but develop over time as start-ups are created and the needs are recognized. We see in the Capitol region that once a critical mass of start-ups in biotechnology and ICT was in place, other organizations were attracted. The organization of the cluster and the entrepreneurial ventures evolved simultaneously and even symbiotically. Having the experience and example of the initial start-ups, the industry became self-sustaining: entrepreneurs attracted physical and human capital to the area, public and private networks built up to support and facilitate the ventures, relevant infrastructure was created through public and private initiatives and services grew up to feed these companies.

One striking fact that emerges is that the history of each cluster is unique, suggesting that the dynamic of cluster development is path or history dependent. In Silicon Valley, for example, we see the strength of the aeronautical and electronics industries being championed into a high-technology conurbation by a small group of people with a vision for the development of the region. While in the Capitol region, we see that changes in government structure and science and innovation policy sparked a frenzy of activity that drew resources to itself as it grew.

Below, the emergence of high technology clusters in biotechnology and ICT is grouped into several phases. The first phase is the emergent phase, where the spark of entrepreneurial innovation is ignited by a confluence of exogenous events. The second phase can be characterized as the self-organization of the cluster and the deepening of the self-reinforcing feedbacks among entrepreneurs, enterprises, institutions and resources. The third phase is the maturation of the industry into the well functioning and rich innovative and entrepreneurial system we observe today.

5. Self-Organization

In the start-up phase, entrepreneurs began by pursuing commercial projects that did not require high levels of investment and which were unlikely to generate the type of large profits that would interest venture capitalists. They started with government contracting, producing rather mundane bread and butter products, such as medical test kits and reagents for biotechnology or services such as computer system integrations and maintenance work in ICT. In addition, the growing number of related firms in the region provided opportunities for subcontracting, work and asset sharing, thus making it easier for the start-up firms to bootstrap and steadily grow without large doses of new capital.

Venture capital seeks opportunity and when there are potentially profitable investment opportunities in a region, venture capital is attracted. Being able to monitor and mentor the entrepreneurial firms in which they invest makes close geographic proximity valuable for venture capitalists (Gompers and Lerner 1999). There are now approximately six venture capital firms headquartered in the region and firms located elsewhere have opened branches. In March 1999, Silicon Valley Bank opened a branch office in Northern Virginia with the comment, "We are trying to be ahead of the curve. As far as new start-ups, this is a real hotbed" (Montgomery and Bacon 1999).

Over time, as start-up companies grew and went public, or were bought out by other companies, the dynamics of the region changed. Most notably, local entrepreneurs who made large fortunes engaged in institution building to support their activities and to encourage further entrepreneurship. Also important was the emergence of networks of supportive social capital that began as membership organizations to promote networking. These activities were primarily private sector initiatives, financed with private funds. By collaborating with state and local government programs, these initiatives resulted in cross-fertilization and a common mission to promote the development of industry in the Capitol region.

Older style quasi-public organizations such as the Washington Board of Trade and the Greater Baltimore Committee have broadened their agendas and spun-off new organizations directed at technology intensive industry. Other government-financed programs found greater success by collaborating with the privately organized networks (Guidera 1996).

6. Critical Mass/System Maturation

There is evidence that the region has achieved a point where it is a self-sustaining system. Many entrepreneurs sold their companies and started private incubators to nurture others. These founders were motivated to share their expertise and to build the region. In addition, at least three angel networks have formed in the last five years. The Private Investors Network (PIN), the Capital Investors Club and the Washington Dinner Club are organizations of experienced entrepreneurs who actively invest in new companies and offer management advice.

Universities in the region have responded to the increased entrepreneurial activity by offering new programs and building branch operations closer to commercial activity. For example, Johns Hopkins University offers a Masters degree in Biotechnology in Silver Spring, Maryland about 50 miles away from the main Baltimore campus. Virginia Tech University opened a branch campus in Northern Virginia about 250 miles from their main campus. The draw has been the number of workers seeking additional training, the opportunities for industry-funded research and the interaction with industry. Notably, the local universities have benefited from the philanthropy of local entrepreneurs. For example, George Mason University began in Fairfax, Virginia in 1950 as a commuter school. It has grown into Virginia's second-largest university with 18 doctorate programs and a focus on technology. Donors have given the university millions of dollars to endow 43 professor chairs, allowing the university to recruit high-profile professors (O'Harrow and Lipton 1996). All of the universities in the area have responded with incubators and other programs to encourage entrepreneurship.

Interestingly too, dedicated university involvement in high-technology enterprise or the promotion of academic-industry transfer occurred only much later in the Capitol region's development. In Silicon Valley, the opposite story could be told; the moulding of Stanford as a producer of entrepreneurs and as an institution that was sensitive to the needs of industry figured prominently in the success of the region and occurred in the initial phase (Kargon, Leslie and Schoenberger 1992).

In the Capitol region, legislative programs also followed the initial phase and addressed the needs of the newly established industry. A group of Virginia businesspeople organized a broad campaign to advocate state tax increases for education in order to address a noted shortage of technology workers and provide greater infrastructure funds (Baker 1995). This initiative attempted to build infrastructure to support the development of local industry. Rather than seeking specific requests for their own business, business leaders were promoting a broader, collectively responsible social agenda (Feldmann 1997). In 2001, the state of Maryland passed twelve legislative acts focusing on providing a supportive environment for technology-based economic development. These cover the full gamut of infrastructure development, training programs and tax incentives.

One more point to consider besides the characteristics of a mature cluster mentioned here, is that an industrial region reaches full maturity when it is stable and able to adapt to adverse shocks (Saxenian 1994). Although the history of high technology in the U.S. is highlighted by brilliantly successful clusters, it is also marked with cluster failures - clusters that were not able to adapt to shocks and whose entrepreneurs and start-up activity folded or went elsewhere when the environment became negative (Leslie and Kargon 1994). The ability to re-configure resources in response to change is an important indicator of maturity (Andersen and Teubal 1999).

In sum, these factors when added together represent a whole that is larger than its parts and system which is self-sustaining and reinforcing. The mature state of the Capitol region looks not much different from Silicon Valley, in terms of venture capital, major research university involvement, university-technology transfer practices, and the types of social capital and entrepreneurial efforts observed. Nevertheless, the genesis of each region is much different - two rich self-sustaining high-technology clusters which appear similar on the outside, but evolved through entirely different paths.

7. Reflective Conclusions and Appreciative Theorizing

The economic success of Silicon Valley in terms of individual wealth creation, corporate profits and job creation, has been so impressive it has pushed government officials in locations across the nation to try to imitate or replicate its success. Government policies aimed at replicating the conditions that exist in the region today are based on the belief that their local areas may also capture the benefits of new high technology firm formation and the attendant economic growth. Much of the prevailing conventional wisdom is drawn from a snapshot of the advanced stage of Silicon Valley's development, that is, on the workings of a fully functioning innovative system. Looking at a successful region in its full maturity, however, may not provide prescriptive information about how such regions do develop. Conditions that we associate with an entrepreneurial environment are the result of a functioning entrepreneurship and do not illuminate the early efforts by which such entrepreneurship first took hold and the cluster developed.

A critical question is how regions change and develop into areas with higher growth potential. Is replication of a mature entrepreneurial environment sufficient to foster entrepreneurship? Notable failures in such places as Dallas and Boston route 128 (Leslie and Kargon 1994) suggest it is not. Saxenian (1994) analyses Silicon Valley from the perspective of how it adapted to restructuring in the semiconductor and computer industry and establishes the importance of social relationships in defining the capacity of the region to evolve, adapt to shocks and accommodate new demands. In this paper, we have examined how one region, initially lacking an entrepreneurial tradition, accomplished the transformation to a fully functioning rich regional system. Such a transformation entails a fundamental shift or phase change from an inert innovative system to an active system. Certainly, the Capitol Region was the site of large government research infrastructure, classified as a State-anchored region using Markusen's (1996) typology. In this regard, the concentrations of resources and highly skilled labour plus access to sophisticated, demanding technology users were pre-existing conditions in the region. The transformation to private sector entrepreneurial growth did not appear to represent movement along a technological trajectory (Kenney and von Burg 1999) but instead was a sustained effort at capacity building that involved human agency, adaptation and evolution. Not only this, a critical point was reached in the development of the region when it jumped from virtually no high technology start-up activity to intense activity with start-ups per year numbering in the hundreds.

In the development of an industrial system of innovation, there are many individual complex stories and personal motivations. It has been said that government employees and contractors could never become successful private businesses - the incentives were very different. Government workers, the logic went, were too removed from the pressures of the market. Government contracts followed a practice of placing a low bid in order to get the job and then making a profit by demonstrating a need for change orders in the absence of competition. This is a very different philosophy from trying to do a job right the first time at the lowest cost. However, the earliest entrepreneurs in biotechnology and ICT were government contractors and employees who proved this logic wrong. What is critical is that the region did provide opportunity for individuals. They began working for the government but then realized that they could adapt their products for dual-use commercial markets. Therefore, they moved to commercial markets and with this development came innovation. Eventually they succeeded. In essence, this was a phase change from latent to active entrepreneurship.

The entrepreneurial event in the Capitol Region was a response to and adaptation to changes that were exogenous to the regional system. In this regard, Federal policies such as downsizing created slack and surplus resources that could find new and more productive uses. Thus, the gales of Schumpeter's creative destruction were unleashed. Policies that created a supply of potential entrepreneurs would not have been sufficient. A complementary set of government policies aimed at creating demand for ICT and biotechnology services through government procurement facilitated the transition. Other exogenous conditions were the policies that provided mechanisms or tools to enable companies to access resources. These affected the supply of new ideas by creating access to intellectual property from government investment.

Both biotechnology and ICT are high opportunity technologies facing growing product demand and therefore are attractive to investors. Firms working in these technologies face favourable market conditions. The degree to which this is exogenous may be debated. Good entrepreneurs may create their own opportunity and thus define the industry. The idea that technology development is endogenous to cluster development warrants more investigation. Through the actions of key individual change agents, the configuration of the cluster and the technological trajectory of the industry may be jointly determined. It does suggest that companies, industries and regions benefit from the same factors and decisions - their evolution may be intricately interwoven.

Currently, a myriad of economic development policies attempt to encourage entrepreneurship. We have shown that rather than being actively promoted and encouraged by economic development policies, these activities had much more humble and pedestrian beginnings. The conditions that we associate with entrepreneurship developed over time. In the early stage of these new technologies, the way in which they would develop was unclear and it would have been difficult to anticipate the types of specific assistance that entrepreneurs needed. Individual entrepreneurs were in the best position to move the technology, the industry and the region forward. This is not to say that there is no role for local government policy in promoting entrepreneurship. Although no early examples presented themselves in the Capitol region, we have not directly examined that question. It is interesting to note that the Silicon Valley success is viewed as an outgrowth of the intense technology transfer and interaction between industry and universities in that region. Local government policies did and do play a role, but these tend to be implemented and effective in the later stages of cluster development. Are there general lessons to be learned from the development of the Capitol region or is this case unique and is every case unique? Certainly, this region benefited from high average household income and higher than average education levels, giving it very different resource endowments from other underdeveloped regions lacking an entrepreneurial culture. The general lesson is that entrepreneurs adapt and when they are successful, they build the types of resources that support their activities.

A distinction should be drawn between the conditions that support innovation and the conditions that support entrepreneurship. The two concepts are related: entrepreneurship facilitates the realization of innovation, as firms are formed to commercialise and advance new ideas. Conducive external environments and resources make innovation activity easier but may not be sufficient to induce new firm formation, which is where the concepts diverge. The critical condition for entrepreneurial enterprise is *opportunity*. Even if the regional conditions do not match those of successful clusters, the incentive to developing locational opportunities leads entrepreneurs to start firms.

Once established, industrial clusters become virtuous, self-reinforcing cycles. Legal and tax frameworks, research institutions and social relationships are definitely areas for public policy intervention in terms of creating a supportive and positive environment for innovation and entrepreneurship. Specifically relevant are the conditions that affect the decisions of individuals to become entrepreneurs, and the ways in which an entrepreneurial culture develops and takes hold. Yet, our understanding of regional economic systems may be enhanced by a consideration of entrepreneurs as economic agents who actively interact with their local environments, adapt to new situations, crises or opportunities using location-specific assets, and finally, build and augment local institutions. Certainly, this is not the last word on this topic. It is my hope that this historically informed appreciative theorizing will inspire others to take a more detailed look. It is only through an appreciation of the nuances of cluster development that we may begin to adequately inform policy.

References

Abetti, P. A. (1992), "Planning and Building the Infrastructure for Technological Entrepreneurship", *International Journal of Technology Management*, 7, pp. 129-139.

Andersen, E. and Teubal, M. (1999), *High Tech Cluster Creation and Cluster Reconfiguration - A Systems and Policy Perspective*, Working Paper.

Anderson, P. W., Arrow, K. J. and Pines, D. (1988), *The Economy as an Evolving Complex System*, Reading, MA, Addison Wesley Publishing.

Appold, S. A. (2000), "The Control of High-Skill Labor and Entrepreneurship in the Early US Semiconductor Industry", *Environment and Planning*, A 32 (12), pp. 2133-60.

Arthur, B. (1994), *Increasing Returns and Path Dependence in the Economy*, Ann Arbor, University of Michigan Press.

Bahrami, H, and Evans S. (1995), "Flexible Re-cycling and High-technology Entrepreneurship", *California Management Review*, 37, pp. 62-89.

Bhide, A.V. (1999), *The Origin and Evolution of New Businesses*, Oxford, Oxford University Press.

Baker, P. (1995), "Giving Allen the Business on Colleges: Executives Say Extra $ 200 Million a Year Is Needed for Education", *The Washington Post*, July 25, p. A1.

Blanchflower, D. G., Oswald, A. and Stutzer, A. (2000), "Latent Entrepreneurship across Nations", Manuscript - August 27.

Blaug, M. (1985), *Economic Theory in Retrospect*, Cambridge, Fourth Edition, Cambridge University Press.

Bygrave, W. and Hofer, C. (1991), "Theorizing about Entrepreneurship", *Entrepreneurship: Theory and Practice*, 16 (2).

Desrochers, P. and Feldman, M. (2000), *The Practical Application of "Truth for Its Own Sake": A Short History of University-Industry Technology Transfer at the Johns Hopkins University*, Working Paper.

Eisenberg, R. (1996), "Public Research and Private Development: Patents and Technology Transfer in Government-Sponsored Research", *Virginia Law Review* 82, pp. 1663-1727.

Feldman, M. P. (forthcoming), "The Entrepreneurial Event Revisited: Firm Formation in a Regional Context", *Corporate and Industrial Change*.

Feldmann, L. (1997), "In Virginia, Businesses Argue For Taxes", *The Christian Science Monitor*, September, 9, p. 4.

Gompers, P. A. and Lerner, J. (1999), *The Venture Capital Cycle*, Cambridge, MA, MIT University Press.

Guidera, M. (1996), "Start-ups Finding it Easier to Get Paired with "Angels"", University of Maryland's Dingman Center joins Investor Network in Matchmaking Effort, *The Baltimore Sun*, August 21, p. 1C.

Haynes, K., Fuller, S. S. and Qiangsheng, L. (1997), "The Northern Virginia Economy: The Changing Role of Federal Spending", *Research in Urban Economics*, 11, pp. 145-161.

Kahn, R. E. and Cerf, V. G. (1999), "What is the Internet (And What Makes It Work)", *Paper prepared for the Internet Policy Institute*, December.

Kargon, R., Leslie, S., and Shoenberger, E. (1992), "Far Beyond Big Science: Science Regions and the Organization of Research and Development", in: *Big Science: The Growth of Large-Scale Research*, Galison, P. and Hevly, B. (eds), Stanford, Stanford University Press.

Kay, N. (2000), "Searching for the Firm: The Role of Decisions in the Economics of Organization", *Industrial and Corporate Change*, 9, pp. 683-707.

Kenny, M. and von Burg, U. (1999), "Technology, Entrepreneurship and Path Dependence: Industrial Clustering in Silicon Valley and Route 128", *Industrial and Corporate Change*, 8, pp. 67-103.
Lazerson, M. H. and Lorenzoni, G. (1999), "The Firms that Feed Industrial Districts: A Return to the Italian Source", *Corporate and Industrial Change*, 8, pp. 235-266.
Lerner, J. (1996), *The Government as Venture Capitalist: The Long-Run Effects of the SBIR Program*, NBER (National Bureau of Economic Research) Working Paper No. W5753.
Leslie, S. and Kargon, R. (1997), "Recreating Silicon Valley", *Business History Review*.
Leslie, S. and Kargon, R. (1994), "Electronics and the Geography of Innovation in Post-War America", *History and Technology* 11, pp. 217-231.
Link, A. N. (1995), *A Generosity of Spirit: The Early History of the Research Triangle Park*, Research Triangle Park, The Research Triangle Foundation of North Carolina.
Malerba, F., Nelson, R., Orsenigo, L. and Winter, S. (1999), "History-Friendly Models of Industry Evolution: The Computer Industry", *Industrial and Corporate Change*, 8, pp. 3-40.
Markusen, A. (1996), "Sticky Places in Slippery Space: A Typology of Industrial Districts", *Economic Geography*, 72, pp. 293-313.
Marshall, A. (1890), *Principles of Economics*, London, Macmillan.
Montgomery, D. and Bacon, J. A. (1999), "How Green is Our Valley"? *Virginia Business*, March, (http://www.virginiabusiness.com/vbmag/yr1999/march99/howgreen.html).
Nelson, R. and Winter, S. (1982), *An Evolutionary Model of Economic Change*, Cambridge MA, Harvard University Press.
Nelson, R. (1993), *National Innovation Systems: a Comparative Analysis*, Oxford, Oxford University Press.
O'Harrow, R., Jr. and Lipton, E. (1996), "George Mason U. Embroiled in Debate about Its Mission", *Washington Post*, 14, p. A1.
Packard, D. (1965), *Stanford Engineering News*, 17 (22) July.
Porter, M. (1990), *The Competitive Advantage of Nations*, New York, Free Press.
PricewaterhouseCoopers (1998), *Toward a New Economy: Merging Heritage with Vision in the Greater Washington Region*, Washington, D.C., Potomac Knowledge Ways Project (http://knowledgeway.org/voice/newecon/homepage.html).
Prevezer, M. (1997), "The Dynamics of Industrial Clustering in Biotechnology", *Small Business Economics*, 9, pp. 255-271.
Roberts, E. B. (1991), *Entrepreneurs in High Technology*, New York, Oxford University Press.
Russell, R. D. (1996), "The Emergence of Entrepreneurship in Eastern Europe: A Self-organizing Perspective", *International Journal of Commerce and Management*, 6, pp. 21-32.
Saxenian, A (1994), *Regional Advantage*, Cambridge, MA, Harvard University Press.
Schumpeter, J. (1942), *Capitalism, Socialism, and Democracy*, New York, Harper Torchbooks, 3rd edition, 1976
Shostak, A. (1987), "History of the Joint Services Electronics Program", in: *Proceedings of the Fortieth Anniversary of the Joint Services Electronics Program*, D. Robb and A. Shostak (eds), Washington, DC, Battelle Laboratories, pp. 15-46.
Stough, R. R. (2000), *The New Generation Technology Economy: Comparative Regional Analysis and the Case of the U.S. National Capital Region*, paper presented at the American Association for the Advancement of Science meeting in Washington, D.C., February 2000.
Stough, R. R., Campbell H., and Haynes, K. E. (1998), "Small Business Entrepreneurship in the High Technology Services Sector: An Assessment of Edge Cities of the U.S. National Capital Region", *Small Business Economics*, V 10, pp. 61-74.

Teubal, M. and Andersen, E. (2000), "Enterprise Restructuring and Embeddedness: A Policy and Systems Perspective", *Industrial and Corporate Change*, 9, pp. 87-111.
Zaltman et al. (1973), *Metatheory in Consumer Research*.

From Exogenous to Endogenous Local Development: The Cases of the Toulouse and Sophia Antipolis Technopoles

Christian Longhi

Abstract. Regional clusters and local organisation of the industry have been proved to be basic elements of economic growth and innovation. The understanding of the processes underlying local development is thus pivotal. The paper attempts to contribute to this aim through the analysis of Toulouse, in the French southwest region Midi-Pyrénées, and Sophia Antipolis, in the southeast French Riviera. Both be considered as the outcomes of the French national system of innovation and the process of decentralisation initiated in the late sixties, and are today well known centres of high technology activity. But, as the numerous failed experiences of top-down reinforcement suggest, the factors the two areas have been endowed cannot explain as such the emergence and development of local innovative milieux. Considering clusters as evolving open complex systems, the paper analyses the reconfigurations of the clusters implied by changes in internal and external conditions that have led to endogenous innovative processes characteristic of technopoles.

1. Introduction

Since the seventies, regional clusters are object of considerable interest. The recognition of their distinctive role on the economic and technological scene became evident with the rise of the industrial districts in the Third Italy. The districts succeeded to perform better than large firms within the same international competitive markets because of the great flexibility allowed by the institutional arrangements set up locally (Becattini 1990). They have represented a kind of localized achievement of the division of labour where the innovative capabilities and competitive efficiency have arisen from organisational designs established by the firms, arranged in networks of specialized sub-contractors in vertically disintegrated processes of production (Brusco 1982) Another important development has been the emergence of high tech clusters based on the knowledge base developed in the universities, as happened in the Silicon Valley or Cambridge, UK (Saxenian 1994; Garnsey 1998).

Despite very different features regarding industries or resource bases, these clusters have in common local endogenous processes of development triggered by innovative SMEs. These models have promoted regional clusters and local organisation of the industry as basic elements of the economic growth. As clearly stated by Porter (1990), growth can be measured at the level of nations, but nations in some sense neither produce nor exchange. Their competitive advantage arises from the strength of regions, cities, districts, which are the locus of production. Likewise, innovation, the capacity of economies to change and adapt to their environments, are deeply rooted in clusters, and strongly related to the organisation of the industry implemented locally.

The understanding of the processes underlying local development is thus pivotal. The paper attempts to contribute to this aim through the analysis of cases very different from the paradigmatic Italian districts or university-based clusters. They are instead the result of industrial and regional development policies, of top down processes implemented in the French economy, that have resulted in the emergence of high tech clusters or technopoles, defined here as regions developed around several interrelated knowledge elements, including but not limited to science parks or research or technology centres (Luger 2001). Indeed, Toulouse, in the southwest region Midi-Pyrénées, and Sophia Antipolis, in the southeast French Riviera, between Nice and Cannes, can both be considered as the outcomes of the French national system of innovation and the process of decentralisation initiated in France in the late sixties, but their evolution was in a sense unpredictable. Obviously, the regional clusters are not naturally given, they are constructed and instituted outcomes of economic processes (Metcalfe 2000). As the numerous failed experiences of top-down reinforcement of local attributes suggest, the factors the two areas have been endowed cannot explain as such the emergence and development of local innovative milieux. The focus has not to be on the factors as such, but on the interactions among them, the feedback effects and cumulative processes they give rise, and the process of development they entail (Garnsey and Longhi 1997). The paper traces these processes in the two cases. A suitable analytical framework has to be adopted to grasp these dynamics and build the history of the innovative milieux.

A technopole, as well as territory in general, can be identified as a spatial cluster of activities, a geographically bounded concentration of interdependent firms and institutions open to its environment, in short an evolving, open complex system. The "system" is a way of conceptualising the set of related elements with a designated boundary which represents economically the territory. Regarding innovation, Andersen and Teubal (1999) identify the business sector (the backbone of the process), the supporting structures (non-business organisations and formal institutions like research centres, universities, business association, formal policy institutions), links and institutions (i.e. user-supplier networks, research-industry relations, labour markets), culture and social structure as basic components. The "system" provides a basis for finding connections between various aspects of the milieu and for tracing its dynamic. At last, territories are enacted systems, individual agents, key actors, or institutions are the basic determinants of their trajectories; as such, despite underlying common processes, their trajectories are unique, unpredictable, and non-replicable (Garnsey 1998).

The relevance of the complexity theory to grasp territorial dynamics has become evident from the recent developments of the literature. The nature of the local development appears as an emergent property of the system, i.e. coherent and unexpected macro behaviour of the whole which is not attribute of the parts, but results from their micro interactions, and which can go with important discontinuities. This analytical framework will allow make clear the development of the local systems, their evolutions following the initial public impetus, adaptations from internal interactions and feedbacks, and from reactions to their environments. The internal as well as external interactions of the local systems are essentials to the definition of the dynamical paths; changes, "small events" (Arthur 1989) regarding these interactions can result in crisis or re-configuration of the systems.

We turn now to the cases of local industrial development. Conceptualising their development in terms of the genesis and growth of complex adaptive systems provides an heuristic device to "organize" their histories, the different reconfigurations that have led to the emergence of technopoles, as well as to understand their fundamental specific features. Diversity is indeed the basic characteristic of local development; both cases involved mechanisms of reinforcement triggered by public policies of regional development, by stimulus to local activity from leading institutions, by strong public - private interdependences, but the combination of these mechanisms of reinforcement differed, and have resulted in self-reinforcing expansion although through highly distinct paths.

2. The Cases

2.1. The Toulouse Technopole

Since thirty years, Toulouse has grown as a major European centre of aeronautics and spatial activities. One distinctive feature of the area is the role of the interdependences between industry, research and training, and services in the development of the local resource and knowledge base, which define it as a technopole. Nevertheless, the contemporaneous system of innovation has not found its source locally, but results from a state-led exogenous process of accumulation of related activities. Its development has been closely related to the construction of the European aerospace industry, in which Toulouse has played a pivotal role. The name of Toulouse is today closely associated with Airbus or Ariane, and symbolized in a thematic park, the "Cité de l'Espace", welcoming numerous visitors. Nevertheless, the development has not followed an unique path. It is obviously gone with ups and downs regarding the cycles of its dominant industries. But above all, the institutional and organizational arrangements qualifying its regime of growth or innovation have deeply evolved. The paper characterizes the different steps of this process, the internal and external circumstances that have induced changes in the local system and led to a technopolitan milieu.

2.1.1. Building of a Productive Capacity

The history of aeronautics in Toulouse begins with World War I, and the location of factories in areas sheltered from the front. Since then, the region has benefited from a strong image effect built on aeronautics, and the further public choices of investment in the area have never been questioned. Key agents explain this development. Latécoère, following a request of the government, created his factory in 1917 to produce fighters, and developed the "Aéropostale" after the war, opening new lines with his planes. The human adventure embodied in the postal airline, with men like Mermoz or Saint-Excupery, gave a strong identity to Toulouse. Dewoitine created his firm as a spin-off of Latécoère in 1920; he developed fighters and employed up to 7,000 people before World War II. The local tradition of artisan industries helped the firms to find the human resources to handle the heterogeneous elements making up a plane. Subcontracting appeared with the industry. As reported by Allinne (1996), Latécoère or Dewoitine has no difficulties to find small subcontracting firms which adapted their activities and endorsed the responsibility of design and development of whole parts of the planes, prefiguring the contemporaneous industrial organization. The specialisation of the subcontractors resulted in a diversified industrial context of interdependent SMEs, whose development has been fostered by the Defence budget.

After the war the state reorganized the industry, merging the numerous manufacturing firms weakened by the war, which failed to adapt to the new competitive context. The nationalization process resulted in the creation of "Sud Aviation", including the Toulouse factories, and "Nord Aviation", the first one specialized in commercial aircrafts and the second one in military. In 1970, these two firms together with SEREB (Société d'Etudes et de Recherches des Engins Balistiques) were merged to form the SNIAS or Aérospatiale, in order to reinforce the capacities of R&D and design, and reach a size compatible with the international standards. The new entity covers the different fields of aerospace activities regarding civil or defence applications. The nationalisations led to the programs Caravelle and Concorde, whose assembling was entrusted to the Sud Aviation factories in Toulouse. This state-led process has resulted according the terms of Muller (1988) in an "arsenal logic", i.e. a highly centralized industrial process driven by the political sphere, the decisions on the programs being made by the ministries of Defence, Air, Finance, and the head of the Air Force. According this logic, the state controlled the definition of the whole characteristics of the programs, the financing of R&D and production and was the main buyer of the products; the budgets allocated were generally exceeded without consequences. Two main basic goals are entrusted to the national firms in charge of the realisation of the programs (Muller 1988; Talbot 2000). First, to reach the technological frontier, without consideration of commercial aspects or airlines needs. Caravelle has been an innovative medium-haul aircraft sold at 278 units, 80 to Air France, but Concorde is remained a prototype. Second, to be the vectors of the regional development policy planned by the government. Aérospatiale played this role in Toulouse, animating a network of subcontractors. The region gained increased productive capacities, but along fordist characteristics; the subcontractors were no more involved in the conception of the elements to be manufactured, and supported the peaks of the market. Only up to 20 % of the subcontracting was localized in the Toulouse area, the bulk of the market, as well as the R&D and design were still concentrated in Paris. The local system of production of this period can be depicted as an hub and spoke (Markusen 1996), the hub itself being driven form the State. This trend was to reverse with the definition of the European program, and the transition towards a "logic" oriented by a market approach.

2.1.2. Take-off of the Technopole

In the late sixties, different public policy decisions have oriented the whole socio-economic development of Toulouse towards a technopole, through the decentralisation from Paris of the main research and education institutions related to aerospace activities, and the induced impetus given to the existing local resources. The emergence of a "collective agent", the CNES (Centre National des Etudes Spatiales) in the local system had a pivotal role to generate positive feedbacks from this accumulation of new resources.

With the development of spatial activities, the French government created the CNES in 1961, a public establishment in charge to define and carry on the spatial programs in collaboration with research, industry and defence. From 1968, the CNES has decentralized in its Space Centre of Toulouse most of its R&D activities; the centre has grown today up to 2,500 researchers. The CNES has developed and coordinated the Ariane program for ESA (European Space Agency), as well as most of the public and defence applications of satellite activities in France and in the European programs. The expertise of the CNES merges different scientific and industrial fields, and it develops thematic R&D on radio communication, observation, orbital sciences and infrastructures, space transport, or generic research on design and development of systems. The Space Centre has been the main catalyst of the development of space activities in Toulouse, and has induced the location in the area of the main actors in the field, public or private. In line with its decentralization, the CNES has prioritised its local relationships and developed synergies between the different actors to feed the needs in pure and applied research as well as industrial development needed by its programs, whose main feature is the innovative content. It has implemented and co-ordinated networks inducing processes of collective learning and the emergence of a local innovative milieu in spatial activities, which cross-different scientific and technological fields. The CNES has also created independent firms based on home technologies to foster the innovative process in the local complex.

The unique feature of Toulouse is that along with the CNES, most of the higher education and research institutions linked to aerospace activities have been decentralized from Paris, making a coherent and interdependent knowledge base. Indeed, the decision to locate in Toulouse the main university level schools ("Grandes Ecoles") of engineers has been conditioned by the simultaneous location of the related institutes of research, because there are closely interwoven institutions. They are both required for the quality of the diplomas delivered, engineers, but also PhD, and for the setting of close links with industry, which is part of the mission of these institutions. This process is gone with the reinforcement of the existing institutes of University and the CNRS, which had already a strong potential in computer sciences, electronics, robotics, materials. Among these main institutions, one can mention ENSEEIHT (Ecole Nationale Supérieure d'Electrotechnique, d'Electronique, d'Informatique et d'Hydraulique de Toulouse), SUPAERO (Ecole Nationale Supérieure de l'Aéronautique et de l'Espace), ENSICA (Ecole Normale Supérieure d'Ingénieurs de Constructions Aéronautiques), ENAC (Ecole Nationale de l'Aviation Civile), ONERA-"Centre de Toulouse" (Office National d'Etudes et de Recherches Aérospatiales), LAAS (Laboratoire d'Architecture des Systèmes, CNRS), LEEI (Laboratoire d'Electrotechnique et d'Electronique Industrielle), CERT (Centre des Etudes sur les Rayonnements), IRIT (Institut de Recherche en Informatique de Toulouse). In addition, the "Ecole Nationale de Météorologie", as well as French National Meteorology offices have been located in Toulouse in a specific area, the "Meteopole". They are backed by the spatial activities or the CNES, using data collected by the satellites Meteosat or Spot, and have developed an important computing expertise needed for forecasts[1].

This education and research complex strongly oriented towards industries or applications has met the tradition of the University Paul Sabatier in Toulouse, as well as the socio-economic and cultural tradition of the city. Indeed, in 1885 already the local government was the main financial support of the new Faculty of Science, and subsidized a course on electricity to meet the industrialization needs of that time; the course has been dedicated from 1902 to the training for local entrepreneurs. Or it provided the premises and the wages of the professors of the "Institut d'Electrotechnique" created in 1907 (Allinne 1996). This tradition shared by the main institutional local bodies has allowed the easily integration of such an amount of new external resources in the area, which have worked as the reinforcement of an existing socio-economic process of development.

The result of the whole process has been the early emergence of a local labour market of high qualification, allowing the "diffusion of embodied and tacit knowledge and technological know-how" (Camagni 1991) and local collective learning. Indeed, the students are vectors of creation of close relationships between education, associate research institutions and firms, through training periods or recruitments, or their involvement through PhD in the problem-solving activities of the firms. Toulouse appears as an attractor for future engineers, technicians studying in its schools or universities, which search for local employment whatever their geographical origin (Grossetti 1990). In 2000, Toulouse gathered more than 107,800 students, 70 % in the four universities and their engineer schools, 5 % in the "Grandes Ecoles"; complemented by vocational

[1]It must be stressed that Toulouse is also one of the main national pole in health, pharmaceuticals, biotechnologies, regrouping an important university and public or private institutes of research. This cluster will not be analysed in this paper, as it does not meet the process of local development induced by strong public policies it focuses on.

trainings for technicians (DRIRE (Direction Régionales de l'Industrie, de la Recherche et de l'Environnement) 2001).

This concentration of education and research is the key of the development of Toulouse. The process of decentralization has often been in France the result of the eviction in regions of standard activities to the benefit of the high value added ones developed in Paris. In Toulouse, it appears that the French National System of Innovation has in some sense been decentralized in some activities. The system works indeed on the basis of the close association of the "Ecoles" of engineers and related research institutes, the large public and private firms, the public agencies or establishments like CNES or ONERA, the whole being articulated by national programs (Grands Programmes). These elements, usually concentrated in Paris, are characteristic of the contemporaneous aeronautics and space activities in Toulouse.

2.1.3. Re-configuration of the Local System

Space Industry. The location of the CNES and the research institutes in Toulouse have been followed by the decentralisation of Matra Espace in 1979 and Alcatel Espace in 1982, which have been two main actors of the development of space industries in France, the third one, Aérospatiale Espace, being located in Cannes near Sophia Antipolis. This process complements the regional development strategy of the government.

The European Space Agency (ESA) has presided over the development of the space industry in Europe. Given the important share of the French government in the ESA budget, the CNES has played a pivotal role in the definition and design of the projects, national but also European, as well as in the choice of firms in charge of the development and design. The technological standards required by the CNES were very high and the satellite programs calibrated for short series, more oriented towards R&D than competitive markets. The growth of the activity has followed the evolutions of the public programs, with some downturns like the Hermes project for example. Given the principles governing ESA, the space industry has been from the beginning organized in cooperative networks of firms (alternatively Aérospatiale, Alcatel, Alenia, DASA, Matra ...) built to complete each program.

Toulouse has been a locus of the construction of the European space industry; obviously, the state control on the space policy and thus the definition of the programs is not questioned. But the knowledge base has grown from the local linkages established between the different actors of the field. Alcatel and Matra have both grown today up to 2,000 employees, drawing on the local labour market to feed their growth, but also to develop new capabilities. The industrial strategy of Matra Espace has been very influential on the constitution of the milieu, and complemented the action the CNES (Dupuy and Gilly 1999). The firm was organized without central laboratory of research and multiplied the local links with the CNES, the institutes of research and the potential subcontractors to solve the complex technological problem-solving activities raised by the space industry. Matra acted as prime contractor of the projects and specialized in integration of systems, implying the coherent merging of the different knowledge bases involved in the definition of highly complex products like satellites. As underlined by Dupuis and Gilly (1999), these problem-solving activities call for multiple skills and confrontations of non-formalized knowledge, which continuously reshape the solution. Proximity appears pivotal: 80 % of the Matra subcontracting has been allocated to local research institutes and firms, essentially in computing services or electronics, but also in firms like Zodiac, for multi-layer insulation equipment for satellites for instance. Matra is also involved in one of the different joint-institutes between firms and CNRS created in Toulouse, which are devoted to fundamental research problems. The relations of Alcatel Space with its local environment have been less intense, its needs in R&D were mostly covered internally from the national integrated group it was part of. The firm specialized on payloads and on board equipments, often in joint programs leaded by Aérospatiale (Cannes), which displayed also important internal potential of research. The situation evolved recently when these entities have been merged, under the control of Alcatel Space in Toulouse. The positive feedback induced by the interactions between research and industry in problem solving activities have rapidly led to the emergence of an innovative milieu in space activities. The coordination of the learning process implemented by the CNES, as well as the pivotal role played by research institutes like LAAS or IRIT involved in research programs with different industrial partners, have been essential in the emergence of localized knowledge.

Aeronautic industry. In the aeronautic activities, the development of the Airbus program, and the induced reorganization of Aérospatiale, had an important role in the reinforcement of the industrial context in Toulouse, and its evolution towards collective learning processes and conception oriented activities. Given the cost of development of programs in aeronautic activities, the cooperation of different companies was necessary in Europe; Concorde was already the result of such cooperation, but, among many other reasons, the institutional arrangements that prevailed in this process worked as constraints upon the commercial success of the project. After many resistances from the defenders of the "arsenal logic" (Muller 1988), the GIE Airbus (Groupement d'Intérêt Economique) was created to design new arrangements, which decreased the prevalence of states in the programs and the inefficiencies it entailed. For example, two assembly lines were working for Concorde, one in France and one in UK, to allow strict equality in the cooperation and equal control over the process. Clearly, aeronautics were considered as strategic activities by the states, and closely controlled by the Defence authorities. The institutional arrangements under the GIE Airbus have been promoted by two key actors, R. Beteille, from Sud Aviation, and F. Kracht, a former engineer of Nord Aviation, from DASA, who had experienced the discrepancies of the past programs. They intended to promote the definition of programs directed by the needs of the clients and the markets. The resulting arrangement has been a GIE owned by the companies associated in the consortium, Aérospatiale, Daimler-Chrysler Aerospace (DASA) and Construcciones Aeronauticas (CASA), but prime contractor for these companies. In 1978, British Aerospace (BAe) joined the consortium. The GIE has been entrusted of the control on design and production, of the relations with the clients, and of test flights. The principle of equality working under the "arsenal logic" is given up, the division of labour obeys the specialisation of the companies for the strategic components of the planes (Aérospatiale is endowed with the integration of systems, the cockpit and the controls), the industrial parts being divided according the financial involvement of the partners (Muller 1988). The same principles are applied in the other European consortium, the GIE ATR (Avions de Transport Régional) linking Aérospatiale and Alenia. The implementation of GIE has allowed to implement an efficient cooperation without merging, any process of privatisation of the aeronautic activities being impossible to consider in France at this time.

The head office, management and design activities of the GIE have been located in Toulouse, giving the area the role of European centre regarding aeronautics. The GIE was an entity without capital and without profit, but it gained a strategic position. There is indeed an asymmetry between Airbus Industry and the partners of the consortium as, through work sharing, the GIE increases the specialization of the partner and holds the control on the whole process and the commercial expertise (Talbot 2000). Airbus has continuously grown, to reach 2,400 persons in 2000.

The development of the Airbus family has induced the restructuring of the activities of Aérospatiale, from an organisation based on largely autonomous sites to a multidivisional organisation (Kechidi 1996). The restructuring has mainly concerned the "Aircraft Division", where all the establishments have been integrated through the creation of a "Direction of Production". The design and conception department have also been merged, and located in Toulouse along with the headquarters. The specialisation process induced by the GIE Airbus in the consortium has been internalised by the firm. It obeys the decomposition of the aircraft on sub-systems, allocated to the partners depending of their specialization, and the process of standardisation induced by the development of the Airbus family, optimised through flexible factories; accordingly, the division of labour between the site obeys their technological specialisation. This process is gone with the complete redefinition of the subcontracting strategy, from its decentralization by sites to its co-ordination from the Toulouse direction. The new internal principles of specialisation, standardisation, automation, have been applied to the subcontractors, and the internal tools of computer-aided technology extended to the network. The share and nature of the subcontracting has also evolved. Aérospatiale internalises only the strategic activities, design, control, cockpit, controls, electronic systems, assembling, engine pylons, assembling, but 10 to 40 % of these activities are planned to be subcontracted. The share of the sub-systems based on diffused technologies is planned from 80 to 90 %, and of 100 % for standard technologies (Kechidi 1996). A further consequence of the specialisation is that subcontractors are entrusted with the whole responsibility of sub-systems, from design to production, meaning the development of internal capacities of conception, involvement in the financing of programs, in maintenance. A multi-layer organisation of sub-contracting is adopted, the first layer firms organising the subsequent levels, with the agreement of the prime contractor; the direct links of the Aérospatiale have fallen from 650 in 1987 to less than 100. The externalisation of sub-systems from design to production has induced new coordination processes between Aérospatiale and the first layer of subcontractors. For example, the Program Meeting Review, i.e. regular bimonthly meetings where the evolution of the plans are discussed and the coherence of the whole process ascertained (Talbot 2000).

The consequence of this reorganisation has been the concentration of the first-layer subcontractors in Toulouse, because of the coordination process going with design activities, and new arrivals of firms, to the detriment of the others regions or Midi-Pyrénées, where volume subcontractors are dispersed. Giving the pressure on costs implied by the market-oriented strategy of Airbus, and thus Aérospatiale, these last are indeed more and more under competitive pressure from foreign firms. For example, Lacétoère, first layer firm, subcontracts activities in Poland, Korea, and has acquired firms in Czech republic. This process is reinforced by the principle of compensation working with the aircraft market. As a consequence, subcontracting covers today the whole countries clients from Airbus. On the contrary, the competitive and technological level of firms located in Toulouse has increased, and they have been able to diversify their markets. Latécoère, Microturbo, Sextant Avionique, Liebherr Aerospace work also for Boeing or Bombardier. Toulouse is thus the node of a global networks characterizing the contemporaneous innovative systems, but with decreasing links with its close periphery. An other consequence of the reorganisation of aeronautic activities is the growth of independent firms in design, conception, consultancies, computing activities in the area of Toulouse, as the market of conception is no more monopolized by an integrated firm. The local aeronautic activities are thus shifting from manufacturing to R&D, design, management, integration of systems, and from internalisation to externalisation, inducing the multiplication of linkages between firms and institutions, and thus the emergence of a technopolitan milieu

On board systems. The existing knowledge base in the university and institutes in electronics, robotics, automation, computing, has induced others related industrial developments, characterized also by the pivotal role initially given to large firms and obeying also to the policy of regional development directed by the government. Motorola for example, which employs 2,500 persons in Toulouse, has specialized the site on on-board systems for cars. Or Siemens Automotive, 1,500 employees, specialized in car safety and comfort technologies, like navigational aids. These locations have resulted in further arrivals, like Bosh, Valeo, Actia on diagnostic apparatus for the car industry. On the whole, an on-board systems cluster has emerged in the Toulouse area, which covers the whole range of applications, from aerospace to car industries, involving design, software development, testing. It results from close interactions between public and private research ranging from electronics to materials, between small and large firms, and it is integrated in international markets. The public or private institutions of research (like LAAS, LEEI, IRIT) are involved in many networks related to different technological domains, and help to the building of specific local competences on the technopolitan area.

2.1.4. From French to European Technopole

The recent period has been highly challenging for Toulouse, as since 2000, the landscape of the aerospace industry has tremendously changed. The first challenge has been the A380, the new long-haul aircraft launched by Airbus Industry. An aspect is the innovative content of the plane. Another aspect has been the location of the assembly line of the A380, which was a strategic element regarding the future centre of the European industry.

From a long time, Toulouse had planned to dedicate industrial wasteland from Dassault-Breguet to the project. But the area resulted undersized for the definitive program, and Hamburg proposed an alternative location open to the sea and adapted to the logistic implied by the A380 assembling. The local public response in the Toulouse area resulted in the creation of a new zone dedicated to aeronautic activities, AéroConstellation, localized in Blagnac. The project has been build by the major of the city, B. Keller, who associated different communes to gather 380 hectares dedicated to the zone. The investment amounts to 4 billions francs: 700 millions from the cities, 300 millions from the "Département" for the development of the zone, 2.5 billions from Airbus to build the assembly line, and 200 millions from Air France, which will locate its maintenance facilities on site. The airport, as well as the existing Airbus assembly lines, the head offices of Airbus, ATR, British Aerospace were already localized in Blagnac. Nevertheless, the city could not stand this new investment alone. The institutional arrangement has been the creation of a grouping of communes around Toulouse (Communauté d'Agglomération), gathering the resources dedicated to local development. AéroConstellation illustrates to socio-economic context of Toulouse. Since the works of Becattini (1990), the understanding of economic successful regions has placed considerable emphasis on the key role of local institutions or on the importance of regional institutional thickness, the interlocking web of institutions including firms, chamber of commerce, local governments (Keeble et al. 1998). Indeed, Keller was consultant in Airbus France, coming from Aérospatiale, J. Diebold, the responsible of the AéroConstellation project in the Toulouse municipality, was engineer in Aérospatiale Airbus, the head of the Chamber of Commerce is also the head of Latécoère. This strong socio-politico-economic context helps to understand the development of the region, its capacity to face changes and to adapt to new conditions; as we have seen, it takes roots in the history of the region, as the development of the university obeyed the same interlocking hundred years ago. But, contrary to the usual district socio-economic context, the regional system of innovation in Toulouse is based on the importance of large firms and the interdependence of public - private institutions.

The "Communauté d'Agglomération" has also promoted the development of an optical regional network to meet the development of the Information and Communications Technologies. This investment complements the activity of France Télécom, which employs 4,000 persons in Toulouse for the development and operating of interurban and pan-European networks. These activities have created more than 12,000 jobs in ten years, and stand as an engine of growth in the region (INSEE 2000). They are closely linked with the space industry, and the satellite telecommunications technologies, leaded by Alcatel. They find also an important regional market, the large firms moving to internet technologies for their control and management systems. The emergence of the internet technologies has reinforced the movement of creation of specialized services SMEs, which emerged with the externalisation of activities in large firms, and sustains the deepening of the technopolitan feature of the local innovation system. The contemporaneous development of the area, traditionally leaded by large firms standing as the "natural" market for the engineers and PhD students, is encouraging entrepreneurship; the recent creation of an incubator by the universities, the institutes of research and the "Grandes Ecoles" to promote the creation of new firms evidences the changes of the local socio-economic context.

The preparation and launch of the A380 program coincides with a major restructuring of the European scene in aerospace industries, from European cooperation networks to European firms. In France, the successive major restructuring of the economy has traditionally gone with nationalisations and re-allocations of assets between firms. This restructuring of the aerospace activities is gone with privatisation, moreover of firms traditionally linked to defence... This process of emergence of European firms has first emerged in the space activities, and covers today the whole range of activities. The France reluctance to abandon the state ownerships in aerospace industry has in fact been overcome by the German defence of pan-European mergers, the new institutional rules resulting from the US-Europe agreements, and need to adapt firms to competitive markets. In space industries, the process began with the merger of Matra Espace with Marconi Space System, and the acquisition of BAe Space Systems by Matra Marconi Space in 1994. But the whole process has been allowed by the surprise Aérospatiale and Matra Hautes Technologies merger set to combine the state-owned company and the Largardère group into a shareholder driven firm. The pan-European strategy promoted by Germany was strongly shared by Lagardère, and the merger has been rapidly followed by the creation of EADS (European Aeronautic Defence and Space), merging Aérospatiale Matra, DASA and CASA. In space activities, two major groups have been constituted. Astrium, from Matra Marconi Space and DASA Space, and Alcatel Space, from Alcated and Aérospatiale Espace. EADS owns Eurocopter, 80 % of Airbus, 75 % of Astrium, 45 % of Dassault Aviation, and a 29.5 % stake in Arianespace. This reorganisation of the industry has been followed by the transformation of the GIE structures into new organisations, for Airbus as well as for ATR. EADS and BAE Systems will hold respectively 80 % and 20 % of the newly created corporate Airbus company, and an equal partnership EADS - Finmeccanica Alenia Aerospazio has been built for ATR Integrated. The GIE structures have allowed to pool the European capabilities in order to build an important resource base and gain an increasing share of the world market, under the constraint maintained national independence. But the growth of the activity has resulted in the need of more management control over the whole industrial process, from design to subcontracting strategy. This process is in some sense the outcome of the growing asymmetry between the GIE and the partners underlined in Talbot (2000).

This reshaping of the European landscape gives Toulouse the central role in the sector. In addition to the assembling of the A380 and the new ATR, it will receive the Airbus and ATR headquarters, activities of design, programme planning, delivery and customer services, and training, confirming the embeddedness of the firms in the local area. The creation of Airbus as an independent corporate organisation will imply a restructuring and rationalisation of all related activities from design, engineering to manufacturing, and in Toulouse, a deepening of the process initiated with the former reorganisation of Aérospatiale. One can expect a reinforcement of the context of first-layer subcontractors, and of design and engineering activities, as well as an increased pressure over the volume subcontractors, deepening the technopolitan nature of the area. Airbus has already announced a planned decrease of 20 % on the cost to allow the commercial viability of the A380 program. The innovative content of the new program has induced, as soon as approved, a strong pressure on the local market on engineers and high qualifications. Large as well as small-specialized firms are adapting their capacities in R&D, conception, to fill the competitive requirement implied by the new institutional arrangements governing Airbus. The trend is an increasing concentration of the innovation process in Toulouse, at the expense of Paris.

The creation of Astrium and Alcatel Space, strongly territorialized in the local milieu, and involved as prime contractors in many new programs, from earth observation to telecommunication and internet, is to reinforce the position of Toulouse in these activities. The restructuring of the firms has allowed their adaptation, from commands of short series, high tech prototypes from thc CNES or ESA, to competition and mass-production going with for example the growing market of telecommunication satellite. The role of the CNES will certainly have to evolve in this new setting, and new institutional arrangements to emerge. Indeed, as a consequence of the restructuring of the aerospace industry, Toulouse leaves in some sense the French national system of innovation, in which it played a pivotal role, to move more deeply towards the European one and to competitive global markets.

2.2. The Sophia Antipolis Technopole

The Sophia Antipolis Park is located in the South of France, on the French Riviera between Nice and Cannes. It is often presented as a project which initiated a new form of local development, the "first technopolis in Europe", built on the idea of cross-fertilization between research and industry championed by its creator, Pierre Laffitte. Indeed, the experiment, built on greenfield, is today an important centre of high technology activities in Europe; 1,193 firms or institutions were located in the park in 2000, offering 21,536 jobs, and some 5,000 researchers and students are present on the park. The experiment is only 30 years old; this short period allows drawing the main characteristics of the history of the project, to trace and understand the different phases of growth and crisis.

2.2.1. *From Vacant Space*

The Sophia-Antipolis experiment of development based on high tech activities is unusual in that the region was one with no industrial or university tradition. Nevertheless, the region was endowed by significant assets, including an international airport, a welcoming climate and a cosmopolitan tradition. The infrastructures were linked to the tourist industry, the main activity of the area. But nothing predisposed the region to economic development based on high technology.

The experience of Sophia-Antipolis is the outcome of "small events" that occurred during the nineteen sixties. It was believed in the first instance that "sun and sea" could attract enterprise in research and leading edge technology, as had occurred in the United States. This view was reinforced by the installation in 1960 of an IBM research centre near Nice, followed by a similar move by Texas Instruments affording credibility to the concept of a new mode of industrialisation based on the sun belt effect. These ideas might well have come to nothing without the influence of one man, Pierre Laffitte, then Director of the renowned "Grande Ecole", "l'Ecole Nationale des Mines de Paris". Laffitte had a utopian vision of a rural "Quartier Latin", a "City of Science and Wisdom" and proceeded to embody these ideas in a specific project. In addition to the initiative of Pierre Laffitte, another "small event" proved to be a crucial impulse. This was the deep involvement of France Télécom in the project and the creation in Sophia-Antipolis of an infrastructure which allowed nodes of a national and international communications networks, giving to the area an important comparative advantage at this time.

The embodiment of Pierre Laffitte's ideas took place rapidly on a dedicated zone located in a forest of pine trees. But the viability of the project, run by a private association, soon came up against financial constraints. From the beginning of the 1970's the direction of the project was transferred to the public sector, with a major role for local authorities, through the creation of a syndicate gathering, the towns on which the project has been settled, the Département des Alpes Maritimes and the Chamber of Commerce. The syndicate is responsible for the general management, the financial policy, the international relations, the promotion and the services to corporations; the control of operations has been delegated to a semi-public company, the SAEM Sophia Antipolis. The state was a leading actor for a period when the project benefited from support from the DATAR, the French regional planning authority, whose involvement allowed directing investments in the area.

2.2.2. *Take-off Stage*

The take-off stage for Sophia-Antipolis was based predominantly on the location of external organisations on the site which were attracted by the quality of the infrastructure made available to them. The implant companies had their own internal resource base but this was but external to Sophia-Antipolis. This form of development resulted in minimal interaction and interdependence between the organisations on the site. Decision centres and industrial strategy making were external to Sophia-Antipolis itself.

However, the participation of the public authorities in the initiative made possible a shift in the scale of the project. After the first period where large public sector companies and laboratories were introduced onto the site, linked to the French decentralisation movement, new locations emerged as the outcome of a commercialisation strategy for the science park based on a major international marketing effort, mainly directed at the USA. Large multinational companies, mostly American, adopted Sophia-Antipolis either as their European administrative base or as an R&D centre charged with adapting their products to European markets. The availability of high-quality telecommunication infrastructures able to support hubs and Nice international airport explain most of these location decisions. These international firms have made major investments, establishing self-contained entities able to operate without local linkages.

In this take-off stage the "Ecole Nationale Supérieure des Mines de Paris à Sophia Antipolis", the "Institut National de Recherche en Informatique et Automatique" (INRIA), institutes from the "Centre National de la Recherche Scientifique" (CNRS), were decentralized from Paris. After the initial impetus of France Telecom, the computer centres of large firms like Air France or group insurance like Organic located in the area, followed by Thomson, Digital Equipment, Rohm and Haas, Dow Corning, Allergan etc.

2.2.3. "Satellite Platform"

The mass effect created by the process of accumulation of external activities has obviously resulted in internal dynamics and positive feedbacks in Sophia-Antipolis, which will be analysed later, but basically the project has grown as a periphery, a "satellite platform" in Markusen (1996) taxonomy, that is, a system directed from outside, rich in external connections but deprived of internal linkages, according to corporate logic independent of any considerations of economic exchange with others economic actors on the site (Longhi and Quéré 1993). Nevertheless, a process reinforcing early expansion set in. As shown in Figure 1, the take-off of the project is emerging in the seventies, and the eighties are characterized by an important process of growth, feed by new locations; this first phase will be followed by a severe slowdown to be discussed further.

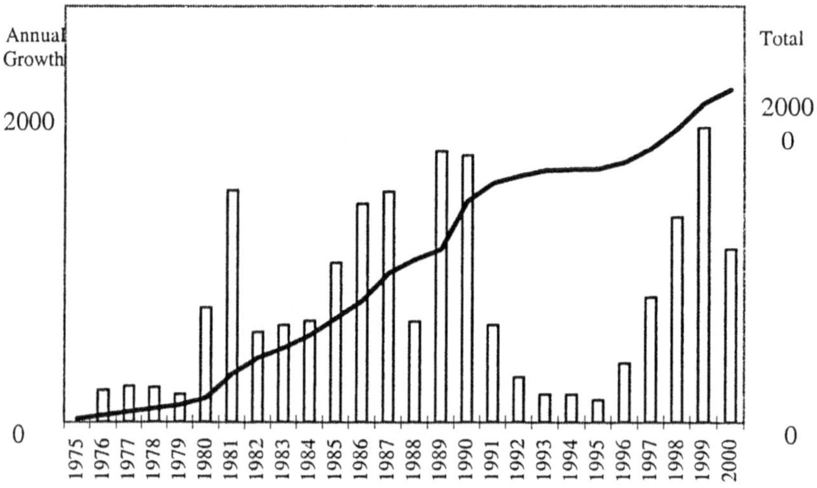

Fig. 1. Sophia Antipolis :growth of employment.
Source: SAEM Sophia Antipolis.

The process of accumulation initially resulted in the emergence of three dominant clusters of technological activity. The first set of activities is made up of the computer science, telecommunications and electronics cluster. This provided the impetus for the growth of the project and structured the development of its industrial context. These activities now represent around 75% of technology jobs in the technopole, with a balance of large French and international (centres of R&D) companies on the one hand and units of large research institutions on the other. These soon reached a critical mass with positive feedback effects.

The second cluster of activities is related to life sciences and health sciences, which gradually built up a significant presence on the site, without actually reaching the coherence of the previous cluster. Mostly large international companies were located in Sophia Antipolis to coordinate their European activities from there and to adapt their products to the "European" market, without no significant local links. As a result the activities and specialities to be found are highly diverse and the critical mass to set off an endogenous growth dynamic has never been reached. The third cluster of activities is linked to natural sciences, environment; historically, it was first to emerge in Sophia Antipolis, but without further growth, and constitutes today a small share of the activity.

2.2.4. Toward a Technopolitan Area

Limits to growth in Sophia Antipolis were closely linked to the initial conditions of the project, i.e. lack of a local labour market, lack of endogenous creation of resources, of SMEs or specialized services. Nevertheless, the accumulation of economic activities on site has triggered positive processes and catch up effects which smoothed progressively these limits until the beginning of the nineties, when a crisis emerged in Sophia Antipolis

A first element concerns the site's potential for training and research. The sitting of the first public institutes on Sophia Antipolis was the result of a decentralisation policy of the major research institutes: the "Ecole des Mines", the CNRS, the INRIA. Their role in the development of the project was very important; INRIA in particular played a fundamental role in structuring local technological development, the emergence of an endogenous innovative capacity and of start-ups on the site. But this potential was not sufficient to train the graduates needed by the developing labour market. A significant development was the sitting in Sophia-Antipolis of research institutes and doctoral studies of the University of Nice in 1986. The university is young and was not associated to the development of the park. Its research institutes and doctoral training began to move from Nice to Sophia from 1986, and progressively a critical mass was reached. The development of university training for engineers has given impetus to the local labour market. In this respect, the process of development of Sophia-Antipolis was the reverse of the "classical" science park processes: the development of the University and the nature of specialist education received their impetus from the accumulation of industrial activity. The large companies arrived first, with decentralised research institutes and only then the university and small and medium high tech enterprises.

A second element concerns the high tech small firms which have grown up in Sophia-Antipolis. The origin of the high tech small firms is very diverse. Many were attracted to Sophia-Antipolis by a site suitable for growth or to build linkages with research institutes (in which their technologies have often been born) or with firms (as subcontractors providing specialized services). Most of the success stories of the park were such implants. Others were spin-offs from the research institutions (INRIA, Ecole des Mines, INSERM in health sciences). The university has also had some successful spin-offs, leaders in their niche. They retain close relationships with research departments, and often try to build local networks around their activities. Pre-production and production of their range of products are often subcontracted to different firms in the region; they are linked with local service activities (such as publishing, translation). These spin-offs were mainly concentrated in computer sciences and telecommunications.

There are also some high tech firms which have located in Sophia-Antipolis to benefit from the prestige of the address without having any links with the local industrial environment. Those that did have linkages were concentrated in the telecommunications sector, in GIS and image processing, software, or biomedical activities. Some acquisitions of high tech SMEs have been made by large firms to renew their technological base, mainly to shift from centralized to decentralized systems of information. On the contrary, spin-offs of large firms have been very scarce on this period. The strategy of internalisation of human resources within large firms and the attractiveness of jobs in these firms have resulted in a "reverse spin-off effect", i.e. their systematic recruitment of high qualified resources from existing SMEs to feed their growth. This characteristic of the project has severely constrained the development of a local labour market and of localized knowledge.

A final element is made up of service activities. This cluster is heterogeneous and its growth has been considerable since 1982 in terms of jobs, so that it makes up 28% of current employment and more than 50% of all firms. This growth is based in part on a catch up effect, due to the absence of service provision on a greenfield site like Sophia-Antipolis; they are basic services required by those living and working on the site and specialized services required by the growing accumulation of firms. The two former categories of firms have made a considerable contribution to the viability of the development process, even if part of service growth has also been make up by a third category of firms without any links with the technological activities of the site. These are essentially the firms linked to BTP or marketing and distribution which have located on in the periphery of Sophia-Antipolis to take advantage of the prestige of the technopolis or of favourable real estate conditions.

These different elements are very important as they contribute basically to the reversal of previous established trends and to the building of an innovative milieu. But the nineties have questioned these positive developments.

The shocks that characterized the early 1990s have indeed been of considerable importance for Sophia-Antipolis. First, it has hit the engine of growth of the project, the computer science activities, implying downsizing in the large firms. Second, it has questioned the strategy of development of the project, based on a marketing strategy aiming at attracting investments of large firms. Indeed, the nineties brought on new forms of globalisation. The 1980s were characterized by multinational developments driven by market processes. The major corporations organized themselves into semi- autonomous regional groups. The attractiveness of Sophia Antipolis for inward investment gave it a comparative advantage over alternative locations. Corporations were not concerned with the specific competencies or capabilities of regions in choosing their location, but were influenced mainly by general conditions in terms of cost and facilities. As we have seen, Sophia-Antipolis was well endowed in this respect.

The nature of this process of globalisation in high tech activities has however dramatically changed in the 1990s (Gordon 1996; Veltz 1993). It is no longer mainly oriented towards manufacturing products or supplying particular markets through adaptation of existing home technologies but towards the process of innovation. As innovation depends on diverse specific capabilities and knowledge, locational differences in these specificities become important. Locations are not easily substitutable, and choices of location are not driven by general advantages but by specific advantages of regions in terms of core competencies. The regime of growth of Sophia Antipolis, based on the attraction of new resources, was clearly a strategy for coping with the 1980s but at odds with the 1990s. It resulted in the location of more conventional activities without any technological content, and in the depreciation of the park's image. A new strategy of development, based on local resources and competencies, had to be generated.

2.2.5. Emergence of an Innovative Milieu

After a period of instability in the beginning nineties, different new trends have emerged in the area. Mainly, the unexpected consequence of the crises has been the change in the regime of growth, from exogenous to endogenous driven processes, and the development of local relations between firms or firms and research institutions involved in locally implemented innovative processes. A main actor of this change has been the professional associations existing on the park, which stood for the absence of clear public strategy to face the crises and to reconfigure the local system.

Various business associations and clubs exist in Sophia-Antipolis. Their role has evolved considerably, from defence of specific interest to more collective purposes. The Telecom Valley Association was the first initiative in this direction from the large firms; created by Texas Instruments and France Telecom, it now includes all local actors in the field of telecommunications, both public and private, large and small. In the face of the dimming of the high-technology image of Sophia-Antipolis, and the risk of relocations away from the park, the Association has attempted, on the basis of the important amount of resources accumulated locally, to label it as the strategic European location for telecommunications. It has brought about a positive dynamic with the large companies beginning to open themselves to the other members of the Association, including smaller firms, to share certain technological problems and resources, to take advantage of complementary expertise and businesses. It can be argued that the park is indeed today effectively a strategic centre for telecommunication activities in Europe, with a specific advantage on wireless technologies. The main European institutions defining the standards for telecommunications (ETSI) and internet (WWW) are located in Sophia-Antipolis, which is thus at the heart of the innovative process and a strategic area to gather information on the future technologies.

Numerous associations and clubs exist in fact in Sophia Antipolis, linking SMEs (Imet, MITSA), research institutes (Persan), specific technologies (Club High Tech), with the explicit aim of generating networking and processes of local collective learning. These local associations and clubs embody in fact the ideas of cross-fertilization developed by P. Laffitte when creating Sophia Antipolis; they are fundamental elements in the coordination of the innovative activities of firms, providing information on technologies as well as on markets, in the building of local ties.

Another important trend to be noticed is an important movement of creation of new SMEs, which was lacking in Sophia-Antipolis. As stressed by Keeble et al. (1998), the creation of new high-technology SMEs and spin-offs is a "major process whereby research ideas, technological innovations and expertise are diffused and shared within a region, rather than being confined and internalised within a single, secretive firm". This process is indeed conducive to collective learning and local positive feedback. The history of Sophia-Antipolis has revealed that in the past opposite processes were in fact at work, namely internalisation of resources within large firms.

Historically, the process of spin-off was restricted to research institutions. With the crisis, it has been extended to large firms. Paradoxically, the most important movement of creation of new SMEs correspond to the major slowdown of the project in terms of employment growth at the beginning of the nineties, and to the process of downsizing and outsourcing induced by large firms. The number of firms increased significantly during the crisis. Downsizing resulted in the diffusion of capabilities up to there internalised in the large firms through the local labour market, and not in a loss of resources. Indeed despite the weakness of linkages between firms or institutions a strong sophipolitan milieu effect has always existed for the people involved in the project, which have tried to find local solution to the crisis. A lot of "forced" spin-offs from large firms have emerged, which have helped to diffuse technological capabilities and competences locally. These firms, created by former engineers of large firms (Thomson, Texas Instruments, DEC) reluctant to leave the project when confronted to "downsizing", have often subcontracting relationships with their parent firms, and important informal networking relations between them, to share problems and solutions. The fact that most of these SMEs have been able to adapt and grow has enhanced the movement.

Recently, numerous local start-ups and spin-offs have emerged in the software industry, multimedia, telecommunication and internet technologies. An important resource base exists locally in these activities. Indeed, many of these new technologies are developed in INRIA or France Telecom. The consequence has been the multiplication of small and micro firms created from local resources, with strong links with research and close interrelationships. This cluster is supported by local business associations, and is embedded within numerous different local formal and informal networks.

This trend towards an endogenous regime of growth could not have existed without the development of the training and research capacities of the area, and their interdependence with the industrial context. The existence of a local labour market, allowing the skilled labour mobility, the diffusion of embodied and tacit expertise and technological know-how is recognized in the literature as the basic element of local development (Camagni 1991; Keeble et al. 1998). As we have stressed, Sophia Antipolis was initially deprived from such processes. The changes initiated by the location of the University has never stopped, and the growth and multiplication of higher education and research institutes is also quantitatively one of the important trend of the recent years. Roughly 3000 students and as many researchers are today in Sophia-Antipolis. The advent of the Institut National de la Recherche Agronomique (INRA) and of the "Institut de Pharmacologie" (CNRS) significantly reinforced the pharmaceutical and health science cluster of activities, even if industry-research linkages were in both cases mainly turned towards outside players. The critical mass effect and the visibility of these developments were insufficient to promote innovative local behaviour. This critical mass has on the contrary been reached in the information technologies with the multiplication of institutes developing specialized doctoral studies, the creation of new schools of engineers from the University (ESSI, ESINSA) or from professional consortia (THESEUS, EURECOM). This potential of higher education and research is continuously growing; with the University, the research institutions (CNRS, INRIA, INRA, INSERM...), the Ecole des Mines, the research and training impulsed by France Télécom, the business schools (CERAM), most of the major French education and research institutions are currently represented in Sophia-Antipolis.

In addition to the qualified resources supplied by the higher education institutions, the process of downsizing and externalisation induced by the large firms has resulted in the reinforcement of this growing labour market, and made qualified resources available to feed the growth of new firms on site, which has been exponential in different cases. But clearly, this local labour market is restricted to the sectors of information and telecommunication technologies. In this sector, the turnover is also very important, showing that new behaviours are nowadays set up in Sophia Antipolis. For example in 1997, the evolution of local information technology employment revealed a net growth of +1,512 (1,460 of which was from existing local firms); this net total is the result of 2,596 recruitments and 1084 resignations, revealing a considerable turnover and mobility of skilled labour. As stressed by Storper (1993) in an other context, these processes of turnover as well as spin-offs are made possible because of the existence of an already institutionalised industry, not only technologically, but also in terms of a specific professional culture, of the existence of human capital specific to an industry. They allow the existence of a local labour market which, together with the local inter-enterprises relationships and spin-offs processes, impulse the emergence of local specific competences belonging to the innovative milieu itself, and not only of individual firms.

The last important change concerns the key actors of the development of Sophia-Antipolis, i.e. the large firms. As we have emphasized, the growth of Sophia Antipolis has resulted from the coincidence of a supply of general advantages (physical infrastructures, climate...) and the needs induced by the multinationalisation of large firms aimed at organizing their European operations. The huge decrease in the arrival of new multinational firms and new investments in the beginning of the nineties has also been the indication of the crisis. The viability of the experience has been dependent on its capability to induce a shift in the regime of growth, from accumulation of external resources to the development of activities based on local specific competences. The attractiveness is no more only dependent of physical infrastructures, but mainly from specific capabilities and human infrastructures. The location of large firms is nevertheless still crucial to the project, but it takes a different sense, it eventually validates its technological and innovative potential and its position of "pole of excellence" in some fields, and constitutes opportunities for the existing local labour market.

Indeed, the arrival of new firms does not now involve neither huge investments nor simply adaptation of existing products or services. The new firms usually settle in the park as tenants with small units and grow locally drawing on local capabilities and on the existing local labour market to develop their innovative activities. For instance, Andersen Consulting located on the park with a small unit of ten people and has subsequently grown locally to employ more than three hundred staff on the base of new markets created from their local activities. The same strategy has been followed by the new comers in Sophia Antipolis, whose location have been clearly induced by the specific capabilities in wireless communication, GSM technologies or internet technologies which have grown locally, from the public research institutes as well as from the R&D developed in the area by large (Texas Instruments, IBM) or medium (VLSI for instance) firms. Most of the main actors in this innovative field have recently located in Sophia Antipolis, with a precise strategy of mobilization of local capabilities and knowledge. This is the case of Siemens, Lucent, Bay networks, Ericsson, whose units have grown on the basis of local resources, or Philips, which acquired VLSI, but also of many small firms which find in Sophia Antipolis an area suitable for growth or for attracting easily the specialized human resources needed in this innovative technological field. The process of outsourcing and the development of information technologies have also considerably changed the local strategy of the large firms; a lot of local relationships linked to the innovative process have emerged locally, through subcontracting with high tech specialized SMEs.

Growth is now quasi-exclusively endogenous, i.e. results from the growth of existing local firms, when before 1990, it resulted mainly from the balance between entry and exit of firms in the project. Besides, the growth is concentrated in the information technologies sector. An efficient local labour market thus exists today for qualified information technology staff. More than half of sophipolitan jobs are highly qualified jobs (executives and engineers); in the information technologies sector, this proportion was 70.8% in 1997.

In the contemporaneous process of innovation, the challenge for Sophia Antipolis was to enter the global networks set up by the high tech firms to sustain their innovative activities, which links large and small and specialized firms as well as public research institutions. The emergence of a new regime of growth and of new institutional arrangements after a crisis of adaptation show that the resources accumulated in the eighties were sufficient to face the shocks and reconfigure the system. But the shocks of the nineties have resulted in an unbalanced local system, in which information and telecom activities have been able to adapt to the new environment and to regenerate growth while other activities (health sciences) are still locked into the previous regime and failing to fulfil their potential. Information technologies have recovered the role of engine of growth they had in the first phase of the project. The engine is no more made of self-contained establishments of large international firms; it results from the positive feedbacks induced by the local interactions of firms involved in innovative activities. This strong movement of specialization of Sophia-Antipolis can raise problems for the balanced growth of the area, as it could increase its sensitivity to asymmetric shocks, despite its specialization in R&D and design activities.

3. Conclusion

Fundamental features of territories, and specifically high tech complexes like Toulouse or Sophia Antipolis can be related to the processes through which resources are created and mobilised. The two areas have many differences and many similarities. They are different in terms of size and scope, of industrial activities, or of socio-economic features. They have in common to be technopolitan areas born from state-led processes, government and local policies based on decentralisation of large public firms and institutions, exogenous processes of growth whose unintended consequences have been the emergence of endogenous processes of creation of resources, of innovative milieux in different clusters of high tech activities. The contemporaneous arrangements of the innovative milieux could not be planned, they are the consequence of adaptations and changes of the local systems, of reconfiguration of their economic working following evolving internal and external circumstances. Diversity, usually considered across different areas, is also a characteristic of the dynamic evolution of the local systems, where reconfigurations have given rise to new internal economic and institutional arrangements, new coherences with evolving environments. Key actors, collective agents play a fundamental role in the bifurcation of the enacted systems that are territories; turning points of evolution of Sophia Antipolis and Toulouse towards technopolitan milieux are examples of these dynamics.

Toulouse had a long university tradition closely associated to economic development and a constituted industrial base in aeronautics, a strong socio-economic system, as well as an identity which explain the further regional and industrial policies that have led to a technopole. Sophia Antipolis on the contrary stood as a vacant space, and the regional policy intended to create a new socio-economic context and change the nature of the development. The reactions to external shocks and internal reorganisations of the local systems have been thus very different, from coordinated initiatives linking the local governments and the public and private institutions in Toulouse, to unexpected self-organized processes in Sophia Antipolis.

The viability of local development based on high tech activities is gone with the steady reinforcement and specialisation of the areas in activities of conception, R&D, problem-solving processes, as well as the increasing necessity of different technological expertises in the economic process. The continuous confrontation of non-formalized knowledge and collective learning feeding the innovative process of high technologies has strengthened the local systems and the decisive role of proximity. The two areas have built in their technological fields specific competences that have attracted new investments and inserted the local systems in global networks of innovation. The structuring of the local and global dimensions of the innovative process is indeed the key of the viability of contemporaneous complex of high technology.

In Toulouse, institutional thickness - i.e. a dense and coherent set of territorial relationships of all the economic actors, private and public, a specific culture, and a shared representation system and development strategy - has been one of the main asset of development, but the role of the State has been permanent, because of the strategic content of the local activities. In turn, the embeddedness of the European aerospace industry in Toulouse, its pivotal role in the European scene, induced by the endogenous process of creation of resources and local specific knowledge, explains certainly the restructuring of the industry allowed by the privatisation processes implemented by the government. This basic change in the institutional arrangements governing the economic activities of the city will certainly deepen its insertion in the global knowledge-based economy. In Sophia Antipolis, the development is the result of the government regional development policy, without the strategic content present in Toulouse. The emergence of an innovative milieu from the accumulation of external resources has been the result of an external shock, a change in the environment the project had to face. The reconfiguration that resulted has allowed the project to go beyond the image effect on which it has grown, towards the emergence of a socio-economic context and the realization of the cross-fertilization concept which presided to its creation.

References

Allinne, J. P. (1996), "Les technopoles sont-elles centenaires? Systèmes innovateurs et création d'entreprises en France", *Revue d'Economie Régionale et Urbaine*, 5, pp. 1870-1960.

Andersen, E. S. and Teubal, M. (1999), *High Tech Cluster Creation and Cluster Re-configuration - A System Policy Perspective*, Mimeo, DRUID, June.

Arthur, W. B. (1989), "Competing Technologies, Increasing Returns and Look-In by Historical Events", *The Economic Journal*, Vol. 99.

Bécattini, G. (1990), "The Marshallian Industrial District as a Socio-economic Notion", in: *Industrial districts and inter-firm cooperation in Italy*, Pyke, F. and al. (eds), Geneve, International Institute for Labour Studies.

Brusco, S. (1982), "The Emilian Model: Productive Decentralisation and Social Integration", *Cambridge Journal of Economics*, vol. 6, pp. 167-184.

Camagni, R. (1991), Local "Milieu", Uncertainty and Innovation Networks : Towards a New Dynamic Theory of Economic Space, in: *Innovation networks: spatial perspective*, Camagni, R. (ed.), London, Belhaven Press.

DRIRE - INSEE (2001), *L'industrie en Midi-Pyrénées*, Dossiers Etudes n°104.

DRIRE (1998), *Les industries de l'aéronautique et du spatial en Midi-Pyrénées*.

Dupuy, C. and Gilly, J. P. (1999), Industrial Groups and Territories: the Case of Matra-Marconi-Space in Toulouse, *Cambridge Journal of Economics*, 23.

Garnsey, E. (1998), "The Genesis of the High Technology Milieu", *International Journal of Urban and Regional Research*, vol. 22 n° 3.

Garnsey, E. and Longhi, C. (1997), *High Technology Locations and Economic Development*, EUNIT (European Network on Industry, Innovation and Territory) International Conference on Industry, Innovation and Territory, Lisbon.

Garnsey, E. and Longhi, C. (1999), "Auto-organisation et émergence des milieux innovateurs", *Revue d'Economie Régionale et Urbaine*, 1.

Gordon, R. (1996), "Industrial Districts and the Globalization of Innovation: Regions and Networks in the new economic space", in: *Wealth from Diversity*, Vence-Deza, X. and Metcalfe, J. S., Kluwer.

Gormand, C. (1993), *L'industrie aéronautique et spatiale*, Paris, L'Harmattan.

Grossetti, M. (1990), "Enseignement supérieur et technopole, l'exemple de l'informatique à Toulouse", *Revue Française de Sociologie*, XXXI.3.

INSEE (1998), *15 ans d'aéronautique et d'espace en Midi-Pyrénées*, Toulouse.

INSEE (2000a), "Aéronautique, espace et sous-traitance", *Repères*, 100.

INSEE (2000b), *Tableaux économiques de Midi-Pyrénées*, Toulouse.

INSEE (2001), "Les NTIC en Midi-Pyrénées", *6 Pages*, 46.

Kechidi, M. (1996), "Coordination inter-entreprises et relations de sous-traitance: le cas d'Aérospatiale", *Revue d'Economie Régionale et Urbaine*, 1.

Keeble, D. et al. (1998), *Collective Learning Processes and Inter-Firm Networking in Innovative High-Technology Regions*, ESRC (Economic & Social Research Council) Centre for Business Research, Working Paper 86, March.

Longhi, C. (1999), "Networks, Collective Learning and Technology Development in Innovative High Technology Regions: the Case of Sophia-Antipolis", *Regional Studies*.

Longhi, C. and Quéré, M. (1993), "Innovative Networks and the Technopolis Phenomenon: the Case of Sophia-Antipolis", *Environment and Planning C: Government and Policy*.

Luger, M. I. (2001), Science and Technology Parks at the Millennium: Concept, History, and Metrics.

Markusen, A. (1996), "Sticky Places in Slippery Spaces: A Typology of Industrial Districts", *Economic Geography*.

Metcalfe, J. S. (2000), "Propspects and Challenge for Research on Innovation", *Co-evolution of Systems of Innovation*, Conference, Berlin, June 8th-9th.

Muller, P. (1988), *Airbus, l'ambition européenne. Logique d'Etat, logique de marché*, Paris, L'Harmattan.

Porter, M. E. (1990), *The Competitive Advantage of Nations*, New York, Free Press.

Saxenian, A. (1994), *Regional Advantage*, Cambridge, MA, Harvard University Press.

Storper, M. (1993), "Regional "Worlds" of Production: Learning and Innovation in the Technology Districts of France, Italy and the USA", *Regional Studies*, vol. 27.5.

Talbot, D. (2000), Institutional Dynamics and Localized Inter-Firms Relations. The Case of Aerospatiale and its Subcontractors in Toulouse, *European Urban and Regional Studies*, 7(3).

Veltz, P. (1993), D'une géographie des coûts à une géographie de l'organisation. Quelques thèses sur l'évolution des rapports entreprises/territoires, *Revue Economique*, n°4.

Biotechnology Development in Germany: The Case of Nordrhein-Westfalen

Francesco Salamini, Anke Sohn and Hartmut Thomas

Abstract. Nordrhein-Westfalen is and has been a centre of excellence for chemical and genetic research for a long time; advanced biology has been added recently. The industrial development of the latter sector is logarithmic here. There is no decline in the demand for research and development either for biomedical or agricultural applications. The forecast of the future of the biotech industry in NRW cannot be included in a pattern containing well-defined goals for the various application fields. In this situation, it is not easy either to estimate the capital requirements in the long term or set priorities for the region. Structural conditions to be kept in view are the enhancement of the "public-research-economy" communication on subjects such as health care, food and agriculture, biotech products and the community, environment and diagnostics. The improved regional credit and functionality of the industrial incubators are also under study. On the other hand, it is a good sign that the supply of *venture capital* for new *start-ups* is rising in NRW; that the public opinion is more favourable than in the past; that the federal and local governments are in favour of the biotech industry in general and of transferring research *know how* to businesses. We can therefore be reasonably certain that the region will be among the biotechnological *hot spots* that will survive in the next fifteen years. The paper reviews the history and the players that have been influencing the development of biotechnologies in NRW: the Bio-Regio project; the role of the Bio-Gen-Tec-NRW agency; the first step of biotech development in NRW; and a short reflection on the future.

1. The BioRegio Initiative

In the 1980s, Germany went through a period of complex - at times fractious - debate on the future of technological societies. Complex because the discussion on the function of science in society, traditionally central to German culture, became more pointed on the heels of the new developments in biotechnology, thus generating a clearer understanding of the meaning and relevance of bioethics (Korff et al. 1998). Fractious because the debate was focused almost exclusively on the risks inherent in genetic manipulations, with a clearly restrictive impact on the laws that were being enacted to regulate this field. Faced with this situation, the scientific and business communities reacted in a predictable manner: scientists and biotechnology investments moved to the United States. However, starting in 1992, small groups of opinion makers, including politicians, businessmen, bankers and scientists, reopened the debate on the risks of biotechnology, broadening it with discussions of the benefits that could be derived from this new wave of innovation. Later on in this presentation, we shall discuss certain aspects of this phase, viewed from the standpoint of the situation that was developing in NRW.

In 1994, the federal government decided to address these issues. The Ministry of Education, Research and Technology (BMBF; Jürgen Rüttgers Minister) sponsored the BioRegio Initiative, which was based on a simple concept: how to develop an integrated proposal for a regional system that would take advantage of a wide range of locally available skills in such fields as advanced molecular research, technology and innovation, and tap investment capital.

The federal government launched a competition, and 17 German states participated. The proposed models were judged by an international panel. Each state offered a different solution, availing itself of the support of the federal government to create the requisite organizational structures (in NRW, the Bio-Gen-Tec-NRW association). The three states selected as winners (NRW, Bayern and Rheinland Pfalz) were each given 50 million German marks that they could use to promote the development of biotechnology (Eichener et al. 2000), encourage the flow of capital toward new fields and, in general, stimulate job creation. The stated goal of Minister Rüttgers was to make Germany the leading European country in the field of biotechnology research and development by the year 2000. In February 2001, H. Thomas, Chief Executive Officer of Bio-Gen-Tec-NRW, could state publicly that Rüttgers's goal had been achieved (Bio-Gen-Tec-NRW, 2001a):

> "A few years ago, we set out to make Germany the European leader in biotechnology. The objective was ambitious and many had their doubts at that time, but today, it has become a reality. Based on the number of start-ups of new biotech companies, Germany is considered worldwide as occupying the number one slot in Europe: the greatly increased availability of venture capital and the keen interest of investors in the shares of young biotech companies, coupled with the set-up boom that is also in progress, are clear signs of a consolidation and international positioning of Germany as a biotechnology location".

2. Biotechnology in NRW: The Function of the Bio-Gen-Tec-NRW Association

Germany got a late start in developing its biotechnology industry but has been catching up. Today, there are 332 biotech companies, which together employ 11,000 people and generate annual gross revenues of US$670 million. In 2000, they invested US$600 million (GFW-NRW 2001).

Many of these companies are located in Nordrhein-Westfalen, the most populated German state (18 million inhabitants), which accounts for 22% of the country's gross domestic product. It is host to a number of multinational companies (Bayer, Henkel, RWE, Krupp-Thyssen, Ford, and many others), and its economy is strongly linked to the chemical industry. The industrial vocation of NRW is supported by a high concentration of public research facilities: 53 universities (500,000 students); 21 scientific organizations, including Max Planck and Fraunhofer Institutes and several National Research Centres; 32 other public agencies financed with government funds; 58 scientific units participating in special research projects (Sonderforschungsbereich, SFB, which are managed by the Agency for University Research, Deutsche Forschung Gemeinschaft, DFG); 68 technology centres and industrial incubators; and 27 technology transfer offices (Anonymous 2001, Bio-Gen-Tec-NRW 2000).

The engine that drives this public research network - the most extensive and important in Europe - is not chemistry, but biology: eight graduate schools of medicine, 12 SFB groups, eight interdisciplinary research centres (IZKF), one Molecular Medicine Centre in Köln (ZMMK), one Biomaterials Centre in Aachen (BIOMAT), eight graduate schools of biology, two biotechnology institutes in Jülich, two Max Planck Institutes in Köln (for plants and neurological research), the Max Planck Molecular Physiology Institute in Dortmund, one Molecular Biotechnology Institute in Aachen (Fraunhofer Foundation), and one Chemio and Biosensor Institute in Münster.

The BioRegio competition ended in 1996. Since then, the state of NRW has been working to turn its biotechnology effort into economic expansion. Today, the state is host to 130 biotech companies located within an area of 13,000 km^2; 220 organizations provide products and services to the biotech companies (GFW-NRW 2001). The rise of this biotech industry has been followed by the Bio-Gen-Tec-NRW association, which assisted the birth of 82 start-ups from 1997 to 2000, requiring a total investment of 635 million German marks (50 from the federal government, 195 from the local government and 390 from private investors). In 2000, the local government invested 100 million German marks (Bio-Gen-Tec-NRW 2001a, Anonymous 2001). Fifty-two of the 80 start-ups created with the support of Bio-Gen-Tec-NRW are located along the Rhine between Köln and Düsseldorf. They engage in a wide variety of activities including purification of nucleic acids, cell separation, molecular carriers, bioreactors, bioprocessing, quality control, molecular chips, bioinformatics, protein design, diagnostics, functional genomics and postgenomics, models of transgenic animals, and molecules that prevent neurodegeneration and cardiovascular diseases.

The main cities in this area, named "Bioriver", are Düsseldorf, with 180 new jobs and the Life Science Centre Incubator; Köln, which accounts for 300 new jobs and is home to Bio-Gen-Tec-NRW, the Max Planck Institute for Plant Improvement, the Institute of genetics of the university and the East Rhine Technology (RTZ) Incubator; and Bonn, where four start-up projects were approved in 2000. The Aachen-Julich region has been the site of interesting interdisciplinary projects financed with private capital and was recently selected as the home of the Fraunhofer Institute for Molecular Biotechnology. In the Ruhr Basin, which specializes in bioprocessing and microstructure technologies, the BioIndustry agency works to encourage the establishment of start-ups in bioindustry and medicine. In addition, the city of Bielefeld has access to a

Genomic Research Centre that has excellent potential for the establishment of start-ups.

The Bio-Gen-Tec-NRW association is a private non-profit organization financed with public funds, private capital and revenues generated by the projects it manages directly. It also manages public funds earmarked for the development of biotechnology in the state of NRW. This association, which has 172 members, provides consulting and support to new start-ups during the establishment and early development phases. New ventures receive financial support in the form of federal and local government funds that are provided through Bio-Gen-Tec-NRW and have access to a wide range of consulting resources. Bio-Gen-Tec-NRW sees itself as a one-stop shop that can help clients define the scope of their endeavour more clearly, prepare a business plan, obtain public financing, enter into contracts with venture capitalists and present their projects at public events (Figure 1).

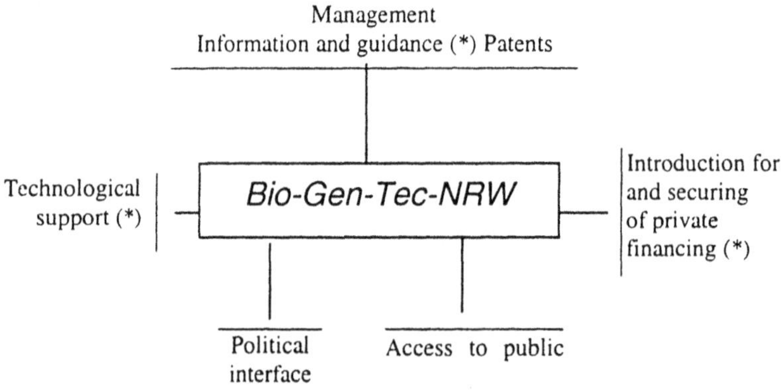

Fig. 1. Services provided by Bio-Gen-Tec-NRW, a non-profit association that helps support the development of biotechnology ventures in the state of Nordrhein-Westfalen. * functions performed by special committees.

The Management, Consulting and Coaching Department makes use of financial, legal and tax consultants (42 in 2000). Clients can also use a voucher program to get help in preparing business plans and receive legal advice. Public funds pay for 75% of the cost of these services. Special attention is paid to securing patent protection or acquiring patents. Bio-Gen-Tec-NRW organizes seminars devoted specifically to patent issues and draws on a network of 26 lawyers who specialize in biotechnology.

The financial network comprises more than 50 German, regional and international investors, including venture capitalists and bank investment funds. At the regional level, five companies with expertise in biotech formed a consortium and invested about 30 million German marks between 1997 and 2000.

Before a start-up project is accepted by Bio-Gen-Tec-NRW, its scientific and business assumptions must be evaluated and approved by the Biotech Competence Network. Special attention is paid to financial considerations, including project risk assessment.

Bio-Gen-Tec-NRW introduces the companies that it sponsors at international trade fairs. It also posts relevant data on Internet in a directory that contains 360 entries (Bio-Gen-Tec-NRW 2001b) on biotech companies, research institutions and venture capital firms, along with information on regional research and development areas. Interaction with the public at large occurs primarily at national trade fairs, at Biomeetings and through press releases (120 in 1998) and press conferences. The association also supports the "Genes We Use Every Day" project, which uses a contest and prizes to disseminate information in the school system.

The most important financial assistance programs used by Bio-Gen-Tec-NRW to support the development of biotech companies are listed in Table 1, which shows the eligible expenditures, the goals of the program and the level of financing provided by financial institutions to support R&D projects or launch start-ups. The most recent program is the Bio-Chip Initiative (16 million German marks), which is funded by two ministries of the local government (Bio-Gen-Tec-NRW 2000, 2001a).

Table 1. Outline of the most important programs available for biotechnology research and startup projects in the state of Nordrhein-Westfalen.

Program	Investments covered	Focus	% of funding from public capital
		FEDERAL GOVERNMENT	
BIOTECHNOLOGY 2000	• R & D • Equipment not included	• Basic research	• Up to 100% (institutions) • Up to 50% (startup)
		NRW REGIONAL GOVERNMENT	
TIP. Support for small and medium-size businesses	• R & D • Equipment included	• New high-risk technologies	• Up to 35% (startup) • Up to 75% (institutions, subcontr.)
INCUBATOR	• Consulting services • Market research • Patents • Investments	• Concept assessment	• Up to 50% of consult. contracts
VOUCHER	• Consulting services • Market research • Patents	• Preseed and seed stages	• Up to 75% of costs

Table 1. (continued)

BIO-CHIP. Support for small and medium-size businesses	• Chips development in: - diagnostics - plant improvement - neurochips - proteins	• Preseed and seed stages	• Up to 75% of costs

3. The Beginning of Biotechnology Development in NRW

Until the 1980s, biology research in Germany was funded by such public agencies as the Max Planck Company and Deutsche Forschung Germishaft (DFG), and by ministries of the federal and local government via research projects. Visible changes in research policy came during the early 1980s, when the federal government decided that it would encourage public agencies to manage the development of cutting-edge technologies, provided their projects involved at least one local industrial enterprise. In the wake of this decision, the Ministry of Scientific and Technological Research (BMBF) created three public Gene Centres (Heidelberg, München and Köln). These Centres were to receive 100% financing provided they met one condition: they would have to show that there was an official relationship linking the institution responsible for setting up the Centre with a privately-owned business. The Köln Gene Centre (Zentrum für Molekulare Gen- und Zelltechologie in Köln, 1986, 1990, 1994) showed that one of its cofounders, the Max Planck Institute for Plant Improvement, received an annual grant from Bayer, Leverkusen, of one million German marks, which was used to fund unspecified basic research programs. The three Gene Centres were financed by the BMBF for three funding periods (1982-1994). The completion of the program for the establishment of the three Centres was followed by a protracted debate as to which projects should then be developed and supported. The result was the decision, supported by Minister Rüttgers, to launch the BioRegio competition among German states for three financing packages of 50 million German marks each. A precondition for entering the competition was the establishment of a regional network of public-sector managers, opinion makers, scientists from universities and other institutions, private-sector managers, union officials, and Chamber of Commerce and Industry representatives. An organizational model was a key part of the competition proposal, which was also required to include a description of the media that would be used to disseminate information about new biotechnology projects, as well as a list of all of the entities engaged in biotechnology work in the project area, showing the scientific and practical significance of their work.

The definition and organization of the networks was not an easy task. It involved balancing the flow of financial support from the federal and regional governments to public - and private - sector institutions with the need to establish quasi-public collaborative superstructures, while at the same time enlisting public support for the selection of a given area as a preferred location for the development of biotechnology businesses. The state of NRW - the city of Köln in particular - was in a privileged position. Between 1988 and 1990, it had been the theatre of a public debate concerning the first release of a GMO in Germany (the case of the transgenic petunia). In Köln, the media had gradually developed a negative bias toward an experiment carried out by the Max Planck Institute for Plant Improvement. In 1992, fearing that as a result of this developing situation local politicians would choose not to pursue opportunities in biotechnology, municipal and state officials, Chamber of Commerce representatives, journalists, scientists, union officials and private-sector managers established a joint committee to coordinate, encourage and channel their efforts to support the transfer of technology from public research to the private sector in a manner that would foster the growth of biotechnology enterprises while addressing the role of science in society. The knowledge gained through this process was instrumental in creating the Bio-Gen-Tec-NRW network in 1994, which was then used in the BioRegio competition as a conduit for the joint efforts of different institutions. It is worth noting that, in this case, the experience gained over the 10 years during which the three Gene Centres had been in operation and the existence of local support, voiced (at least in Köln) by a committee of people who shared the same clearly-focused objectives, turned out to be determining factors of success: the three states, where the Köln, München and Heidelberg Gene Centres were located, were chosen to receive the financing packages, in preference to 14 other states.

The less immediate reasons for the success achieved by the state of NRW - even though initially the environment was particularly challenging and local industrial development priorities were nebulous at best - are basically two: a high and widespread level of scientific expertise and the organizational ability of a group of people representing diverse professional interests who were sustained by their belief that they were contributing to the transfer of scientific knowledge.

4. It Could Happen Tomorrow

The state of Nordrhein-Westfalen has been and remains a centre of excellence for research in chemistry and genetics, to which it recently added advanced biology. Industrial development in advanced biology is currently in a logarithmic phase. There appears to be no slackening of demand for research and development for biomedical and agricultural applications. It is clear, however, that current forecasts of the future of the biotech industry cannot be pigeonholed within a model that sets well-defined objectives for each separate field. As a result, it is difficult to forecast long-term capital requirements and to define priorities at the state level. However, close attention should be paid to key structural developments, including an intensification of the debate involving the public, the research community and businesses on such issues as medicine and health, food and farming, biotech products and society at large, and the environment and diagnostics. The increased availability of local credit and the success of industrial incubators should also be taken into account. Positive developments in the state of NRW include an increase in the availability of venture capital for start-ups, a more favourable public opinion and a more constructive attitude by the federal and local government toward the biotechnology industry in general and the transfer of research to businesses in particular. Therefore, it is reasonable to expect that this German state will become one of the biotechnology hot spots (among perhaps 20 or 30) that will survive worldwide over the next 15 years.

References

Anonymous (2001), "Spotlight on NRW", *Nature*, supplement March 22, Vol. 410.
Bio-Gen-Tec-NRW (2000), *Info.* no. 1 + 2, Bio-Gen-Tec-NRW.
Bio-Gen-Tec-NRW (2001a), *Annual Report 2000*, Bio-Gen-Tec-NRW.
Bio-Gen-Tec-NRW (2001b), *Bio-Gen-Tec Atlas 2001*, Bio-Gen-Tec-NRW.
Eichener, V., Schaaf, S., Schulte, F. and Weingarten, J. (2000), *Erfolgsfaktoren für die Biotechnologie-Regionen. Sozialwissenshaftliche Begleitforschung zu den BioRegios*, Hans Böckler Stiftung.
GFW-NRW (2001), *Welcome spezial. Biotech-NRW 2001*, GFW-NRW.
Korff, W., Beck, L. and Mikat, P. (eds) (1998), *Lexikon der Bioethik. Im Auftrag der Goerres-Gesellschaft*, Gütersloher Verlagshaus.
Zentrum für Molekulare Gen- und Zelltechnologie in Köln (1986), *Abschlussbericht 1982-1986. Förderungsvorhaben BCTO365 2 (BMBF)*, Max-Planck-Institut für Züchtungsforschung, Institut für Genetik der Universität zu Köln.
Zentrum für Molekulare Gen- und Zelltechnologie in Köln (1990), *Abschlussbericht 1987-1990. Förderungsvorhaben BCTO365 2 (BMBF)*, Max-Planck-Institut für Züchtungsforschung, Institut für Genetik der Universität zu Köln.
Zentrum für Molekulare Gen- und Zelltechnologie in Köln (1994), *Abschlussbericht 1991-1994. Förderungsvorhaben BCTO365 2 (BMBF)*, Max-Planck-Institut für Züchtungsforschung, Institut für Genetik der Universität zu Köln.

Support for Technology-Based Firms: The Role Played by Property-Based Science Parks

Don Siegel, Paul Westhead and Mike Wright

Abstract. To encourage the formation and development of independent technology-based firms property-based Science Park initiatives in the United Kingdom have been established. The performance of Science Park and off-Park firms was compared with regard to several indicators. Also, the role played by the Science Park manager/director in the development of firms located on "managed" and "non-managed" Science Parks was explored.

1. Introduction

Policy-makers and practitioners are investing in people and are encouraging locally based solutions to create "learning regions" (Keane and Allison 1999). They are particularly interested in encouraging R&D and innovative intensive areas. Policy-makers and practitioners believe new technology-based firms (NTBFs) play a pivotal role in competitiveness, job generation, innovation, wealth creation and regional development Organisation for Economic Co-Operation and Development (OECD), 1993; Department for Trade and Industry (DTI), 1998; (Tether and Storey 1998).

The phenomenal growth in the number of NTBFs (and resultant employment) around, for example, university technology-based incubators such as Stanford University in Palo Alto and Massachusetts Institute of Technology (MIT) in Boston (Roberts 1991; Saxenian 1994) in the United States provided a role model for the development of Science Parks in the United Kingdom. Adherents of the "linear model" assume a chain of successive interrelated activities. These begin with basic scientific research, pass through applied and more developmental research activities, the development of new product and process ideas, the evolution and testing of prototypes, to commercial production and finally to diffusion. As a result, the local technological environment can attract and encourage the formation and development of technology-based firms in localities. This view has, however, been questioned. Massey et al. (1992) have asserted that there is no simple linear relationship between science and production, and no single model of the innovation process. Nevertheless, the linear model commands widespread support and continues to guide the strategies selected by policy-makers and practitioners.

Cooke and Morgan (1993) have asserted that "industrial districts" of firms (Castells and Hall 1994) can be created by public policy. To stimulate NTBFs, successive British Governments have encouraged (in conjunction with universities, higher education institutes (HEIs), local authorities and various financial institutions) the development of various property-based incubator initiatives. Science Parks represent one form of property-based incubator initiative. After a period of enthusiastic appreciation, scepticism is growing surrounding the value of sponsored environments (Flynn 1993) and property-based initiatives (Massey et al. 1992; Felsenstein 1994; Jones 1996; Oakey and Mukhtar 1999). There is still insufficient knowledge about the formation of "hub" (i.e., a Science Park acting as the generator of externalities for tenant firms) and "spoke" (i.e., firms located on Science Parks) districts. Despite considerable investment in the Science Park movement, limited empirical research has focused upon the contextual and process issues associated with the behaviour and attitudes reported by entrepreneurs who own independent firms located on property-based Science Park initiatives.

This paper makes a contribution to this debate. To assess the benefits of property-based Science Park initiatives in Great Britain, a demand-side perspective was utilised. Information was gathered surrounding the perceived demand-side advantages of a Science Park location. The performance of independent technology-based firms located on Science Parks was compared with the performance of a "control group" of similar firms located off-Park. Performance contrasts between the two groups of firms were explored with regard to several indicators reported by owners of independent Science Park and off-Park firms. The role played by the Science Park manager/director in the development of firms located on "managed" and "non-managed" Science Parks was explored.

2. Science Parks in the United Kingdom

Property-based initiatives that can be regarded as Science Parks have been identified and defined by the United Kingdom Science Park Association (UKSPA). According to UKSPA (1999), a Science Park as a property-based initiative includes the following features: it encourages and supports the start-up and incubation of innovation-led, high-growth, knowledge-based businesses; it provides an environment where larger and international businesses can develop specific and close interactions with a particular centre of knowledge creation for their mutual benefit; and has formal and operational links with centres of knowledge creation such as universities, higher education institutes and research organisations. Within this context, Science Parks in the United Kingdom may provide an intellectual and physical infrastructure as well as management support that is actively engaged in the transfer of both technology and business skills towards their clients.

Cambridge and Heriot-Watt Science Parks were established in the United Kingdom in 1972. In 1983, seven Science Parks were fully operational in the

United Kingdom. By 1989, there were 32 Science Parks in the United Kingdom and this plateau was maintained until 1993 (UKSPA 1999). In 1999, the United Kingdom had 46 fully operational Science Parks. This "new wave" growth in the Science Park movement is mainly due to the fact that the "new universities", previously the polytechnics, began to establish Science Parks.

Carter (1989) identified three types of management strategies leading to the development of Science Parks in the United Kingdom. First, is a university led and funded strategy (which is the least common model) in which a university or HEI establishes a Science Park on its own and continues to be responsible for its development and/or management. Second, is a joint venture company strategy, with the Science Park being run by a separate legal entity. This mechanism ensures a continuing positive role for the HEI and other investors in the joint venture company. Third, and most common is a co-operative venture strategy where partners work together within a flexible and informal framework. This is the strategy with the least input from the HEI and the academic involvement tends to be fairly limited, often via the HEIs industrial liaison officer. Services available to Science Park firms can be delivered through three main types of management agreements (Carter 1989; Grayson 1993). The first is the informal team. This is the cheapest and most flexible approach in which the Science Park partners divide the various management tasks between themselves but there is no full-time, on-site management presence. The second is the single on-site manager. This has the advantage of allowing the build-up of expertise relating to the specific needs of Science Park firms. Background and experience of the manager are important, especially the balance of industrially related scientific skills, promotional skills and skills in accessing finance. The third is the on-site management company. Here, a formally integrated management structure, including representatives from all investors in the Science Park, may provide a more secure basis for its long-term development.

UKSPA has made a distinction between "managed" and "non-managed" Science Parks in the United Kingdom. In 1992, UKSPA suggested that a "managed" Science Park had a generally full time manager on site whose principal task was to manage the Park (Westhead and Storey 1994). Science Park managers (i.e., gate-keepers) can legitimise the activities of entrepreneurs and can increase and diversify an entrepreneur's commercial, professional and social networks. In this study, the role played by Science Park managers is explored. The following Science Parks were regarded by UKSPA, in 1992, as being "non-managed": Aberdeen, Aberystwyth, Cambridge, Highfields (Nottingham), Listerhills (Bradford) and Loughborough (Westhead and Storey 1994, p. 170).

3. Research Questions

Science Parks have various objectives. These objectives include: the promotion of HEI/industry linkages and the transfer of technology from HEIs to Science Park firms; the promotion of the formation of new technology-based firms; the encouragement of spin-off firms started by academics; the encouragement of the growth of existing technology-based firms; attracting firms involved in leading-edge technologies; creating synergy between firms; improving the performance of the local economy; improving the image of the location, particularly for areas of industrial decline; creating new jobs directly as well as indirectly; and enhancing the competitiveness of new as well as existing firms in the region.

Developers of Science Parks assume that the owner of an independent technology-based firm faces barriers to commercialising their knowledge into profitable businesses due to their liabilities of small size and youthfulness. They have assumed that these issues can be addressed by providing owners of NTBFs and established firms with high specification and flexible premises located in landscaped environments (although some tenants have to pay a rent premium); shared resources and business services; links with local universities or HEIs; and opportunities for co-operation with other tenants. Further, a Science Park manager (i.e., a gate-keeper) can be a mechanism reducing uncertainty (and fixed costs) for owner-managers of NTBFs. A gate-keeper can enhance the reputation of new and small firms (i.e., those with limited social and business networks) and enable some owner-managers of firms to gain access to an uninterrupted supply of crucial resources from a variety of organisations including skills from HEIs and finance from banks, venture capitalists and development and enterprise agencies. Science Park management can proactively stimulate transfer and promote links between local HEIs and tenant firms through informal contacts and visits from an HEI's Industrial Liaison Officer, and thus enable firms to tap into important technical information from HEIs (Westhead and Batstone 1998, 1999). Proximity to similar firms and organisations with a technology focus (i.e., HEIs) can, in addition, provide opportunities for business development through informal and formal linkages. Network theory (Johannisson et al. 1994) suggests that entrepreneurs (and firms) with dense and varied social and business networks of contacts can gain information and resources to overcome business blind spots. Universities and HEIs can act as a vehicle for the sponsorship of firms through the use of grants and contracts and industry sponsored research. Moreover, firms utilising the resources of HEIs (i.e., testing equipment, laboratories, the technical expertise of academics and students, etc.) can control their own fixed costs.

The above arguments suggest that the performance of independent firms will be enhanced if their owner-managers select a Science Park location. To guide policy, the following research questions are explored in this review:

R1. Were independent technology-based firms located on Science Parks in 1986 more likely to survive over the 1986 to 1992 period than off-Park firms?

R2. Were independent technology-based firms more likely to survive than firms engaged in more "conventional industries"?
R3. Were independent technology-based firms, reporting a link with a local HEI in 1986 more likely to survive over the 1986 to 1992 period?
R4. Were independent technology-based firms located on "managed" Science Parks in 1986 more likely to survive over the 1986 to 1992 period than firms located on "non-managed" Science Parks?
R5. Did independent Science Park firms report higher levels of absolute employment growth over the 1986 to 1992 period than off-Park firms?
R6. Was absolute employment growth over the 1986 to 1992 period concentrated in a small number of independent technology-based firms?
R7. Did independent technology-based firms report higher levels of absolute employment growth than firms engaged in "conventional" activities?

4. Research Methodology

Considerable debate remains over the appropriate research methodology to measure the effectiveness of property-based initiatives. The above discussion has highlighted the supply-side rationale leading to the development of Science Parks in the United Kingdom. A demand-side perspective (i.e., a bottom up rather than top down) was used in this study to assess the benefits of a Science Park location. Researchers assessing the outcomes associated with programmes and initiatives to assist NTBF formation and development rely on asking participants (e.g., owner-managers of firms) their views. This procedure has several limitations:

- there is a clear risk that respondents may be unrepresentative of the population of businesses located on Science Parks;
- respondents may be tempted to give the answers they think the questioner expects to hear;
- the impact of a property-based Science Park initiative can only be judged by comparing the performance of Science Park firms with comparable firms that have not selected a Science Park environment;
- the effectiveness of property-based Science Park initiatives could be exaggerated (or undervalued) by failing to take account of the demographic characteristics of firms that have selected a Science Park environment; and
- whilst reported opinions of owner-managers of Science Park firms are of interest, there is a need to collect "hard" indicators of actual subsequent firm behaviour and performance.

To overcome the first limitation, the scale and nature of firms located on Science Parks at specific points in time need to be ascertained. A longitudinal comparative static methodology was used to assess the benefits associated with a Science Park location. To gauge the benefits associated with a Science Park location in Great Britain, the owner-managers of firms located on Parks were contacted. In 1986, data was gathered from Science Park and off-Park firms (i.e., firms not located on a Science Park) in Great Britain by Monck et al. (1988)[1]. A "follow-on" questionnaire survey of Science Park and off-Park firms previously interviewed by Monck et al., was conducted in 1992/93 by Westhead and Storey (1994).

It is difficult to generalise results from a single Science Park environment. To overcome this problem, interviews were conducted, in 1986, with a representative sample of owner-managers of Science Park firms located in diverse range of Science Park environments. Due to time and resource constraints, interviews could not be conducted with the owner-managers of all firms located on Science Parks. A random sample of Science Park firms was drawn, in 1986, by Monck et al., from the population of all firms located on Science Parks in Great Britain. Also, structured questionnaires (i.e., developed with the advice from Steering Committees) were used in 1986 and 1992 to gather information from owner-managers of Science Park firms. Face-to-face interviews with the owner-managers of Science Park firms were conducted (in 1986 and 1992) to ensure high response rates to the surveys. Face-to-face interviews ensured that the owners of the businesses responded and all appropriate information was collected.

To minimise the impact of the second limitation, a variety of objective as well as more subjective outcomes associated with a Science Park environment were monitored. The remaining limitations were overcome by collecting data (in 1986 and 1992) from a matched control group of owners-managers whose firms had never been located on a property-based Science Park initiative. Using trade directories and the Yellow Pages business directory, Monck et al., identified a random sample of technology-based firms not located on a Science Park in 1986. Samples of Science Park and off-Park firms were matched, in 1986, with regard to factors that influence business performance (i.e., business age, main industrial activity, ownership status and standard region location; Storey 1994). To overcome the final limitation, the Monck et al., study conducted in 1986 and the "follow-on" sample survey conducted in 1992/93 by Westhead and Storey (1994) gathered employment data from Science Park and off-Park firms. The Monck et al., (1988) surveys as well as the Westhead and Storey (1994) "follow-on" surveys are summarised below.

Monck et al. (1988) conducted a questionnaire survey of independent and subsidiary organisations located on Science Parks in Great Britain. In total, 284 direct face-to-face interviews were conducted in 1986 by Monck et al., with the owner-managers of 183 Science Park firms (row 1 in Table 1).

---

[1]The 1986 survey of Science Park firms conducted by Monck et al. (1988, p.64) and the 1992 survey conducted by Westhead and Storey (1994) adopted the following UKSPA definition of a Science Park as a property-based initiative which included the following features: has formal and operational links with a University, other Higher Education Institution or Research Centre; is designed to encourage the formation and growth of knowledge based businesses and other organisations normally resident on site; and has a management function which is actively engaged in the transfer of technology and business skills to the organisations on site.

Table 1. Survival and closure of surveyed science park organisations over the 1986 to 1992 Period (a).

	Science Park sample (b)			Off-Park sample		
	Total	Ind	Sub	Total	Ind	Sub
1. Number of surveyed organisations in 1986	183	135	48	101	92	9
Sample of 35 organisations selected for interview in 1990 during the pilot follow-on survey but excluded from the 1992/93 follow-on survey						
2. Survived - either original or new address by the end of 1992	24	17	7			
3. Confirmed closure by the end of 1992	11	6	5			
4. Survived - interview completed during 1992/93	59	49	10	50	44	6
5. Survived - either original or new address, 1992	31	26	5	19	18	1
6. Confirmed closure by the end of 1992	46	25	21	21	19	2
7. Organization has no telephone listing / not recorded in telephone book or trade directory by the end of 1992	12	12	0	11	11	0
8. Total sample closure rate (including no telephone listings) over the 1986 to 1992 period		43 (32%)			30 (33%)	

Table 1. (continued)

9. Number of organisations in the valid 1992/93 follow on sample (excluding those 23 independent Science Park organisations re-interviewed in 1990 and the organisations that had closed over the 1986 to 1992 period)	75	62
10. Valid response rate to interview survey, 1992/93	65%	71%

Notes: (a) The tracking of Science Park organisations was successfully achieved because information was collected from Science Park managers surrounding organisation name changes and / or organisation relocations.
(b) Nine surveyed firms located on a Science Park in 1986 have subsequently survived and moved to an off-Park location.
(Ind) Independent organisation in 1986.
(Sub) Subsidiary organisation in 1986.

Source: Westhead, 1997, Table 1, p.49.

This sample constituted 53% of all tenants on Science Parks in Great Britain at that time. The coverage, however, varied from one Science Park to another (for example, only 35% of firms located on the Cambridge Science Park were interviewed) and no firms were surveyed at Aberystwyth, Heriot-Watt, Kent and St. Andrews. Monck et al. (1988) acknowledged that the 1986 survey contained an over-representation of "new wave" Science Parks but argued "... that the firms in this survey do provide an adequate sample of Britain's new high technology industries, providing adequate geographical, technological, sectorial and ownership coverage" (pp. 110-111).

In 1990, Storey and Strange (1992a) conducted a follow-on pilot study of 35 Science Park organisations (rows 2 and 3 in Table 1). During 1992/93, empirical evidence from a second, and much more extensive, longitudinal "follow-on" survey of independent and subsidiary organisations located on Science Parks in 1986 was collected (Westhead and Storey 1994). It was decided not to re-survey those organisations contacted in 1990.

Table 1 shows that out of the 135 independent Science Park organisations interviewed in 1986, 92 (68%) remained in business in 1992 (rows 2, 4 and 5)[2]. Rows 3 and 6 in Table 1 show 31 independent Science Park organisations had closed. A further 12 independent Science Park organisations could not be traced (row 7), and were regarded as closures. Hence, the total number of independent Science Park organisation closures was 43 (32%; row 8).

Row 9 in Table 1 shows there were 75 independent surviving Science Park organisations in the valid 1992/93 "follow-on" sample (excluding the 17 surviving independent organisations re-interviewed in 1990 and 43 independent organisation closures). Structured questionnaire interviews were collected from surviving independent organisations (in 1986) during late 1992 and early 1993. Row 4 in Table 1 indicates 49 "follow-on" interviews were conducted with independent Science Park organisations (65% valid response rate; row 10)[3]. These interviews explored whether a Science Park environment provided technology-based firms with a variety of benefits. The management function on site and the role-played by the Science Park manager/director was also explored.

---

[2]The following widely used definition of organisation closure was used: "An independent business is regarded as a closure if, in 1992, it is no longer identifiable as a trading business. An independent business which moves locations but continues as a trading business is *not* regarded as a closure. If the business is a subsidiary or a branch plant then it is regarded as having ceased if it no longer trades at its previous location" (Westhead and Storey 1994, p. 25).

[3]Responses to the "follow-on" survey were examined for non-response bias (Westhead and Storey 1994, pp. 33-39). No marked differences between the 1986 and 1992 "follow-on" survey returns were recorded with regard to ownership characteristics, sectorial or geographical coverage or the age of the surveyed organisations.

5. Results from the Longitudinal Comparative Static Study

R1. Were independent technology-based firms located on Science Parks in 1986 more likely to survive over the 1986 to 1992 period than off-Park firms?

Westhead and Storey (1994) conducted a survival analysis of the independent and subsidiary organisations located on and off Science Parks in Great Britain interviewed by Monck et al. (1988). They monitored business survival over the six-year period between the end of 1986 and 1992. In 1992/93, Westhead and Storey attempted to contact all the 183 organisations interviewed in 1986 (Table 1). Using a widely used definition of business closure (Garnsey and Cannon-Brookes 1993), they noted that 43 independent Science Park organisations had closed. A further 30 independent off-Park organisation closures were identified. No statistically significant difference in closure rates was detected between the Science Park and off-Park samples (32% compared with 33%). A chi-square test confirmed that this difference was not significant at the 0.05 level. Westhead et al. (1995) concluded that sponsored Science Park environments had not significantly improved the probability of business survival. It, however, must be appreciated that several off-Park firms were located in sponsored Government designated "assisted areas" for regional development assistance.

R2. Were independent technology-based firms more likely to survive than firms engaged in more "conventional industries"?

It was difficult to compare the closure rates reported by independent Science Park and off-Park firms over the 1986 to 1992 period with the closure rates recorded by independent firms engaged in more "conventional industries". From the outset, it needs to be appreciated that any comparison would be imperfect. It was difficult to obtain any official statistics surrounding the survival of independent businesses located in Great Britain. Data was, however, gathered from a database that focused upon the number of businesses deregistered for Value-Added-Tax (VAT) in the United Kingdom. Between 1980 and 1990, about 11% of the stock of firms de-registered each year (Daly 1991). Based on an annual 11% deregistration rate over the six year period (i.e., 1986 to 1992), it was "expected" that 115 total sample independent firms (i.e., the independent firms in the Science Park and off-Park samples combined in 1986) would have closed. The "observed" number of businesses closing in the total sample of independent firms was 42 firms fewer than "expected" (73 "observed" closures compared with 115 "expected" closures). This suggests that, subject to provisos, the closure rate of independent essentially technology-based firm was lower that that recorded by United Kingdom businesses in general.

R3. Were independent technology-based firms, reporting a link with a local HEI in 1986 more likely to survive over the 1986 to 1992 period?

In 1986, Monck et al. (1988) made a distinction between independent firms reporting at least one link with a local HEI and those reporting no links. With reference to the combined sample of independent Science Park and off-Park firms, Westhead et al. (1995) charted the survival of firms with and without HEI links in 1986 to the end of 1992. The combined on and off-Park "follow-on sample", however, cannot be regarded as a representative sample of all technology-based firms in the United Kingdom in 1992/93. They found that a larger proportion of firms (131 firms: 72%) that had reported a HEI link in 1986 were survivors in 1992 than firms that had not reported a link (19 firms: 53%). A chi-square test confirmed that this difference was significant at the 0.05 level. Westhead et al., inferred that links with a HEI had enabled some firms to overcome some of the liabilities of "small size" and "newness"[4].

R4. Were independent technology-based firms located on "managed" Science Parks in 1986 more likely to survive over the 1986 to 1992 period than firms located on "non-managed" Science Parks?

Westhead and Batstone (1999) compared the closure rates between independent firms (in 1986) located on "non-managed" (a stock of 41 firms in 1986) and "managed" (a stock of 85 firms in 1986) Science Parks[5]. Over the 1986 to 1992 period, 10 firms (out of 41 firms) on "non-managed" Science Parks had closed compared with 28 firms (out of 85 firms) located on "managed" Parks. It is interesting to note that the closure rate was lower on the "non-managed" rather than the "managed" Science Parks (24% compared with 33%). However, a chi-square test confirmed that this difference was not significant at the 0.05 level.

Studies have found that older and larger firms are more likely to survive (Storey 1994). The lower closure rate on "non-managed" Science Parks may, therefore, be simply due to the fact that more firms on "non-managed" rather than "managed" Parks were older and larger in size. Further, the demographic differences between firms on the two types of Parks may be due to the fact that many "non-managed"

[4]Westhead et al. (1995) used multivariate logistic regression analysis to detect the combination of factors associated with the survival of technology-based firms in the Monck et al. (1988) samples over the 1986 to 1992 period. A Science Park location, in 1986, was not found to be a significant determinant of business survival. They found that younger businesses and those competing against small firms were significantly less likely to survive. However, "team start" businesses which were owned by more than one shareholder and those that had a founder with a bachelor's degree or more were significantly more likely to survive.

[5]In 1986, 135 independent Science Park firms were interviewed on Science Parks defined by UKSPA. The 9 independent firms surveyed, in 1986, located on the Bolton, East Anglia and Springfield House (Leeds) Science Parks were regarded by UKSPA as Science Park firms. However, in 1992, these locations did not conform to UKSPAs Science Park definition. Firms surveyed in Bolton, East Anglia and Leeds were no longer regarded as Science Park firms. Westhead and Batstone (1999) excluded these firms from their study that compared the closure rates between independent firms located on "managed" and "non-managed" Science Parks.

Science Park managers have exhibited a reluctance to attract and incubate new and small weak firms. The closure rate difference may, in part, be due to deficiencies in the management of some "managed" Science Parks. Moreover, the difference may be due to "managed" (i.e., those generally located in depressed and peripheral regions) Park managers being less selective (and more prepared to take risks) surrounding their choice of tenants. This, in itself, may stem from a desire to let the highest appropriate proportion of property in order to maximise rental income flow. Additional multivariate statistical analysis controlling for demographic sample differences between firms in the two groups is, therefore, required to explain this aggregate closure rate difference.

R5. Did independent Science Park firms report higher levels of absolute employment growth over the 1986 to 1992 period than off-Park firms?

Flynn (1993) warned that sponsored environments (i.e., Science Parks) may encourage negative selection (i.e., non-survival attributes being transferred to the next generation of firms) and the survival of competitively weak organisations. The direct employment contributions reported by surviving independent Science Park and off-Park firms in the "follow-on" samples were compared by Westhead and Cowling (1995). Employment data for independent firms in 1986 was available for two points in time for 46 surviving Science Park firms (out of 49 firms) and a further 31 surviving off-Park firms (out of 44 firms). The mean employment size, in 1986, was 11.3 employees for Science Park firms compared with 21.4 employees for off-Park firms. This difference between the two groups of firms was not statistically significant at the 0.05 level. By 1992/93, the Science Park firms had grown to employ on average 26.8 people whilst the mean employment size for off-Park firms had grown to 37.8 employees. Again, this difference was not statistically significant. The mean absolute employment growth reported by Science Park and off-Park firms over the 1986 to 1992 period was virtually identical (means of 15.5 employees and 16.4 employees). No statistically significant difference was reported Science Park and off-Park firms. Westhead and Cowling (1995) concluded that both types of firms had made contributions to job generation. Moreover, there was no real evidence that the Science Park sample included a large proportion of weak organisations that had been propped up by a sponsored environment[6]. Nevertheless, additional multivariate statistical analysis

---

[6]Westhead and Cowling (1995) used multiple correlation and regression analysis to detect the combination of factors associated with log transformed employment change over the 1986 to 1992 period reported by surviving firms in the Monck et al. (1988) Science Park and off-Park samples. A Science Park location, in 1986, was not found to be a significant factor explaining subsequent employment change. With respect to factors influencing high levels of employment growth, a high firm size (in 1986) was found to act positively on employment growth, as was a graduate level education for the key founder. On the finance side, firms that had access to and used a multiplicity of sources of start-up finance tended to report superior levels of employment growth.

controlling for sample selection bias is still urgently required to confirm this suggestion.

R6. Was absolute employment growth over the 1986 to 1992 period concentrated in a small number of independent technology-based firms?

New and small firms are not a homogeneous group and only a small proportion of NTBFs have the enthusiasm, ability, resources or inclination to grow (Oakey 1991). In aggregate, new and small firms make a significant contribution to the total employment pool but this aggregate data masks a highly skewed distribution. In short, whilst most firms employ a few people, only a few firms provide significant employment. This assertion was conformed by evidence from the "follow-on" samples of Science Park and off-Park firms.

Two firms in the Science Park sample reported no change in employment size over the six-year period between 1986 and 1992. A further 9 Science Park firms had reduced their employment size in total by 40 employees. The 35 employment growing Science Park firms had generated an additional 753 jobs. Interestingly, the 5 fastest employment growing Science Park firms (11% of this sample) accounted for 57% of these gross new jobs (i.e., 428 jobs). The skewed distribution of employment was even more apparent in the off-Park sample. Four off-Park firms reported no change in their employment size but a further eight firms indicated that they had reduced their employment size in total by 115 employees. At the other extreme, the 19 employment-growing firms had generated an additional 623 gross new jobs but the 5 fastest employment growers (i.e., 16% of this sample) had generated 84% of these gross new jobs (i.e., 521 jobs).

Presented evidence suggests that few technology-based firms are significant direct employment creators. Not surprisingly, a case for targeting support toward firms (and entrepreneurs) with growth potential has been proposed (Westhead 1995). Considerable debate has, however, been generated surrounding the benefits of targeting support and "picking winners" (Storey 1994).

R7. Did independent technology-based firms report higher levels of absolute employment growth than firms engaged in "conventional" activities?

Responses from the 46 Science Park firms and the 31 off-Park firms in the "follow-on" sample were combined by Westhead and Cowling (1995) to generate a "combined sample" containing 77 technology-based firms. They detected that firms in the "combined sample" were larger and had recorded more absolute employment growth than monitored firms engaged in technology-based (Oakey et al. 1988) as well as "conventional" activities (Jones 1991; Storey and Strange 1992b).

6. Conclusions and Implications

At first sight, evidence presented in this review appears to be disappointing. This, in part, is due to the unrealistic expectations surrounding the benefits associated with property-based initiatives. Very few statistically significant performance differences were detected between Science Park and off-Park firms with regard to several indicators (i.e., business closure rates and absolute employment change)[7]. It was, however, noted that Science Park and off-Park firms reporting at least one link with a local HEI in 1986 were significantly more likely to survive over the 1986 to 1992 period than their counterparts that had failed to develop links with a local HEI in 1986. Presented aggregate empirical evidence also confirmed that technology-based firms were more likely to survive and report superior levels of direct absolute employment growth than firms engaged in more "conventional" industries. Policy-makers and practitioners concerned with encouraging wealth creation and job generation need to appreciate that relatively few owners of independent technology-based firms have the inclination or ability to grow the employment size of their ventures. Whilst adding to the employment pool, presented evidence suggests that relatively few technology-based firms are significant direct employment generators in the short-term. Supporting the case for the targeting of public policy support, a skewed pattern of absolute employment growth, particularly by off-Park firms, was detected. As found elsewhere, relatively few firms had generated the vast majority of absolute employment growth. This evidence again warns that assessments of public policy interventions should not solely focus upon business survival rates. Additional multivariate statistical research is required that makes a distinction between low and high performing surviving firms in terms of a broad range of "hard" (i.e., employment growth, sales growth, profitability, ability to export, R&D inputs and outputs, etc.) and "soft" (i.e., entrepreneur satisfaction, contribution to a local community, number of links with a local HEI, etc.) performance indicators.

Several Science Parks are encouraging NTBF in depressed and turbulent environments that do not have a culture of technology-based firm activity. Some owner-managers located in these environments have selected "managed" Science Parks to gain access to critical resources as well as to increase the legitimacy of their ventures. This study has highlighted that different types of Science Parks exist. Although the difference was not statistically significant, it was noted that business closure rates on the "non-managed" Science Parks were lower than those reported on "managed" Parks. It can be inferred here that Science Parks placing a greater emphasis on increasing the supply of entrepreneurs who own technology-based firms in depressed environments without a strong culture of technology-based enterprise will be associated with high levels of business turbulence. In

[7]No statistically significant differences were detected between independent Science Park and off-Park firms with regard to "inputs" into R&D and R&D "outputs" (Westhead 1997). However, Science Park firms in the follow-on sample had reported in 1992/93 significantly more links with local HEIs than off-Park firms (Westhead and Storey 1995).

contrast, owners of "non-managed" Science Parks who place a greater emphasis on developing rather than incubating new firms may be inevitably associated with lower business closure rates. We believe it is unwise to assess the benefits associated with "managed" Science Parks with regard to a single performance indicator. A more balanced perspective of the benefits associated with the heterogeneous Science Park movement needs to be appreciated.

One of the implications of these findings is that more effective ways of linking tenant firms to the facilities and resources provided by a local HEI need to be given greater attention by Science Park managers. For example, HEI industrial liaison officers and Science Park managers need to become more proactive in setting up systems which continually encourage the development of more formal technical links between HEIs and industry located on and off Science Parks (Westhead and Storey 1995). Areas that need to be considered by Park managers include the need to enhance their technology and related market information gathering as well as ensuring the dissemination of this information to their tenants.

Additional research evidence is needed to provide policy-makers (House of Lords Select Committee on Science and Technology, 1997) with more accurate assessments of the direct and indirect benefits as well as costs of future policy decisions towards encouraging the supply of academic entrepreneurs and technology-based firms. Large and representative surveys of entrepreneurs who have selected a Science Park location for their businesses in the United Kingdom (and other countries) need to be collected. Responses from the owners of NTBFs should be selected to ascertain the benefits associated with different types of Science Park initiatives. Several "hard" (i.e., business closure rates, employment size, absolute employment change "inputs" into R&D and "outputs" from R&D, etc.) and "soft" (i.e., social and business networks, trust, co-operation and satisfaction, etc.) outcomes need to be monitored. Future studies need to carefully monitor the deadweight and displacement outcomes associated with different types of Science Parks. Most notably, how many owners of Science Park firms would have gone into business anyway without the provision of supportive Science Park environments.

The effectiveness of intermediaries (Perry 1996), such as Science Park managers, in the commercialisation process and the linking of Science Park firms with HEIs, other tenants on Parks a well as other firms located off-Park needs to be carefully monitored. To promote mutual learning and the sharing of expertise between technology-based firms, additional research is warranted surrounding the role-played by gate-keepers in maximising goodwill trust between tenants located on Science Parks and off-Park firms. Superior gate-keeper practice needs to be highlighted and disseminated to other practitioners.

References

Carter, N. (1989), *Science Parks Development and Management*, London, The Estates Gazette Limited.

Castells, M., and Hall, P. (1994), *Technopoles of the World: The Making of 21st Century Industrial Complexes*, London, Routledge.

Cooke, P., and Morgan, K. (1993), "The Network Paradigm: New Departures in Corporate and Regional Development", *Environment and Planning C: Society and Space*, 11, pp. 543-564.

Daly, M. (1991), "VAT Registrations and Deregistrations in 1990", *Employment Gazette*, 99, pp. 579-588.

Department for Trade and Industry (DTI) (1998), *Our Competitive Future: Building the Knowledge Driven Economy*, Competitiveness White Paper, London, HMSO, Cm 4176.

Felsenstein, D. (1994), "University-Related Science Parks - 'Seedbeds' or 'Enclaves' of Innovation?", *Technovation*, 14, pp. 93-110.

Flynn, D. M. (1993), "A Critical Exploration of Sponsorship, Infrastructure, and New Organizations", *Small Business Economics*, 5, pp. 129-156.

Garnsey, E. W., and Cannon-Brookes, A. (1993), "The "Cambridge Phenomenon" Revisited: Aggregate Change Among Cambridge High-Technology Companies Since 1985", *Entrepreneurship and Regional Development*, 5, pp. 179-207.

Grayson, L. (1993), *Science Parks: An Experiment in High Technology Transfer*, London, The British Library.

House of Lords Select Committee on Science and Technology (1997), *The Innovation Exploitation Barrier*, London, House of Lords, Paper 62.

Johannisson, B., Alexanderson, O., Nowicki, K., and Senneseth, K. (1994), "Beyond Anarchy and Organization: Entrepreneurs in Contextual Networks", *Entrepreneurship and Regional Development*, 6, pp. 329-356.

Jones, C. (1996), "The Theory of Property-Led Local Economic Development Policies", *Regional Studies*, 30, pp. 797-801.

Jones, M. (1991), *Small Firms Study 1985, 1988 and 1991*, Coventry, University of Warwick MBA, Unpublished Dissertation.

Keane, J., and Allison, J. (1999), "The Intersection of the Learning Region and Local and Regional Economic Development: Analysing the Role of Higher Education", *Regional Studies*, 33, pp. 896-901.

Massey, D., Quintas, P., and Wield, D. (1992), *High Tech Fantasies: Science Parks in Society, Science and Space*, London, Routledge.

Monck, C. S. P., Porter, R. B., Quintas, P., Storey, D. J., and Wynarczyk, P. (1988), *Science Parks and the Growth of High Technology Firms*, London, Croom Helm.

Oakey, R. (1991), "High Technology Small Firms: Their Potential for Rapid Industrial Growth", *International Small Business Journal*, 9, pp. 30-42.

Oakey, R. P., and Mukhtar, S-M. (1999), "United Kingdom High-Technology Small Firms in Theory and Practice: A Review of Recent Trends", *International Small Business Journal*, 17, pp. 48-64.

Oakey, R., Rothwell, R. and Cooper, S. Y. (1988), *The Management of Innovation in High Technology Small Firms*, London, Pinter Publishers.

Organisation for Economic Co-Operation and Economic Development (OECD) (1993), *Small and Medium-Sized Enterprises: Technology and Competitiveness*, Paris, Organisation for Economic Co-Operation and Development.

Perry, M. (1996), "Network Intermediaries and Their Effectiveness", *International Small Business Journal*, 14, pp. 72-80.

Roberts, E. B. (1991), *Entrepreneurs in High Technology*, New York, Oxford University Press.

Saxenian, A. (1994), *Regional Advantage: Culture and Competition in Silicon Valley and Route 128*, Cambridge, MA, Harvard University Press.

Storey, D. J. (1994), *Understanding the Small Business Sector*, London, Routledge.

Storey, D. J., and Strange, A. (1992a), "Where Are They Now? Some Changes in Firms Located on UK Science Parks in 1986", *New Technology, Work and Employment*, 7, pp. 15-28.

Storey, D. J., and Strange, A. (1992b), "New Players in the "Enterprise Culture"?", in: *Small Enterprise Development: Policy and Practice in Action*, Caley, K., Chell, E., Chittenden, F. and Mason, C. (eds), London, Paul Chapman Publishing, pp. 85-95.

Tether, B., and Storey, D. J. (1998), "Smaller Firms and Europe's High Technology Sectors: A Framework for Analysis and Some Statistical Evidence", *Research Policy*, 26, pp. 947-971.

United Kingdom Science Park Association (UKSPA) (1999), *15th Anniversary 1984-1999*, Birmingham, The United Kingdom Science Park Association.

Westhead, P. (1995), "Survival and Employment Growth Contrasts Between Types of Owner-Managed High-Technology Firms", *Entrepreneurship Theory and Practice*, 20, pp. 5-27.

Westhead, P. (1997), "R & D "Inputs" and "Outputs" of Technology-Based Firms Located On and Off Science Parks", *R&D Management Journal*, 27, pp. 45-62.

Westhead, P. and Batstone, S. (1998), "Independent Technology-Based Firms: The Perceived Added Value of a Science Park Location", *Urban Studies*, 35, pp. 2197-2219.

Westhead, P. and Batstone, S. (1999), "Perceived Benefits of a Managed Science Park Location for Independent Technology-Based Firms", *Entrepreneurship and Regional Development*, 11, pp. 129-154.

Westhead, P. and Cowling, M. (1995), "Employment Change in Independent Owner-Managed High-Technology Firms in Great Britain", *Small Business Economics*, 7, pp. 111-140.

Westhead, P. and Storey, D. J. (1994), *An Assessment of Firms Located On and Off Science Parks in the United Kingdom*, London, HMSO.

Westhead, P. and Storey, D. J. (1995), "Links Between Higher Education Institutions and High Technology Firms", *International Journal of Management Science*, Omega, 23, pp. 345-360.

Westhead, P., Storey, D. J., and Cowling, M. (1995), "An Exploratory Analysis of the Factors Associated with the Survival of Independent High-Technology Firms in Great Britain", in: *Small Firms: Partnerships for Growth*, Chittenden, F., Robertson, M. and Marshall, I. (eds), London, Paul Chapman, pp. 63-99.

High-Tech Industrial Clusters in the Asian Countries: The Case of the Electronics Industry in the Hsinchu-Taipei Region

Kung Wang

Abstract. This article mainly discusses Taiwan innovation system (TIS), that is, the operation system by enterprises, government, universities and public research laboratories of science and technology (S&T) industry development, and also the most representative characteristics that can explain TIS, that is, 1) the technology development program (TDP) of the Ministry of Economic Affairs (MOEA) for instance, to illustrate the industry innovation policy by Taiwanese government; 2) the Industry Technology Research Institute (ITRI) for instance, to explain R&D, and technology diffusion strategy of research institutes; 3) the Hsinchu Science-based Industrial Park (HSIP) for example, to show the way of enterprises S&T commercialisation; 4) bringing back overseas scholars and also the talents cultivation, to introduce supply of human resources in the academic circle. Finally, it introduces the practical achievements of TIS by the cluster effect formed for the development of Taiwan semiconductor industry.

1. Introduction

Taiwan is an island country, with limited natural resources and small market scale. After the agricultural society in 1940s, there was the light industry of the people's livelihood in 1950s, and then the petrifaction heavy industry in the 1960s. At the beginning of 1970s, both the government and domestic and foreign scholars and experts thought that, in order to maintain economic development, Taiwan should develop towards hi-tech industry. Looking back at the industry situation of Taiwan in the 70s, however, it was universally family business, and capital market has not taken shape yet, with weak equipments and capacities in terms of basic researches by universities. In order to enable domestic industry circle to obtain foundation for hi-tech development, government, enterprises, and experts and scholars from universities and public research laboratories all worked together, trying to build an innovation system for the development of hi-tech industry. Important S&T organizations, S&T policies and measures, and research institutes are as shown in Figure 1.

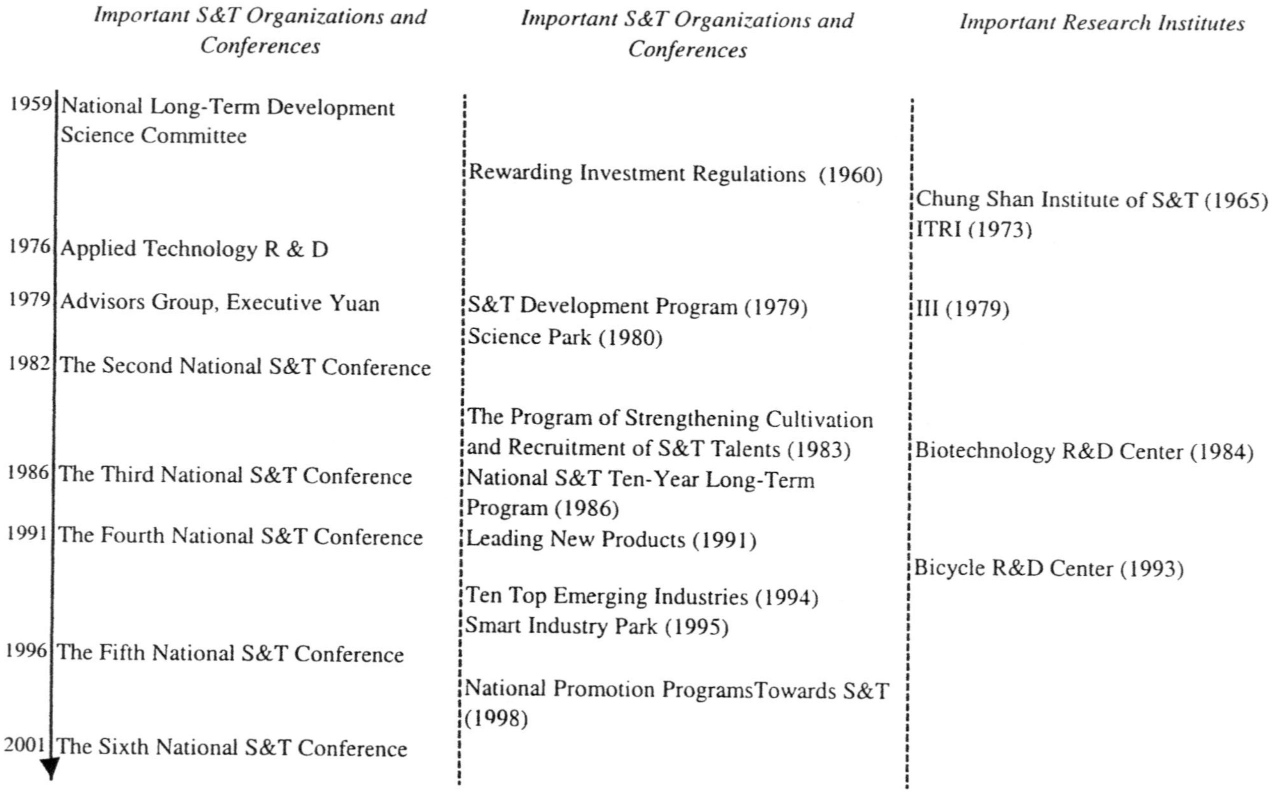

Fig. 1. Brief history of Taiwan S&T policy.

Under its unique innovation system, after efforts for almost 30 years, Taiwan has outstanding performance concerning number of patents acquired in USA and papers published on Science Citation Index (SCI) and (Engineering Index (EI), comparing with other rising countries in Asia. Its overall development achievements of S&T industry are quite good, as shown in Table 1.

The main objective of this article is to introduce the innovation system with Taiwan characteristics, with the main structure as follows: first, it discusses about TIS, that is, the operation system by enterprises, government, universities and public research laboratories in S&T industry development, and also the most representative characteristics that can explain TIS, that is, 1) take TDP of the MOEA for instance, to illustrates the industry innovation policy by Taiwanese government; 2) Take ITRI for instance, to explain R&D, and technology diffusion strategy of research institutes; 3) Take the HSIP for example, to show the way of enterprises S&T commercialisation; 4) take the instance of bringing back overseas scholars and also the talents cultivation, to introduce supply of human resources in the academic circle. Finally, it introduces the practical achievements of TIS by the cluster effect formed for the development of Taiwan semiconductor industry.

2. Documentation Discussions on National Innovation System (NIS)

2.1. A Survey of Academic Thought: Definitions of, and Views on, a National Innovation System

The concept of a national innovation system basically follows Schumpeter's (1939) discussion on technology innovation. Among scholars who study this topic, some of the more renowned academics are Freeman, Lundvall, and Nelson. They each define a national innovation system from different research angles.

Freeman (1987) defines a national innovation system as "...the network of institutions in the public and private sectors whose activities and interactions initiate, import, modify and diffuse new technologies". *He discusses the interplay between activities that lead to advances in technology and the organizations in the public and private sector that comprise the national innovation system.*

Lundvall (1922, p. 2) says of the national innovation system that "... a system of innovation is constituted by the elements and relationships which interact in the production, diffusion and use of new, and economically useful, knowledge... and are either located within or rooted inside the borders of a nation state", and studies the system from the perspective of the institutions that, taken together, make up the national innovation system, as well as from the microcosmic point of view of the individual.

Nelson (1993) calls the national innovation system a "... a set of institutions whose interactions determine the innovative performance... of national firms".

Table 1. Comparison of R&D achievements by the four small dragons of Asia from 1993 to 1999.

	Year	1993	1994	1995	1996	1997	1998	1999
R.O.C.	Name of patents approved in USA	1,510	1,814	2,087	2,419	2,597	3,805	4,526
	Ranking	7	7	7	7	7	5	4
	Number of reports published on SCI	4,793	5,852	6,682	7,500	7,758	8,601	8,931
	Ranking	22	20	19	18	19	19	19
	Number of reports published on EI	2,687	3,543	3,560	4,220	4,839	4,026	4,376
	Ranking	11	12	11	11	11	9	11
South Korea	Name of patents approved in USA	830	1,008	1,240	1,567	1,965	3,537	3,679
	Ranking	11	10	9	8	8	8	7
	Number of reports published on SCI	3,014	4,041	5,384	6,422	7,807	9,503	10,992
	Ranking	29	24	23	21	18	16	16
	Number of reports published on EI	1,342	2,135	2,528	3,479	4,911	3,645	4,777
	Ranking	18	15	13	13	10	11	10
Singapore	Name of patents approved in USA	44	59	61	97	100	136	152
	Ranking	30	24	24	24	25	24	22
	Number of reports published on SCI	--	--	--	--	--	--	--
	Ranking	--	--	--	--	--	--	--
	Number of reports published on EI	770	965	1,190	1,152	1,603	1,213	1,854
	Ranking	22	23	21	25	24	23	19
H.K.	Name of patents approved in USA	182	220	248	247	261	--	--
	Ranking	20	19	19	19	19	--	--
	Number of reports published on SCI	1,539	1,779	2,915	3,090	3,703	--	--
	Ranking	38	38	35	35	28		
	Number of reports published on EI	448	789	978	1,310	1,719	1,483	1,688
	Ranking	31	28	24	23	21	20	21

Source: NSC, 2000b.

Basically, the focus of his research is to compare the innovation systems adopted by various countries.

2.2. The NIS and the Knowledge Economy

In today's era where business is moving towards the concept of a knowledge economy, an understanding of how the national innovation system operates, and a realization that the successful operation of the knowledge economy depends on the smooth flow of knowledge among enterprises, universities and research institutions, enables policy-makers to develop policies and measures that promote successful innovation.

With this in mind, and following these academics' research, OECD (1997) attempted to apply to the national innovation system the indices used to evaluate the development of the knowledge economy. The aim was to discuss and compare interactions among the various players in the national innovation system, in order to gauge how effectively those interactions were at facilitating the flow of knowledge in the greater technology sector. Four types of interactions between different constituents of the innovation system were compared:

1. Interactions among Enterprises.
2. Interactions among Enterprises, Universities and Public Research Laboratories.
3. Diffusion of Knowledge and Technology to Firms.
4. Movement of Personnel.

2.3. Framework for a NIS Used in this Article

In this section, all the components of Taiwan's policy towards the NIS will first be examined in broad outline. Then the focus shifts to a microcosmic analysis of the way in which the individual institutions making up the innovation system interact with one another. Interactions between manufacturers in the industry, co-operation among enterprises, universities and public research laboratories, interactions that lead to technology diffusion, movements in the labour force, international knowledge exchanges, and other kinds of interactions will be examined to facilitate discussion on these five aspects of communication within the NIS (Figure 2).

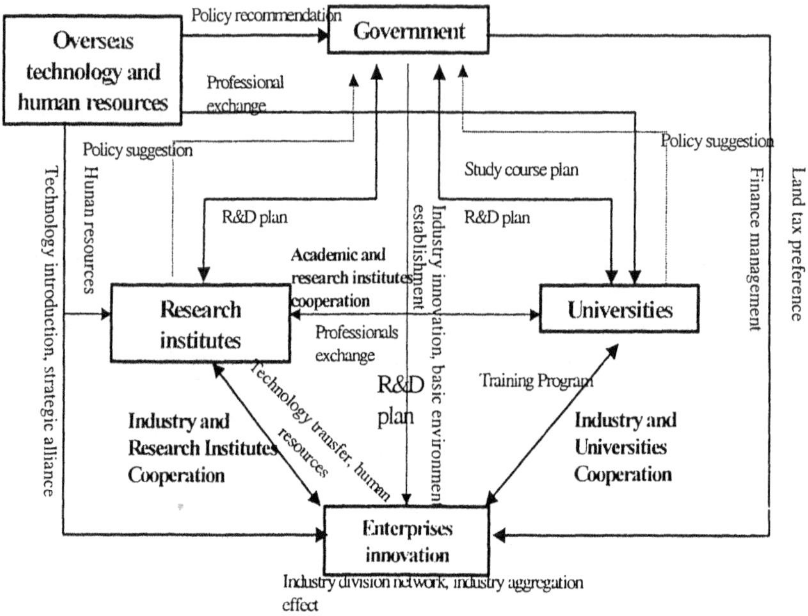

Fig. 2. Taiwan innovation system illustration.
Source: This article.

3. The Framework of Taiwan S&T Innovation

Taiwan technology development adopts the principle of overall planning and execution based on different labour division. For the industry technology development, the MOEA takes charge of the technology development work of applied researches. In addition to research units directly under its jurisdiction, the R&D departments in state-run enterprises, and research institutions appointed by Technology Developed Program (TDP), mainly engage in application and development of industry technology, transferring then to enterprises circle to carry out products technology development and commercialisation activities. Technology of the enterprises circle can be originated from technology transfer or research participation of research institutions and academic circle etc., as well as from technology introduction and strategic alliance with overseas companies or research institutions. It is as shown in Figure 2.

Research institutions interact with the enterprises circle by the method of technology transfer and cooperation with enterprises, and also assist in technology development of enterprises by technology guidance, information supply, and talents training etc. The special-topic research plans by NSC, promotes the academic circle to assist enterprises to develop technology-by-technology guidance and talents training etc. In the respect of development measures for technology of applied research by the government, besides NSC's cooperation program between enterprises and universities, research program of innovation technology, and target-oriented research program, there are also the program of introducing overseas technology, S&T development program and TDP by enterprises mainly led by the MOEA. In terms of commercialisation and application measures, there are the key components and parts by NSC and the MOEA's leading new product program, enterprises cooperated research program, and the R&D program of state-run enterprises assisting Small and Medium-Sized Enterprises (SMEs) etc.

For the interaction among enterprises, government, universities and public research laboratories, research institutions and academic circle transfer their R&D results to enterprises by all means, and then relying on each research institution, have interaction with overseas companies of enterprises or overseas research institutions through technology introduction, cooperated research, overseas investment and strategic alliance. The MOEA also implements subsidy funds, cooperation funds, and investment offset etc. measures, to encourage enterprises to engage in R&D. In conclusion, Taiwan industry technology development at present mainly focuses on independent R&D, assisted by technology introduction, and also stresses malignant interaction among enterprises, government, universities and public research laboratories, with the final goal of promoting the technology upgrade of the industrial circles.

4. TDP of the MOEA

In accordance with the "S&T Development Program" issued by the Executive Yuan in 1979, TDP becomes an important policy tool helping the MOEA to drive development of enterprises technology. TDP is used to research and develop key components and parts or common key technology, and then transfers to industrial circles for commercialisation and even quantity production investment, to reach the goal of using enterprises technology to raise enterprises competitiveness. Thus, the idea of TDP is to promote development of enterprises technology and stable economic growth, and strengthen competitiveness of enterprises and national life quality, through the R&D work of TDP.

The Technology Section of the MOEA is responsible for the promotion of TDP. In order to help the upgrade of enterprises technology, technology that has the common features of advantages in market, technology intensive, enterprises connection and verifiability etc., and also high-benefit technology with the capability to increase public welfare and public investment etc. are chosen as the main R&D items, therefore, the promotion of TDP plays a decisive role in the development of overall industry and economy.

The positioning of TDP is mainly to carry out the development of planned technology based on requirement by the development of enterprises. Enterprises development factors include capital, human resources, technology, market, resources, environmental protection, infrastructure and related management system etc., therefore, TDP selects technology needed for promoting enterprises development as the prerequisite, adopts the development strategy of independent technology development and bringing in overseas hi-tech industry, and engage in product development of important components and parts, important examination technology, decide technology standards and establish management system of intellectual property, to further raise technology of enterprises.

In order to help forward industry upgrade, strengthen effects to industrial circles by TDP, various forms and channels are used to implement the R&D outcomes of TDP on enterprises. During the program execution period, enterprises can make use of R&D outputs of research institutions by participation in advance, cooperated research and authorization transfer etc. After the program is finished, research institutions put into effect the R&D outcomes in the industrial circles by technology transfer and patent authorization etc. Information exchange, seminar, professionals diffusion and all kinds of industrial services can all help to establish the exchange channels between research institutions and enterprises, and this kind of channel also helps to diffuse R&D outcomes of TDP to enterprises without taking a form, professionals diffusion is especially a method that applies technology directly and practically in enterprises.

In order to fulfil the important mission of upgrading industry technology, the Technology Section selects and adopts measures based on industry change appropriately, furthermore, it has to have an insight into present and future industry requirement from a foresighted and macroscopic point of view, to raise proper measures during different stages (MOEA 2001). Looking back, in addition to constantly appointing mainly ITRI, Development Centre for Biotechnology (DCB), institute for Information Industry (III) etc. non-profit research institutions to carry out technology development starting from 1979, and adopting all kinds of channels like technology transfer, technology seminar, industrial services, and derivative companies establishment etc. to implement R&D outcomes extensively in the industrial circles, to really help S&T industries to become main force of Taiwan industries today, it also uses proprietors as the direct object of measures to promote enterprises to join in R&D in an all-round manner, to fulfil the measures of upgrading enterprises technology. This can be traced back to 1994 when promoting enterprises cooperation, proprietors participated in since the planning stage of TDP, so that the R&D program by research institutions can be more suitable to requirement by enterprises, and the participation by proprietors during the research procedures can help to accelerate the commercialisation procedures of technology as an answer to the shortening of technology life cycle. In order to accelerate in helping the industrial circles to set up S&T type enterprises, using research institution capabilities, facilities, talents and knowledge etc. resources accumulated by TDP, it provides "public laboratories" as the place for the industrial circles to engage in R&D and business establishment. In order to correspond to the gradual upgrade of industry technology and extensive requirement, under the situation of appointment to juridical person research institutions being insufficient for all these demands, the Technology Section tried in 1997 to appoint the industrial circle to carry out TDP, namely the "enterprises TDP". Following the development trends in the knowledge economy era, strengthening the creation of knowledge is its important factor, so is the development of human resources. Consequently, the first consideration is to introduce abundant R&D and human resources in universities systematically into the innovation of enterprises, the Technology Section tried the "Universities TDP" of appointing universities to carry out TDP in 2000.

5. R&D and Technology Diffusion Strategy of Research Institutions

In order to develop towards high added value and technology-intensive industry, promote the establishment of emerging hi-tech industry, and improve the present industry structure to reach sustainable development, government helps the establishment of research institutions in phases and also takes the responsibility to improve industry technology based on requirement by industry development. For instance, ITRI was established in 1973, with a predominant role in innovation technology and as the pilot who started up domestic industry technology; III was established in 1979, to drive development of information software/hardware industry in Taiwan. In addition, for the significance of biotechnology, government also sets up DCB, with the expectation of improving our biotechnology. Since industry structure of Taiwan is still with SMEs at the most, in order to assist technology development of SMEs, seven non-profit research institutions were helped to be set up in 1992 and 1993 successively, including printing, plastics, shoemaking, bicycle, precision machinery, stone materials processing, and pharmaceutical technology research centre.

Among them, ITRI is the largest non-profit industry technology application research institute in Taiwan, and has engaged in R&D until now for over 20 years. Starting from TDP execution appointed by the MOEA, it builds up its own R&D ability, and then pushes enterprises to engage in technology development, further drives our industry structure to advance towards the technology-intensive development direction. There are around 6,000 employees in ITRI during these recent years, with more than half owning master or Ph.D. degree. They cover extremely extensive fields of technology specialty, and are distributed in seven research laboratories and four research centres. Including Electronic Research & Service Organization (ERSO), Computer & Communication Research Laboratories (CCL), Opto-Electronic & System Laboratories (OES), Mechanical Industry & Research Laboratories (MIRL), Union Chemical Laboratories (UCL),

Material Research Laboratories (MRL), Energy & Resources Laboratories (ERL), and Centre for Measurement Standards (CMS), Centre for Aviation & Space Technology (CAST), Biomedical Engineering Centre (BMEC), Centre for Industrial Safety and Health Technology (CISH).

ITRI involves actively in the development of industry technology, to assist the industrial circle to improve technology, further to establish oncoming S&T industry and upgrade conventional industries. In order to reach the goal of improving industry technology, the research and development management of ITRI adjusts duly with the requirement by industries, for instance, it emphasized R&D planning since early stages, selected technology items suitable to be developed in Taiwan, and focused on bringing in technology. Following the ardent demand by the industrial circles, it made great efforts in diffusing R&D achievements, and also transferred experts appropriately to equally emphasize R&D of emerging technology industry and putting it into effect. At present, it drives the industrial circle to engage in R&D activities to actively create economic benefits, and also cooperates with balanced industry development of government to participate in newly risen fields. For the task of improving industry technology, ITRI will still actively implement R&D outcomes on enterprises in the future, besides bringing economic benefits, but also will open laboratories to accelerate establishment of S&T innovation enterprises. On the other hand, it will also engage in R&D of foresighted technology, to open up new technology and foresighted fields for enterprises, and also keep an unfailing supply of motive power of technology to create technology required by enterprises during the opportune time. At the same time, it will accelerate internationalisation of R&D and combine with the development of science-based industrial park (SIP).

The technology development strategy of ITRI is the R&D strategy oriented by the target of establishing core technology, environment construction, R&D innovation, and also improving R&D energy of enterprises. In the respect of establishing core technology, it is to integrate each technological field and create the maximum benefits; in the respect of environment construction, it is to build up information network environment and promote exchange of specialists from different fields; in the respect of R&D innovation, it is to strengthen foresighted technology R&D, promote internationalisation and also recreation of innovation culture, to enable ITRI to own an unfailing supply of motive power of technology; in the respect of improving R&D energy of enterprises, it is to use active methods of putting R&D outcomes into use by enterprises such as cooperation among enterprises, technology transfer, R&D strategic alliance, and public laboratories etc., to accelerate the improvement of R&D energy by enterprises.

6. The Technology Commercialisation Method in HSIP

Taiwan government adopts the policy of establishing Science Park to develop hi-tech industry of Taiwan. Considering that the main conditions for developing hi-tech industry are S&T professionals, the most advanced technology and fast acquirement and digestion of information, Hsinchu was chosen as the place to establish the first SIP of our country, the reasons are that, there are Tsing Hua University, Chiao Tung University and ITRI etc. academic and research institutions in Hsinchu, they provide excellent human resources abundantly, who can assist manufacturers to break through technology bottleneck. There is also convenient traffic condition in Hsinchu. The establishment of HSIP is an important measure taken by the Taiwanese government for promoting development of hi-tech industry. After development for 20 years, it is indeed the representative for Taiwan hi-tech industry acknowledged universally by home and abroad. There are six main industries formed already in HSIP, namely (Science-based Industrial Park Administration 1998):

1. integrated circuits: this industry is the one with the fastest growth in HSIP during the recent ten years, including integrated circuits design, manufacture, materials, assembly and testing, manufacturing process equipments etc.
2. Computer and peripherals: computer and peripherals industry is the main industry of HSIP, including computer system, storage equipments, input/output equipments, network equipments, special software and related mechanical and electrical components and parts etc.
3. Communication: communication industry has introduced the following four categories of products in HSIP, electronics communication system (including telephone set, modem, and exchange board), microwave system and components, optical fiber system and components, and satellite communication system.
4. Opto-electronics: this industry has brought in opto-electronics system components (such as thin-film transistor flat display, color display tube, CD-ROM drive, digital static picture camera), touch image sensor, opto-electronics semiconductor, and photodiode etc.) and optical system components (such as instruments and lens etc.).
5. Precision machinery: precision machinery industry has brought in automation system in HSIP (e.g. Computer Numerical Control (CNC) tool machinery, industry used controller, robot, water knife, vacuum machine, factory information automation etc.) and automatic components (such as casting, surface processing etc.).
6. Bioengineering: bioengineering technology has already brought in vaccine reagent, medical appliances and seed vaccine etc. in HSIP.

Ever since it was established in 1980, HSIP has introduced 292 hi-tech manufacturers until the end of 1999, altogether 82,778 employees and investment at NT$6,183. After development for 20 years, HSIP has grown up to become a place with significance for Taiwan hi-tech industry development.

Since the main objectives for establishing SIP in Taiwan is to promote development of high technology and industry, and the government uses the essential factors of hi-tech industry development as the object for carrying out its policies, this section explains the ways of innovation by HSIP based on different measures by the government to promote each essential factor to benefit manufacturers or entrepreneurs who engage in development of hi-tech industry. There are mainly technology sources, human resources, funds requirement, preferential land tax and other levies, and management environment five items of important factors, to be used to illustrate the way of bringing technology into products by hi-tech manufacturers in HSIP.

6.1 Technological Resources

HSIP had a large external demand in early days, ITRI could be regarded as important technological sources for HSIP, and manufacturers could also appoint ITRI to develop product technology or solve technological problems. For technological demand, government appointing applied research institutions such as ITRI near HSIP to carry out R&D and also put R&D outcomes into use by the industrial circles, has brought obvious results. With the growth of HSIP, the technological sources have already included overseas or technology transfer and cooperation, or from nearby Tsing Hua and Chiao Tung University. However, manufacturers in HSIP presently are still searching for technology development related information from ITRI, and also looking for new field, so its role has changed from the predominant one to supporting functions like encouraging to innovate, information, and assisting settlement of technological problems, only in some innovation fields, it is still an important technological source for HSIP.

In order to encourage manufacturers to engage in R&D of innovation technology, government provides grants for R&D of innovation technology to manufacturers in HSIP. It has been 14 years since it was implemented in 1986, and has become an important resource for helping HSIP to increase technology competitiveness. The government has provided grants to 24 research programs in 1999, totalling NT$67 million, at 28.1% of the total invested amount of NT$241 million. Up to 1999, altogether 478 programs received grants, at the amount of NT$896 million. Furthermore, in order to encourage manufacturers to research and develop key components & parts and products, the government has provided key components & parts and products subsidy for R&D by manufacturers in HSIP. From July 1992 to June 1999, there are 88 that have passed examination, and for the total R&D funds of NT$6.33 billion, NT$2.472 billion was approved to receive subsidy (Science-based Industrial Park Administration 1999).

6.2. Human Resources

For the development of high technology, talents are an important factor, and the aggregation of high-level human resources is the most significant characteristics of HSIP, and also an important resource promoting the gestation of hi-tech industry. Concerning the supply of human resources, manufacturers can easily recruit human resources needed, and can provide training for on-the-job personnel and raise personnel quality, at the same time, it attracts talents, including bringing in hi-tech entrepreneurs, that is why government chooses the place near universities and research institutions to set up HSIP.

6.2.1. Bring in High-Level Technology Entrepreneurs

Taiwan has started to engage in the development of hi-tech industry since 1980s. Since local people were lack of R&D ability, overseas scholars were recruited, especially those who have worked in Silicon Valley. They brought technology ability, marketing information, equipments purchase information etc. to Taiwanese, and based on their experience overseas, technology transfer and cooperation were expanded and even branch companies were set up, and Taiwanese also went to join in overseas technology seminar to master the information about advanced technology. In order to create newly risen enterprises, overseas scholars were recruited to bring in technology. The companies established by Taiwanese scholars who went to study overseas, from its start-up to healthy growth, all made significant contribution. There are 109 these kind of companies in HSIP by the end of 1998, with nearly 3,056 people. The technology and ideology brought back by these scholars have deeply rooted in the SIP and developed, which shows the performance of this measure obviously.

In addition, many high-level technology talents from research institutions were attracted to start enterprises in SIP, for example, the first non-governmental integrated design company in Taiwan - Syntek Design Technology Co. Ltd. was established in HSIP in 1981. The General Manager of the Company was formerly the Manager of Circuit Design Department of Electronics Bureau, ITRI, and the R&D Manager and Layout Engineer were also originally from ITRI. In 1983, it established Greatek Electronics, and among its main shareholders, seven are from ITRI, each takes on its main technology post.

6.2.2. Nearby Universities and Research Institutions

During the early days, HSIP manufacturers were in great demand of technological professionals for starting new business. In addition to recruiting overseas scholars, there are derived companies from ITRI, such as United Microelectronics Corporation (UMC), Taiwan Semiconductor Manufacturing Corporation (TSMC) etc., all bring human resources to HSIP. Moreover, there are also quite a lot who resigned and started business in HSIP or joined in manufacturers, and technology diffusion is also one of R&D results that government uses to evaluate ITRI. In S&T circles, all consider the transfer of technology people is the best way for technology transfer, thus, for electronics and information talents of HSIP, ITRI is an important source.

The demand on human resources by HSIP has expanded gradually, including those located at basic level and middle level. ITRI is still an important source for middle-level manpower of manufacturers in HSIP, and for the gradual expansion of basic-level manpower required, graduates from Tsing Hua and Chiao Tung University provide the supply. In addition, the manpower demand by HSIP during early phases required experience, therefore, students who just graduate regard ITRI as the "Shaolin Temple", a place for studying R&D ability, and advance to HSIP after three to five years, even manufacturers in HSIP regarded this as a reference criterion when evaluating employees' value. After the demand increases gradually and companies reach certain growth stage, by the method of hens leading chickens, graduating students become the very important new force. In particular, for the characteristics, prospect, payment and welfare of hi-tech industries in SIP all have very strong attraction, there is no special difficulty in the respect of recruiting good staff. Only for the change on industry technology, the semiconductor industry in HSIP has greatly attracted many talents, thus brought some squeezing phenomenon, consequently, what manufacturers recruit may not necessarily be from the right department and can master the work within short time, which increases the cost for on-the-job training and short-term training by enterprises in HSIP, and also is a waste of education resources.

In order to cooperate with technology development and manpower demand by enterprises, improve manpower quality, and shorten the time for new entrants' before-the-job training, the Administration Bureau of HSIP has cooperated with research institutions like Chiao Tung and Tsing Hua University, ITRI etc. to help all kinds of training for professional technology human resources, such as professionals training for deep sub-micrometer and liquid crystal display, integration between communication and computer, computer and automation etc. eight items of specialized technology, and also management, a second skill, and evening on-the-job training etc. From 1993 to 1998, there are altogether 24,390 people undergoing training, and 18,271 participating in management course and special topic discussion (MOEA 1999), which indicates that people who work in HSIP are taking an active attitude towards participation in training courses when facing fast development of technology, to keep their competitiveness. Recently, the Ministry of Education has opened the on-the-job class for master degree, which will be a catalyst to further promote improvement on manpower quality. This is the convenience on manpower recruitment and training of human resources brought by the government choosing places close to academic and research institutions to set up science park.

6.2.3. Employees' Profit Sharing System

In order to attract participation by more professionals, manufacturers in HSIP adopt the system of employees drawing dividend and becoming shareholders, which indeed bring significant influence on HSIP. Basically, this system can enable both the labour and capital to become rich, simplify the relationship between the labour and capital, and reduce significantly the inner transaction cost of enterprises. Its positive effect lies in keeping employees, so it can attract talents for individual company, from the overall point of view, however, professionals flow in this market, difficult to be deeply rooted (pulling force from the market). On the other hand, since talents are easily attracted, the speed of enterprises creation and experience exchange is even faster.

6.3. Source for Capital

During the early stages, there was SIP providing land needed for setting up plants, but difficulty in acquiring capital, especially for technology industry with high risks. For the gradual establishment of capital market in Taiwan, capital changes to mainly come from capital market, and the following gives an analysis on influence of HSIP hi-tech manufacturers' capital demand concerning policies like government providing funds for starting business and working capital etc.

6.3.1. Government Provides Funds for Starting Enterprises

During the early stages of development, the environment for HSIP to acquire funds was not abundant. The government coordinates Bank of Communications to provide financing for S&T enterprises or carry out investment on S&T enterprises, and the Development Funds of Executive Yuan was established for investing in S&T enterprises. In order to put its technology into practice by enterprises, ITRI also establishes innovation investment company, besides assisting ITRI to transfer technology to enterprise, it also assists manufacturers to acquire funds or get investment from the company to accelerate the commercialisation of high-risk and high-market potential technology of manufacturers. Furthermore, it rewards establishment of entrepreneurial and investment enterprises, and related hi-tech companies invested, which all have significant influence on the start-up funds for manufacturers in HSIP. For instance, among the 32 manufacturers brought in Taiwan in 1996, there were 9 Taiwan entrepreneurial investment companies, having carried out investment to ten newly entered manufacturers among them (IEK 2001).

6.3.2. Assist Manufacturers to Raise Working Capital

With the gradual establishment of Taiwan capital market, the stocks of manufacturers in HSIP were listed, and for their proper management, professional managers run the companies, and staff become shareholders, so that the acquirement of capital for company operation or expansion also comes from employees' investment, as well as capital from many entrepreneurs, venture capital, bank investment, and investment transferred from related enterprises. Since manufacturers in HSIP have close relationship in terms of personnel exchange, information exchange, and technology transfer or cooperation, for enterprises with great potential, when expanding factories and increasing investment, all are very easy to obtain funds from non-official relationship.

Transferred investment is also one of the main methods to acquire funds in SIP, especially the electronics industry. When manufacturers have very good operation, for the upper, middle and lower-reach key components and parts and products, they will transfer investment to set up new companies, thus to take shape within short time in terms of funds, market, product, important high-level managers, management system etc., shorten the preparation period for starting new business and reduce cost. Therefore, very closely connected network is formed, as long as the economy is in good cycle, these factors helping to start business can almost create another manufacturer at any time.

In order to assist manufacturers to acquire working capital, the government has adopted many measures, such as the measure of accelerated depreciation of equipments, so that those with very large investment on equipments can delay in paying taxes for accelerated depreciation, and also get working capital without interest, which have quite significant influence on semiconductor and optoelectronics mainly for manufacture. The government also looses its regulation on retaining surplus and compulsory distribution. Following the trends of getting funds easily by HSIP manufacturers, the influence on them is also relatively reduced. Furthermore, the government provides middle and long-term low-interest bank loan, and invests in manufacture of key products for important S&T enterprises, this policy has the function of leading manufacturers to invest in products with foresight and also risks. So it suits relatively more to the demand of expanding funds by manufacturers of HSIP.

6.4. Preferential Land Tax and Other Levies

There was insufficient investment on high technology in Taiwan during the 1980s. In order to encourage investment, the government has adopted many measures, for instance, the income tax for profit business can enjoy 5 (4) years of exemption from taxation, accelerated depreciation, paying taxes within limited amount, investment offset, loosened expenses etc. Preferential taxes, which can really be considered as great achievements in reducing and exempting from taxes.

Among all measures of reducing and exempting from taxes in HSIP, that the income tax for profit business can enjoy 5 (4) years of exemption from taxation brings the most favourable treatment for manufacturers, since it has significant influence on manufacturers' profit, reduce payment of taxes. For the business income tax of Taiwan enterprises, it is averagely 25%, in order to encourage hi-tech manufacturers' investment in HSIP, the government has adopted preferential measures on their business income tax. The first several years for hi-tech new businesses are normally in deficit, however, it still encourages investment, and its achievements can be detected from the fast accumulation of capitals in HSIP.

The government gives five-year free business income tax (exemption five) for businesses belonging to emerging and important S&T enterprises, and four-year free business income tax for locating in HSIP (exemption four). Since almost all enterprises in HSIP can enjoy five-year exemption and also four-year preference, they enjoy more favourable treatment comparing with manufacturers outside HSIP. The main goal of exemption five is to provide free tax during the initial stage of establishment to S&T type enterprises with relatively great risks, to encourage entrepreneurial spirit and start emerging S&T industries.

In order to reduce the cost for instruments and equipments of hi-tech enterprises and raw material, and further to increase business profit, the government has adopted import duties and goods tax reduction and exemption measures, to promote manufacturers in HSIP to mainly purchase machines, raw materials, components and parts from overseas, which are more advanced than local technology and therefore can raise domestic industry.

6.5. Operation Environment

In order to provide manufacturers with excellent hi-tech enterprises operation environment, the government has adopted many measures such as single service window, simplified administration procedures, and keeping improving infrastructure etc. to build an excellent development environment for manufacturers. The government also adopts cargo automation system, to create competitive advantages for HSIP manufacturers and also grab the business opportunities. HSIP has taken the lead to adopt cargo automation system since Nov. 1992. This cargo automation system connects through SIP information network (SPNET), to carry out goods import/export examination operations (visa, hi-tech products control, software export examination), warehouse operation (enter/exit storehouse) for related units (Administration Bureau and warehouse and transportation centre, manufacturers, customs declaring office, post office, III software examination station inside HSIP), and also transform applications after signature and take storehouse into electronics data interchange (EDI) message, to transmit specially via trade-van, which not only provides fast and convenient services to SIP enterprises, increases products competitiveness, but also improves the administrational efficiency and image of government organs.

Since the Administration Bureau of HSIP provides single window administrational service, which saves much time and documents travelling, so that manufacturers can devote themselves to business, expanding product market, and overseas and Taiwan hi-tech manufacturers all regard permission into SIP as a great honour and pride. All operational activities needed after manufacturers enter the HSIP can be obtained inside HSIP, so that enterprises operation can reduce operational cost, and also prepare excellent living environment for employees, reduce external cost, and also increase exchange of people from different companies.

7. Supply of Human Resources from Universities

In order to cooperate with the demand on high-level manpower by the industrial circles and development of Asia-Pacific Regional Operation Centre, the Executive Yuan has specially combined enterprises, government, universities and public research laboratories to organize employment-oriented short-term professional training for high-level manpower, after the training, they can be directly employed by enterprises, so that people with above master degree and related skill but insufficient experience in practice, can serve enterprises after the short-term training (MOEA 1999). For example, (1) assist talents from home and abroad above master degree to engage in short-term research work: in order to assist those from home and abroad at above master degree to get a job, before they decide whom to work for, arrange them to engage in short-term research work to store up talents. There are funds for subsidizing around 100 short-term researchers, to employ by agreement, with the agreement term at most one year; (2) assist talents at Ph.D. to engage in short-term research work: assist talents with Ph.D. degree to work in non-governmental organizations, encourage non-governmental organizations to use Ph.D. level talents, to cultivate talents needed for important national construction plan, promote implementation of high-level human resources, and also raise quality of non-governmental talents. The National Youth Commission subsidizes research remuneration at 50 researchers every year, with the longest subsidization term at most one year.

In the respect of bringing in R&D talents, to accelerate the upgrade of domestic industries, the government brings in specialists lacked in Taiwan as well as trains up actively high-level R&D talents, to improve S&T R&D ability, accelerate the development of supporting industry to the most advanced S&T, and expand recruitment of overseas high-level S&T talents, for example, NSC provides subsidy to S&T figures from China or from China to the third region to come to Taiwan to participate in S&T research program for public and private universities, public research institutions and national laboratories. For the MOEA, in order to assist domestic non-governmental enterprises to expand recruiting overseas technology talents to return home to serve, any specialist beneficial to raise technology and added value of enterprises can raise work program together with their employer, after examination and approval by "Examination Committee for Recruiting Overseas Industry Specialists", they can receive subsidy for the specialists' salary part, at the limit of three years.

In order to strengthen the implementation of overseas high-level technology talents and domestic talents above master degree, the government has adopted many recruitment measures (NSC 2000a), such as: (1) set up service window of "strengthen implementing high-level S&T talents program", to provide introduction of this program and also information needed by each domestic institution, and the execution unit includes each unit under the Executive Yuan, state-run enterprises and ITRI etc.; (2) hold high-level human resources seminar and exhibition, to assist talents above master degree to understand working environment of Taiwan, and encourage enterprises to recruit high-level talents; (3) starts to hold seminars for introducing overseas new S&T since May 1994, to discuss the ten top emerging industries, and also assist scholars travelling overseas to come home to give lecture and pay visits.

In the respect of training industry technology talents, in order to cooperate with industry upgrade and train talents required by industry development, the MOEA has advanced comprehensively the "Phase I five-year plan for training of industry technology talents" from July 1990 to June 1995, including raising product quality, industry design, industry automation, industry manufacture process technology etc. talents training, and started carrying out phase II five-year plan since 1996. It carried on advancing "Five-year plan for training of industry technology talents" in 2000, to strengthen combining academic, research institutions and non-governmental enterprises teachers, facilities and teaching materials to raise substantially the percentage for graduates from the science and engineering department of junior college to engage in manufacturer through on-the-job training and professional switching training, to strengthen the technological level of workers presently in conventional industries, and also solve the problems of generally insufficient manpower in conventional industries (MOEA 1999).

Corresponding to the serious insufficiency in hi-tech engineer-level talents of emerging S&T key industries, the MOEA and Employment and Vocational Training Administration Council of Labour Affairs have advanced the "Hi-tech key industries talents training program" in July 1998, lasting three years. The MOEA also cooperates with the "S&T talents training and implementation program" by the Executive Yuan to establish the "Cultivating high-level cross-industry talents with S&T background program" in 2000. It is estimated that around 60 R&D or management talents with S&T background will be selected during three years from Taiwan emerging hi-tech fields related enterprises and research institutions, recommended by the institution they hold a post, subsidized by the government, mainly to participate in overseas practice and assisted by short-term training, to have six to nine-month investment evaluation, technology transfer, and intellectual property etc. professional training and practice. In 2000, among these 220 people recommended, 50 have passed examination, and after around two-month training in Taiwan, it is planned that 20 will be chosen to have further practice overseas.

In addition, in order to cooperate with the development rends of 3C integrated S&T, the government subsidizes each circle to engage in R&D of 3C integrated S&T and products, and the Ministry of Education will also started to encourage all universities to have courses that teach 3C integrated S&T since 2000. It is estimated that 750 3C S&T integrated talents can be training each year, and also 150 high-level talents with master and Ph.D. degrees. Corresponding to the highly development of electronics, information and electric machinery in Taiwan, there is serious insufficiency in related talents, especially wireless communication, Internet, and software talents, and also for the demand on Business to Business (B to B). E-commerce talents, electronics industry technology talents cultivation program, industry computer talents cultivation program, and strengthening information electronics hi-tech talents cultivation programs etc. were implemented, by on-the-job class and cultivation class, having trained up more than 13 thousand people from 2000 to 2002, which solves the lack of talents in domestic S&T industries. Among them, the demand for communication and network talents is even more pressing, so the cultivation for information electronics hi-tech talents will be expanded.

8. Innovation System Forms Cluster Effect

An important case of Taiwan hi-tech industry development - the innovation course of the semiconductor industry, mainly includes the function played by the government, research institutions, universities and enterprises. During the development courses of Taiwan semiconductor industry, the government sponsored R&D funds of ITRI, and ITRI plays the sower of planning industry development and also the cultivator who engages in R&D, to assist the establishment of emerging industries, and create the prosperity for Taiwan semiconductor industry, so that it becomes the representative of domestic hi-tech industries.

The formation of overall integrated circuits industry can be described as that, the government chooses integrated circuits technology, ITRI brings in 7-micrometer integrated circuits technology from USA, after digestion and R&D, transfers to industrial circles to carry out commercialisation. Further combines with enterprises to develop sub-micron technology, and develop towards independent memory products. During the industry development course for as long as 30 years, the government subsidizes funds for R&D of ITRI, and ITRI plays the sower of planning industry development and also the cultivator who engages in R&D, to assist the establishment of emerging industries, and create the prosperity for Taiwan semiconductor industry.

Our semiconductor industry started to establish factory by Kaohsiung Electronics Company in 1966 to assemble transistor. From 1969 to 1973, there was Philips, TI, RCA etc. foreign companies establishing assemble factories of integrated circuits in Taiwan. Among Taiwan capital factories, there are Lingsen, Orient, Siliconix etc. During this period, our semiconductor industry has been in the assemble stage of integrated circuits. In the respect of academic field, only Chiao Tung University has semiconductor lab for cultivating semiconductor talents. After introduction by ITRI and establishment of HSIP, the cluster effect was incubated. For the efforts made for more than ten years, it finally helps Taiwan semiconductor industry to take a place on global stage.

From the cluster effect of Taiwan information electronics industry, from Taipei, Taoyuan County to Hsinchu County, in such a 80-Kilometer long and narrow geographic area, it is almost the gathering place for main manufacturers. If just checking integrated circuits industry, the obvious cluster situation can be found. Except for Nanya Technology Corp., the left 15 integrated circuits wafer factories are all situated in HSIP, and integrated circuits design industry is distributed in Hsinchu and Taipei, package manufacturers base in Hsinchu, Taichung, and Kaohsiung. Under the situation of insufficient land in HSIP, integrated circuits manufacture develops towards Tainan Science Park, and package manufacturers change to Hukou. Integrated circuits design industry still stays in Hsinchu and Taipei for its small land demand and high requirement on professionals and information.

For peripheral supporting industries, such as mask, conducting wire frame, chemicals, equipment agents etc. related proprietors, under the consideration of serving customers without having to go far, most establish their base inside HSIP or in its vicinity, so that the upper, middle, and lower-reach system of integrated circuits industry are almost all concentrated in a closed geographic area. We can regard this kind of industries cluster situation as a gigantic production satellite system or imagine it as a "virtual" large company with vertical integration of upper and lower reach. Almost without geographic and transportation limitation, more economical ways can be used to carry out mutual communication or products transportation. Comparing with large enterprises system of USA, Japan and Korea, for this large "visual" company of this mode can have the effect of each "enterprise system" concentrating on their own specialized fields, to develop the flexibility and fast emergency reaction ability of small enterprises into full, thus their competitiveness is greatly raised.

The unique innovation system of Taiwan causes the cluster effect of HSIP without the intention. For semiconductor industry, more than 90% of its upper and lower-reach enterprises are concentrated in Taipei, Taoyuan, and Hsinchu etc. County and City, so it is the most obvious one, as shown in Figure 3. This kind of cluster effect helps Taiwan semiconductor industry to take a place globally. In 2000, Taiwan has a market share as high as 70% of global Original Equipment Manufactor (OEM) market, and around 20% share of Dynamic Random Access Memory (DRAM) production.

The cluster effect of semiconductor industry in HSIP has quite evident effect on related businesses establishment. Take integrated circuits design for instance, there were 3 integrated circuits design companies in Taiwan during the first half of the 1980s, only one was in HSIP. For the 84 companies established after 1995, 40 were established in HSIP, and as high as 62 if calculating together with Hsinchu County and City. Please refer to Table 2.

Table 2. Relation table of Taiwan integrated circuits (IC) design industry establishment time and geographic location.

Number	1980~	1985~	1990~	1995~
Taipei City	1	7	9	8
Taipei County	1	6	4	3
HSIP	1	4	12	40
Hsinchu City			3	12
Hsinchu County				10
Tainan County				1

Source: Semiconductor Industry Yearbook, June, 2001.

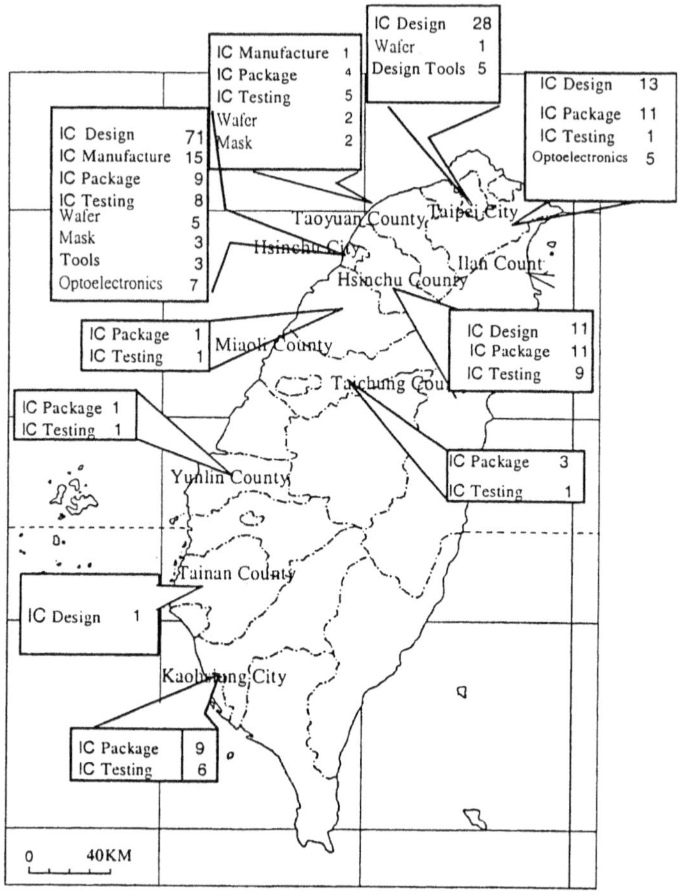

Fig. 3. Taiwan semiconductor industry manufactures cluster.
Source: Industrial Economics & Knowledge Centre (IEK), June 2001.

9. Conclusions

In order to promote the development of Taiwan hi-tech industry, based on the development environment in 1980s, aiming at the factors needed for developing hi-tech industries, all kinds of measures as many as seeds have been implemented, to advance the germination and development of high technology. After development for 20 years, these seeds have grown up to become large trees, and today's hi-tech forest. For this forest, these seeds can bring new vitality but their function has gradually withdrawn.

Taiwan's policy didn't pre-select electronics information industry, but all factors pushed the growth of electronics information industry in Taiwan at the fastest speed. The global development environment and electronics information industry, especially semiconductor industry, have driven the development of overall high technology. It can be further expected from the phenomenon of communication and Internet deepening into every level of mankind, that they will bring limitless opportunities for Taiwan to develop high technology.

For the innovation system that is built in Taiwan after 20 years, semiconductor industry is a successful case from null to prosperous development. In the future, the innovation system will put forth efforts on established system. Facing the knowledge economy era and fast development of S&T, knowledge has become the driving force for economic development, and gradually replacing the leading positions of traditional manufacturer factors of land, labour, and resources etc. Industry based on S&T knowledge, therefore, has become the main power for economic growth of advanced countries in technology. For example, the knowledge industry output value of OECD countries, more than half are related with creation and application of knowledge, especially electronics information, communication, and biotechnology industry with gene as the principal part, will become important S&T industries for the development of knowledge economy.

References

Freeman, C. (1987), *Technology and Economic Preference: Lesson from Japan*, London, Printer.

Industrial Economics & Knowledge Centre (June 2001), *2001 Semiconductor Industry Yearbook*, Taiwan, IEK.

Lundvall, B-A. (ed.) (1922), *National Innovation Systems: Towards a Theory of Innovation and Interactive Learning*, London, Printer.

Ministry of Economic Affairs (1999), *White Book of Industrial Technology 1998* Taiwan, R.O.C., MOEA.

Ministry of Economic Affairs (2001), *MOEA Technology Development Program 2000*, Taiwan, MOEA.

National Science Council (2000a), *The Annual Report of Science and Technology in R.O.C.*, Executive Yuan, Taiwan, NSC.

National Science Council (2000b), *Indicators of Science and Technology R.O.C. 2000*, Executive Yuan, Taiwan, NSC.

Nelson, R. (ed.) (1993), *National Innovation Systems: a Comparative Analysis*, New York/Oxford, Oxford University Press.

OECD (1997), *National Innovation Systems*, OECD.

Science-based Industrial Park Administration (1998), *The Annual Report of Hsinchu Science-based Industrial Park*, Taiwan, NSC.

Science-based Industrial Park Administration (1999), *The Statistics of Quarterly Report of Science-based Industrial Park*, Taiwan, NSC.

Financial Markets, Industrial Clusters and Small and Medium-Size Enterprises

Angelo Tantazzi[1]

Abstract. While the quantitative importance of Italian industrial clusters is indisputable, the structural weaknesses that are inherent in the cluster system have been pointed out repeatedly in recent years. The study identifies the economic conditions where the Stock Market can help strengthen and energize Italy's industrial clusters. The cluster companies already listed on the Stock Market are identified, and their importance for capitalisation and turnover of the market are calculated. The contribution of cluster companies seems to be particularly relevant in the group of industrial small and medium-size listed companies.
The new segmentation of the Stock Exchange is finally described. The recent introduction of the STAR Segment (Segmento Titoli con Alti Requisiti) in the stock market is designed to create a new regulatory, organizational and microstructural framework that is better suited to enhance the value of small and medium-size industrial enterprises. This solution provides a source of diversification and offers retail and institutional investors attractive investment opportunities in areas that previously had been neglected due to problems of visibility, transparency and liquidity.

1. Italian Industrial Clusters: Definitions

A study of Italian industrial clusters entails an analysis of a segment of the economy the importance of which is comparable to the contribution that large multinational groups or capital-intensive industries provide to the national economy in Italy and in other countries. The fact that industrial clusters have also been called "spontaneous multinationals" (Fortis 1998) is indicative of their significance. It has been estimated that, in absolute terms, the value added generated in the early 1990s by the products that are emblematic of Italian industry, a large portion of which is manufactured in industrial clusters, was greater than the value added created in the United States by the pharmaceutical, oil refining, office machines and beverage industries combined. Italian textile, garment and footwear manufacturers generate more value added than the German automobile industry. Industrial clusters and specialized manufacturing systems

[1]My thanks to Alessandra Franzosi and Enrico Pellizzoni (Market Research and Analysis, Borsa Italiana) for their invaluable contribution to the analytical portion of this presentation.

account for an estimated two million jobs, and industrial clusters contribute one-third of Italy's exports.

While the quantitative importance of industrial clusters is indisputable, the structural weaknesses that are inherent in the cluster system have been pointed out repeatedly in recent years. The main weaknesses include the vulnerability of clusters to innovations that originate outside the cluster, the lack of economies of scale at the individual company level (offset only in part by system-wide economies of scale) and the inability of industrial clusters to formulate a well-defined strategy in the absence of a system leader. It has been suggested often that these problems could be overcome if the individual manufacturers can be made to grow to a large enough size while, at the same time, retaining their entrepreneurial spirit, flexibility and ties to local traditions. One of the obstacles to growth for these enterprises (as well as for the Italian economy as a whole) is the absence of a financial system that can provide adequate support to companies in their efforts to expand.

In order to understand how the stock exchange and the financial markets in general can help strengthen and energize Italy's industrial clusters, one must first clearly define the concept of industrial cluster, with the goal of identifying the different configurations that exist within clusters and of understanding, given these different internal organizations, the various problems and specific challenges they face. Studies of industrial economics usually make reference to "small business systems" the main structural characteristics of which are summarized below:

- There is a strong manufacturing specialization at the local level, which is identifiable not just by the preponderance of one industry or area of manufacturing, but also by a production system that links different subsystems. These systems are all tied to a single product that is characteristic of the local economy.
- The production of the cluster is sufficiently large to account for a sizable percentage of the national and/or international output for the industry or specific product that the cluster specializes in.
- There is a very clear division of labour among the companies in the local system, which produces a high level of horizontal integration with low transaction costs.
- There is a plurality of small and medium-size enterprises instead of one leading company that dominates the local system.
- An information system has evolved spontaneously within the cluster, transforming the knowledge of the individual entrepreneurs into a common asset.
- The employees are highly skilled.
- There is a high rate of new business creation.
- There is an abundance of entrepreneurial spirit that fuels the growth of the manufacturing system.

This definition is broad and applies to a number of possible system configurations. In order to measure the importance of cluster companies quantitatively at the stock market level and determine whether they have a different relevance within the context of the overall financial system, according to the characteristics that identify each area and its different growth potential, we must first analyse these complex systems systematically, classifying them into three types of small and medium-size business systems (Garofoli 1983):

- In the so-called specialized manufacturing areas, companies compete in the same markets, since they produce the same goods or do the same processing work. The system's structure is horizontal.
- In the case of "local manufacturing systems", small enterprises operate within the same industry, as is the case for "specialized manufacturing areas", but intra-industry transactions between companies are quite extensive (quality subcontracting). There is no production integration among the different areas of manufacturing, the structure of the industrial system is adequate and technology is generally imported from outside the system.
- In the so-called system districts (for example, Biella, Vigevano, Prato, Carpi, Sassuolo), the division of labour among the different enterprises is quite extensive, resulting in the gradual development of a local manufacturing system. There are intra - and inter - industry transactions that lead to the creation of a system that produces capital goods used to manufacture finished products. The development of the district is driven by the utilization of local resources (entrepreneurs, capital, skilled workforce, local technology).

The different situations described above, while characterizing different production systems, also provide an outline that could explain the evolution of industrial clusters. The classic Italian industrial cluster, in its original incarnation, meets the criteria of the third definition provided above - i.e., the system district. The distinctive features of industrial clusters are small individual companies, a fragmented manufacturing system and the existence of a deeply rooted individual and/or family-based entrepreneurial culture. Under these circumstances, the growth of cluster companies to a size large enough to permit access to financing as a means of further expansion - including, ultimately, access to the capital markets - generally occurs over the long term.

However, the world of traditional industrial clusters is particularly vulnerable to structural change caused by the occurrence of extraordinary conditions resulting from the shock of innovations and/or external competition.

When this occurs, industrial clusters generally are forced to alter their "natural" growth path, changing their traditional configuration and morphing into "local manufacturing systems" or "specialized manufacturing areas". This process of transformation is driven by the development of new, dynamic organizations that are better suited to grow rapidly. Obviously, the emergence of companies of unequal size can undermine the specificity of a traditional industrial cluster, because large enterprises lack the operating flexibility and adaptability that make a cluster organization inherently efficient. However, in the presence of an external shock, substantial changes, including the adoption of a new organizational structure, are essential to the very survival of a manufacturing system.

The availability of adequate financial resources is crucial, particularly during these transitional phases, and is of vital importance in allowing districts to reorganize their manufacturing systems and remain competitive. Let us now review a few examples:

- The creation of large companies such as Benetton, Diadora, Dolomite and Tecnica in the sports apparel cluster located in the province of Belluno followed the arrival of synthetic materials, which required the attainment of economies of scale, and, during the last 20 years, the conversion from leather-working to plastics technologies, as well as the movement of production facilities to countries with lower labour costs and the ability to control multiple brands in order to compete successfully in international markets. As these changes occurred, the cluster responded to external technological and competitive pressures by shedding its artisan structure to become an organization in which small, highly specialized subcontractors handle the simpler aspects of the manufacturing process and support the leading companies in the cluster.
- In the case of the industrial cluster that produces packaging equipment in the province of Bologna, the turning point in its evolution into a system that gravitates around a few companies that are world leaders in their respective markets (Sasib, Ima, Nuova Fima Imballaggi, Mg2 and Officine Meccaniche Corazza) occurred with the integration of mechanical and data processing technologies in the manufacturing process and the need to develop international sales and marketing strategies to counter increasing pressure from German and Japanese competitors.

When the definition of industrial cluster, according to which the common denominator of all cluster companies is a strong local character and membership in a manufacturing system with well defined territorial boundaries and a high degree of similarity (in terms of number of employees, valued added produced and propensity to export), is broadened to include a local economic system, it is conceivable that competition in the same product segment could trigger a natural process whereby the most nimble and dynamic companies grow the fastest. The development of a sizable number of medium-size enterprises occurs more easily in the absence of the strong system-wide inter- and intra-industry relationships that normally exist in industrial clusters characterized by transactions "between equals".

The financial markets can play a different function in each of the business systems described above, which can also be viewed as different phases of the same phenomenon as it evolves over time. Traditional industrial clusters obtain new resources by forging close links with the banking system. Other types of financing become necessary during the phase when companies grow from local enterprises to cluster leaders with a strong international bias. Financing for such companies first takes the form of private equity financing and can evolve into direct access to the capital markets, provided the cultural environment is favourable and the companies have developed sufficiently structured management organizations.

2. Industrial Clusters and the Stock Market

Owing in part to the differences in the way industrial clusters evolve (as outlined above), the number and size of cluster companies that are publicly traded in Italy can vary widely, depending on which definition of "industrial cluster" is chosen.

If the most comprehensive (but also the most restrictive) definition of "system district" is used - i.e., the definition that specifies a high degree of intra - and inter-industry integration achieved as a result of local growth that is made possible by the optimum utilization of local resources only, the percentage of publicly traded companies is relatively small. Based on the classification of industrial clusters developed by Ceris (Istituto di ricerca sull'impresa e lo sviluppo) – CNR (Consiglio Nazionale delle Ricerche) (1997), only four cluster leader companies are listed on Borsa Italiana's capital markets. They are: Luxottica and Safilo, which are the leaders of the Cadore (Veneto region) eyeglass cluster; Ratti for the Como silk cluster; and Olivetti for the Canavese (Piemonte region) information technology cluster.

Table 1 lists publicly traded companies that operate in Italy within districts characterized by a high degree of manufacturing specialization, identified on the basis of the percentage of the total national exports of a specific product provided by the respective provinces. The definition used in this case is that of "specialized manufacturing area".

Using this definition, the number of publicly traded companies increases to 25. The most important companies include Benetton and Stefanel (province of Treviso) and Marzotto (province of Vicenza) for the garment industry; Tod's (province of Ascoli Piceno) for the footwear industry; and Merloni (province of Ancona) for the household appliance industry. These large companies are the product of manufacturing systems - the Northeast cluster and the Adriatic backbone cluster - that started at the local level and, during the last ten years, became important players in the national economy. Other examples include smaller but very dynamic companies such as Gewiss, which produces electrical equipment in the province of Bergamo; I.M.A. Industrie Meccaniche, which manufactures packaging equipment in Bologna; and Emak, a producer of agricultural equipment in the province of Reggio Emilia.

The relative importance of this group of "publicly traded cluster companies" is significant both economically and in terms of its impact on stock market indicators. Based on aggregate data taken from corporate annual reports for 1999 (source: BUREAU van Dijk 2001), this group of companies employed more than 48,000 people and had revenues of 15,941 billion lire and invested capital of 22,554 billion lire, including 6,705 billion lire in stockholders' equity.

With respect to stock market statistics, at the end of May 2001, this group of companies accounted for 2.7% of the total capitalization of the Italian stock market. If the focus is narrowed to the macro group of industrial companies (to which all publicly traded cluster companies belong), the relative weight of these companies is greater, since they account for 21.7% of the group's total capitalization.

Table 1. Quantitative impact of cluster companies on the Italian Stock Market*.

	COMPANY	AREA	CAPITALIZATION (millions of euros)	% TOT. CAP.	JAN-MAY TRADING VOLUME (millions of euros)	% TOT. TRADING VOLUME
	STOCK MARKET – BLUE CHIP SEGMENT					
1	BENETTON GROUP	Treviso – clothing	3,241.0	0.47%	561.3	0.195%
2	LUXOTTICA	Belluno – eyeglasses	8,025.8	1.16%	144.8	0.050%
3	MARZOTTO	Vicenza – clothing	1,050.3	0.15%	210.0	0.073%
4	SAFILO	Padua – clothing	1,268.2	0.18%	216.9	0.075%
5	TOD'S	Ascoli P. – footwear	1,452.6	0.21%	92.6	0.032%
	TOTAL CLUSTER COMPANIES		**15,037.9**	**2.17%**	**1,225.6**	**0.426%**
	TOTAL BLUE CHIP SEGMENT		*660,876,2*	*95.47%*	*283,654.3*	*98.486%*
	STOCK MARKET – STAR SEGMENT					
6	CSP INTERNATIONAL	Mantua – hosiery	83.8	0.01%	11.1	0.004%
7	INTERPUMP GROUP	R. Emilia – pumps	325.5	0.05%	59.6	0.021%
8	MARIELLA BURANI	R. Emilia – clothing	220.0	0.03%	62.5	0.022%
9	SABAF	Brescia – plumbing fixtures	147.0	0.02%	30.6	0.011%
	TOTAL CLUSTER COMPANIES		**776.3**	**0.11%**	**163.8**	**0.057%**
	TOTAL STAR SEGMENT		*5,288,9*	*0.76%*	*911.7***	*0.317%*
	STOCK MARKET – REGULAR SEGMENT					
10	DANIELI & C	Udine – wood products	270.4	0.04%	14.5	0.005%
11	FILATURA POLLONE	Biella – wool	15.7	0.00%	2.3	0.001%
12	GEFRAN	Brescia – electr. equipment	84.5	0.01%	29.9	0.010%

Table 1. (continued)

13	GEWISS	Bergamo – electr. equipment	592.1	0.09%	51.8	0.018%
14	GILDEMEISTER	Bergamo – metalwork. equip.	119.5	0.02%	1.7	0.001%
15	CERAMICHE RICCHETTI	Modena – floor tiles	203.1	0.03%	12.7	0.004%
16	IDRA PRESSE	Brescia – metalwork. equip.	30.3	0.00%	1.3	0.000%
17	IMA	Bologna – packaging equip.	316.3	0.05%	37.1	0.013%
18	MARCOLIN	Belluno – eyeglasses	74.9	0.01%	4.6	0.002%
19	MERLONI	Ancona – appliances	471.7	0.07%	56.5	0.020%
20	SIMINT	Modena – clothing	293.2	0.04%	109.4	0.038%
21	STEFANEL	Treviso – clothing	141.3	0.02%	6.6	0.002%
22	CEMBRE	Brescia – metalwork. equip.	46.2	0.01%	1.6	0.001%
23	EMAK	R. Emilia – farm equipment	60.6	0.01%	4.8	0.002%
24	LINIFICIO	Bergamo – cotton textiles	28.7	0.00%	2.7	0.001%
25	RATTI	Como – silk	40.4	0.01%	1.1	0.000%
	TOTAL CLUSTER COMPANIES		**2,788.8**	**0.41%**	**338.6**	**0.118%**
	TOTAL REGULAR SEGMENT		*26,043.0*	*3.76%*	*3,448.9*	*1.198%*
	TOTAL CLUSTER COMPANIES		**18,603.1**	**2.69%**	**1,728.0**	**0.600%**
	TOTAL STOCK MARKET		*692,208.1*	*100.00%*	*288,014.8*	*100.00%*

* This list includes only companies that operate in the "specialized manufacturing areas" as defined by Fortis in Il Made in Italy, Editrice Il Mulino.

** Trading volume is also for the January–March period, during which the STAR Segment was not yet operational.

Source: Figures based on data provided by Borsa Italiana Spa.

When stated as a percentage of the 2001 trading volume (the Blue Chip Segment accounts for 95% of the total), the contribution of publicly traded cluster companies is relatively small (0.195% of the total). However, when the blue chip shares are excluded, the relative weight of the cluster companies rises significantly, reaching 11.4% of capitalization and 11.5% of the trading volume.

It is important to remember that cluster companies represent a considerable portion of all publicly traded small and medium-size companies and that the potential impact of unlisted cluster companies is highly significant when viewed against the need to expand, both quantitatively and qualitatively, the roster of Italian listed companies and bring the Italian stock market on a par with those of other major European countries. At this point, about 2,000 small and medium-size companies operating in traditional industries (internal estimate of Borsa Italiana) are large enough to be considered for listing. More than one-third of these companies belong to manufacturing sectors that are concentrated in industrial clusters operating in the following fields: mechanical engineering (200 companies), textiles/garments (191 companies), food (134 companies), electrical equipment (65 companies), furniture and furnishings (57 companies), ceramics (31 companies) and jewellery (9 companies).

3. New Borsa Italiana Initiatives for Small and Medium-Size Italian Enterprises

The recent introduction of the STAR Segment in the stock market is designed to create a new regulatory, organizational and microstructural framework that is better suited to enhancing the value of small and medium-size industrial enterprises while providing a source of diversification and offering retail and institutional investors attractive investment opportunities in areas that previously had been neglected due to problems of visibility, transparency and liquidity. In keeping with this approach, the stock market listing of cluster companies should meet with a more favourable reception than in the past.

Borsa Italiana has designated liquidity and the transparency of information as the two key areas in which STAR Segment companies will be required to make a major effort.

The presence of a sponsor in the primary market, with responsibility for performing the requisite due diligence work and ensuring periodic follow-up research after the initial listing, and of a specialist in the secondary market (who can be the same as the sponsor), coupled with regulatory requirements that ensure a steady flow of information (prompt release of quarterly financial data in Italian and English for STAR companies), helped solve the problems of limited direct and indirect information, which is typical of small companies, and lack of liquidity. This approach removes the danger of an inefficient pricing mechanism and eliminates cost of capital distortions that occur when the market demands a liquidity premium at the time of offering. For the same reason, the specialist is required to ensure an orderly market, setting bid and asked quotes within a limited spread while at the same time doing research that is designed to support the issuer company through the publication of information on its performance and the performance of the industry within which it operates.

STAR companies must pay special attention to establishing effective corporate governance systems so as to be able to provide the market with adequate disclosure on their organization and on their general and operating performance. These companies must comply with the Code of Conduct for Listed Companies. Accordingly, they must appoint a sufficient number of independent directors to their boards, establish an audit committee to oversee internal control systems and appoint an "investor relator" to ensure continuity in their relations with investors and intermediaries. The online services that Borsa Italiana has developed to meet the special needs of institutional investors complete the framework of programs designed with the specific purpose of increasing the visibility of new issuers and creating the conditions necessary to trigger a virtuous circle that will be beneficial for:

- Issuers, who will be able to successfully float secondary offerings with less burdensome costs.
- Institutional investors, who will operate in a market that will be more supportive of their interest in small and medium-size enterprises.
- Intermediaries, who will have an opportunity to specialize in origination and research in the area of small and medium-size companies and in the establishment of small-cap funds.
- The quantitative and qualitative expansion of the market, since a listing system that ensures maximum transparency and orderly trading conditions will help foster interest among companies that are potential candidates for listing.

Even though the STAR Segment was created only recently, making the analysis period relatively short, early evidence on trading in STAR securities has already shown positive results:

- During the April–May period, the average number of contracts for STAR securities executed daily was 14.6% higher than in the first three months of the year, while the overall market declined by 2.3%.
- A similar trend was observed in the average daily value of the transactions (+9.3% for STAR, -2.0% for the overall market).
- Volatility decreased by 17% for STAR securities, as against a drop of just 10.9% for the overall market.

At present, four of the 25 listed companies from the clusters shown in the preceding table have agreed to be included in the STAR Segment. They are: CSP International (a company based in the Castel Goffredo cluster, in the province of Mantova, that controls some of the top women's hosiery brands), Interpump Group (hydraulic pumps in the province of Reggio Emilia), Mariella Burani (garments and fashion, also in the province of Reggio Emilia) and Sabaf (kitchen furniture in the province of Brescia). At present, they represent 20% of the listed companies included in the STAR Segment, 14.7% of the Segment capitalization and 35.9% of the value of all trades executed in 2001.

The fact that these companies chose to join STAR during the start-up phase (trading in the Segment's securities began on April 2, 2001) shows that the Segment's organizational and structural characteristics are being well received by enterprises. The STAR Segment was launched to fill the relationship gaps that existed until now in certain areas between the financial markets and medium-size companies, despite these companies' excellent fundamentals and strong competitive positions.

Economic literature and the case histories of individual companies concur in showing that the financial markets play a decisive role in the development of economic systems. The entities that oversee the operations of these markets are responsible for stimulating the growth of their national economies by providing enterprises with the tools necessary to seize the opportunities created by changing competitive conditions. In pursuit of this goal, Borsa Italiana has focused its efforts on restructuring the Italian financial markets. Easier access to the financial markets and, consequently, to new sources of capital for small and medium-size companies means making available to industrial clusters - the very heart of Italian industry - a vital tool in their effort to manage the competitive repositioning required by an accelerating process of globalisation.

References

BUREAU van Dijk (2001), CDRom AIDA, march.

Ceris-CNR (1997), *Innovazione, piccole imprese e distretti industriali*, Roma, Documenti CNEL III, Quaderno 7.

Fortis, M. (1998), *Il "made in Italy" – Quando stile e creatività non sono solo moda*, Bologna, Il Mulino.

Garofoli, G. (1983), "Le aree-sistema in Italia", *Politica ed economia*, No. 11.

Garofoli, G. (1984), *Modelli locali di sviluppo*, Milano, Franco Angeli.

List of Authors

Giacomo Becattini. Professor of Political Economics at Firenze University. Member of the Accademia Nazionale dei Lincei (Roma), member of Accademia Toscana di Scienze e Lettere "La Colombaria" (Firenze) and of the Accademia dei Georgofili (Firenze). Chairman (1992-1995) of the Società Italiana degli Economisti. Received a number of awards, both scientific (Scanno Award for economics) and literary (Prato and Firenze Awards). Since 1985 honorary member of Trinity Hall college, Cambridge, England, and honorary citizen of Prato since 2001. Chief editor of *Sviluppo Locale* (Torino, Rosenberg & Sellier) and co-editor of *Il Ponte* (Firenze) since 1957; member of the Boards and Scientific Advisory Committee of the following journals: *Economia e politica industriale* (Milano, IEFE), *Economia Marche* (Fabriano, Fondazione Merloni), *Economia e società* (Lecce University), *Marshall Studies Bulletin* (Firenze University), *Storia del pensiero economico* (Firenze University), *History of Economic Ideas* (Pisa University), *Ricerche di storia sociale e religiosa*, (Vicenza, Istituto di Storia Sociale e Religiosa). He is also a member of the Scientific Committee of a number of research institutions such as IRPET, Firenze, and IRIS, Prato. He has been coordinating the "Prato weeks on local development", known as "Artimino's Weeks", since the foundation in 1991. Member of the Firenze's Municipality Government from 1980 to 1985.

Giovanni Dosi. He attended a post-graduate course at ISTAO, Ancona, 1977-78, and holds a Ph.D. in Economics, 1983, Sussex University, GB. From 1984 to 1987 he was researcher at IUAU, Venezia. From 1987 to 1998 he was Professor of Applied Economy at the Faculty of Statistics of Roma's Università "La Sapienza". Since 1998 he is Professor of Economic politics at the Scuola Superiore Sant'Anna, Pisa, Italy. From 1992 to 1997 he was head of the international research project "Technological and Economic Dynamics" at the International Institute of Applied System Analysis (IIASA), Laxenburg, Austria.
Editor for Europe for the journal *"Industrial and Corporate Change"*. In the last two decades he was a Visiting Scholar/Visiting Professor at a number of Universities, including UC Berkeley, Rio de Janeiro, Strasbourg, Stanford, Sussex, Columbia University.

Maryann P. Feldman. Johns Hopkins University, Baltimore, Maryland Research Professor, Department of Mathematical Sciences Director, Centre for Policy Research, Whiting School of Engineering Adjunct Associate Professor, Economics. Dr. Feldman holds a Ph.D. in economics and management and an M.S. in policy analysis and management from Carnegie Mellon University. She received a Bachelors of Arts in economics and geography from Ohio State University.

Her book, *The Geography of Innovation* (Kluwer 1994), is in its second printing. She is also editor of *The Oxford University Handbook of Economic Geography* (with Gordon Clark and Meric Gertler, Oxford University Press, 2000), *Innovation Policy in the Knowledge-Based Economy* (with Al Link, Kluwer, 2000) and the forthcoming *Institutions and Systems in the Geography of Innovation* (with Nadine Massard, Kluwer, 2001).

Marco Fortis. Director of the Department of Economic Studies of Edison, contract Professor of Industrial Economics at the Faculty of Political Sciences of Università Cattolica (Milano) since 1989. Member of the Scientific Committee of the Research Centre in Economic Analysis, International Economics and Economic Development (CRANEC) of Università Cattolica (Milano), member of the Scientific Committee and Vice-President of the Fondazione Comunità e Innovazione, Vice-President of the Fondazione Guido Donegani, member of the Board of Directors of the Fondazione Carlo Erba and of the Università Carlo Cattaneo, Castellanza.
Publications: numerous books and articles in books, journals and newspapers on the subjects of Italian economy, industry and local production systems, technology, development and international trade. Among his main books: *"Prodotti di base e cicli economici"* (Il Mulino, 1988, Iglesias Prize winner) and *"Il made in Italy"* (Il Mulino 1998). With Alberto Quadrio Curzio he edited *"Il made in Italy oltre il 2000"* (Il Mulino, 2000), belonging to the Fondazione Comunità e Innovazione series of books.

Murray Gell-Mann. He received a Bachelor of Science from Yale, 1948 and a PhD from MIT, 1951. Married poet Marcia Southwick in 1992. Distinguished Fellow, founding member, and former Chairman of the Board, Santa Fe Institute, Santa Fe, New Mexico, USA. His research interests are on topics related to simplicity and complexity. He is author of *The Quark and the Jaguar* (translation *Il quark e il Giaguaro*) and co-editor of *The Evolution of Human Languages* and *Understanding Complexity in the Prehistoric Southwest.* R. A. Millikan Professor Emeritus of Theoretical Physics, California Institute of Technology, where he taught from 1955 to 1993. He worked on weak interaction theory, "Eightfold Way" scheme for particle classification, quarks, quantum chromodynamics, renormalization group. He received the Nobel Prize in physics, 1969. Member of Board, J. D. and C. T. MacArthur Foundation, helped start World Resources Institute and MacArthur grants for nature conservation, was co-inventor of Rapid Assessment Program for timely evaluation of natural areas considered for protection. On UN Environment Program's Global Roll of Honour for Environmental Achievement. He shared Ettore Majorana Science for Peace Prize. He was Member of the US President's Science Advisory Committee, from 1969 to 1972 and of the Council of Scientific Advisors from 1994 to 2001.

John H. Holland. Professor of Computer Science and Electrical Engineering and Professor of Psychology at the University of Michigan; member of the External Faculty of the Santa Fe Institute and member of the Executive Committee of its Board of Trustees.
Holland has received numerous awards and honours, including the Levy Medal of the Franklin Institute, the Henry Russel Lectureship of the University of Michigan (the highest honour the University can confer upon a member of the faculty), the prestigious MacArthur Fellowship, and the Pender Award of the Moore School of Engineering at the University of Pennsylvania. Holland is known internationally as "the father of genetic algorithms". His current interests include the theory of complex adaptive systems and cognitive processes. His most recent book is *Emergence*, Perseus Books (1998).

David A. Lane. He is Professor of Statistics in the Faculty of Economics at the University of Modena and Reggio Emilia. In 2002 he will join the new Faculty of Communication, Economics and Information in that University's Reggio Emilia campus. Lane is also a member of the External Faculty and the Science Board of the Santa Fe Institute. Among other academic honours, Lane has been elected a fellow of the Guggenheim Foundation, the Institute of Mathematical Statistics, the American Statistical Association and the International Statistical Association. He has published articles and books in a variety of fields, including probability theory, statistical inference, epidemiology, medical decision-making, economics and business. He was Chairman of one of the sessions of the conference whose proceedings are collected in this book.

Christian Longhi. Professor at L.A.T.A.P.S.E.S. – Laboratoire "Trasformations de l'Appareil Productif et Stratégies Economiques Sectorielles" (Centre National de la Recherche Scientifique), Valbonne (France) and IDEFI – Institut de droit et d'Economie de la Firme et de l'Industrie (CNRS).
His publications include: "Technopolises and Technological Development, in: Y. M. Rabkin (ed.), *Diffusion of New Technologies in the Post-Communist World*, Dordrecht, Kluwer Academic Publishers, 1998; "Networks, Collective Learning and Technology Development in Innovative High-technology Areas: The case of Sophia-Antipolis", *Regional Studies*, 1999; Regional Evolutionary Trends in the 1990s, in: D. Keeble and F. Wilkinson (eds), *High-Technology Clusters, Networking and Collective Learning in Europe*, 2000 (with D. Keeble); Intégration européenne et dynamiques régionales, in: D. Torre, E. Tosi (eds), *Intégration européenne et institutions économiques*, Paris, De Boeck, 2001.

Mario A. Maggioni. He holds a Master of Science and a Ph.D. from the Warwick University, Coventry (UK). Now Associate Professor of Economics at the Faculty of Political Sciences at Università Cattolica (Milano). Member of the Steering

Committee of AISRe (Associazione Italiana di Scienze Regionali) and Associate Editor of "Network and Spatial Economics".
He is the author of a number of articles and books on the subjects of economic development, innovation and technological change, interactions between educational processes and labour market and economic structure and local production specialization at the local, national and European levels, and interactions between "old" and "new" economy. Member of a number of Italian and foreign inter-University research groups. His methodological interests span a broad range from reticular analysis to ecologic-population patterns and multisectorial analysis.

Franco Malerba. He received a Ph.D. in Economics at Yale University (USA), 1983. He is now full Professor of Industrial Economics, Università Bocconi and Director of CESPRI (Centre on Innovation and Internationalisation Processes); Eeditor of *Journal of Industrial and Corporate Change*; Associate Editor of *Journal of Evolutionary Economics*, Advisory Editor of *Research Policy*. Director of the European Summer School of Industrial Dynamics. President-elect of the International Joseph Schumpeter Society.
His main research interests are: Industrial Economics, Economics of Innovation, Industrial dynamics. Recent publications: S. Breschi, F. Malerba, L. Orsenigo, "Technological Regimes and Schumpeterian Patterns of Innovation", *Economic Journal*, vol. 110 (463), April 2000, pp. 388-410; F. Malerba (ed.), *Economia dell'Innovazione*, Roma, Carrocci, 2000; A. Gambardella, F. Malerba (ed.), *The Organization of Economic Innovation in Europe*, Cambridge University Press, 1999; F. Malerba, L. Orsenigo, *Technological Entry, Exit and Survival: An Empirical Analysis of Patent Data Research Policy*, vol. 28 (6), August 1999, pp. 643-60. He was chairman of one of the sessions of the conference whose proceedings are collected in this book.

Alberto Quadrio Curzio. Professor of Political Economy, Dean of the Faculty of Political Sciences and Director of the Research Centre in Economic Analysis, International Economics and Economic Development (CRANEC), Università Cattolica, Milano. He is member of Istituto Lombardo and Socio Corrispondente of the Accademia Nazionale dei Lincei. Chief editor of the journal "Economia Politica", Il Mulino and member of the Scientific Committee of other journals among which "Structural Change and Economic Dynamics"; "Journal of Policy Modelling".
He has been Chairman of Società Italiana degli Economisti (1995-98) and elected member of CNR for ten years. He received the St. Vincent Award for economics and the international Cortina Ulisse Award. He received from the President of the Italian Republic the "gold medal Award". He wrote approximately 300 scientific publications, about 30 of which in English; one has been translated into Chinese. His main subjects of study are: economic theory of scarce resources, income distribution, education, economic development and economic-institutional forms

thereof. He is Chairman of the Scientific Committee and Vice-President of the Fondazione Comunità e Innovazione.

Enzo Rullani. He is Professor of Business Strategies at the Faculty of Economics, Venezia University, after a period spent at the Udine and Verona Universities. Professor of Economies of Industrial Enterprises and Technological Processes at the Ph.D. in Business Economics, Bocconi University, Milano, Italy. He was a Visiting Scholar at the MIT, Boston, in 2000. Chairman of the Technical-Scientific Committee of the CUOA Foundation, Vicenza Business School, set up by the North-East University. He is Chairman of COSES, the Venezia consortium specialized in territorial analyses, and of TEDIS, a research centre on new communications technologies at the Venezia International University.
He is author of numerous works on management and economics of industrial businesses, among which *Sistemi ed evoluzione nel management* (edited with S. Vicari), Milano, Etas Libri, 1999; "Complessità e informazione nella scienza economica", *Pluriverso* 2 , 1996.

Charles F. Sabel. He received an A.B. from Harvard in 1969 and a Ph.D. in 1978. Currently he is Professor of Law and Social Science, Columbia Law School.
Publications include: "Quale volano per gli standard internazionali di protezione sociale? Proposta per un progressivo miglioramento delle condizioni di lavoro nell'economia globalizzata" with Archon Fung and Dara O'Rourke, in: *Il diritto del Mercato del Lavoro*, n.1/2001; "Treatment Courts and Emergent Experimentalist Government" (with Michael Dorf), *Vanderbilt Law Review* (April 2000); "An Unlikely Democracy: The US at the Millennium", *Internationale Politik und Gesellschaft* 1/2000 and presented at the OECD Forum for the Future Conference, Berlin, December 1999; *Design, Deliberation and Democracy*, (1998); *The Second Industrial Divide: Possibilities for Prosperity* (with Michael Priore, 1984).

Francesco Salamini. In 1971 he received the degree of "Professor of Plant Genetics", Roma University (libera docenza). He now is Director of the Max-Planck-Institut für Züchtungsforschung in Köln, Department of Plant Breeding and Yield Physiology, Honorary Professor at he University of Köln, Professor (on leave) in the University of Milano. From 1971 to 1974 he was Professor of Botany and Plant Physiology, Università Cattolica, Piacenza, Italy. In 1969 he was Research Associate, Dept. of Botany and Plant Pathology, Purdue University, West Lafayette, Indiana, USA. In 1988 he received the Medaglia dei XL, Accademia delle Scienze, Roma. Since 1990 he is member of the Accademia Nazionale dei Lincei, Roma. In 1992 he was member of the Accademia Europea, London. In 1995 he received the Fondazione Invernizzi Price for the career. From 1985 to 1999 he was National Coordinator Program "Tecnologie Avanzate Applicate alle Piante", Ministry of Agriculture, Roma, Italy. Since 1998 he is

Member of the Accademia dei Georgofili, Firenze. Since 1998 he is member, of the Accademia Nazionale di Agricoltura, Bologna. Since 2000 he became Chairman of Scientific Committee, Plant Genome Program, Genoplante, France.
He published more than 480 articles in specialized journals and is editor of 4 books. Member of the editorial board of *Theoretical and Applied Genetics*, *J. Gen. & Breed.*, *J. Plant Breeding, Maydica, Plant Molecular Breeding, Genetical Research, Atti Lincei, Plant Biology.*

Angelo Tantazzi. He is Chairman and Economist at Prometeia Calcolo, Bologna. He is also Chairman of Borsa Italiana SpA and Vice-president of the publishing company Il Mulino. He is member of several government's study commissions and Associate Professor of Economic Politics at the Faculty of Political Sciences of the Bologna University.

Kung Wang. He is Professor at the Graduate Institute of Industrial Economics, National Central University, and Adviser at the Industrial Technology Research Institute.
Among his past experience, he has been Adviser of the National Science Council; Director General of the Science Based Industrial Park Administration; Commissioner of the Fair Trade Commission The Executive Yuan. He has also been visiting Assistant Professor at the City University of New York; Researcher at the M.I.T. in the International Motor Vehicle Program; National Expert of the Republic of China in the Asian Productivity Organization; National Expert in the Pacific Economic Cooperation Committee; Director of the Board of Directors, Asia Pacific Science and Technology Association; Adviser of the Ministry of Economic Affairs of the Council for Economic Planning and Development, Ministry of Transportation and Communication.

Paul Westhead. He is Professor of Entrepreneurship in the Institute for Enterprise Development at the University of Nottingham. He was formerly Professor of Entrepreneurship at the University of Stirling after being a research fellow at Cranfield University, Imperial College and Warwick University. He has conducted numerous research projects sponsored by agencies such as the Department for Trade and Industry, the Department for Education and Employment, the Leverhulme Trust, the Confederation of British Industry, Lloyds Bank, NatWest Bank, KPMG Peat Marwick, PricewaterhouseCoopers, Shell UK Limited and the United Kingdom Science Park Association. His research interests include habitual entrepreneurs, technology-based firms, Science Parks and the internationalisation of small firms. He has published over 60 refereed journal articles. In collaboration with Professor Mike Wright, he recently edited "Advances in Entrepreneurship: 3 Volumes" (2000) published by Edward Elgar. He is on the editorial boards of *Entrepreneurship Theory and Practice*,

International Small Business Journal and the *International Journal of Entrepreneurial Behaviour and Research.*

Contributions to Economics

Michele Bagella/
Leonardo Becchetti (Eds.)
The Competitive Advantage of Industrial Districts
2000. ISBN 3-7908-1254-4

Frank Bohn
Monetary Union and Fiscal Stability
2000. ISBN 3-7908-1266-8

Jaime Behar
Cooperation and Competition in a Common Market
2000. ISBN 3-7908-1280-3

Michael Malakellis
Integrated Macro-Micro-Modelling Under Rational Expectations
2000. ISBN 3-7908-1274-9

Stefan Baumgärtner
Ambivalent Joint Production and the Natural Environment
2000. ISBN 3-7908-1290-0

Henri Capron, Wim Meeusen (Eds.)
The National Innovation System of Belgium
2000. ISBN 3-7908-1308-7

Tobias Miarka
Financial Intermediation and Deregulation
2000. ISBN 3-7908-1307-9

Chisato Yoshida
Illegal Immigration and Economic Welfare
2000. ISBN 3-7908-1315-X

Nikolaus Thumm
Intellectual Property Rights
2000. ISBN 3-7908-1329-X

Max Keilbach
Spatial Knowledge Spillovers and the Dynamics of Agglomeration and Regional Growth
2000. ISBN 3-7908-1321-4

Alexander Karmann (Ed.)
Financial Structure and Stability
2000. ISBN 3-7908-1332-X

Joos P.A. van Vugt/Jan M. Peet (Eds.)
Social Security and Solidarity in the European Union
2000. ISBN 3-7908-1334-6

Johannes Bröcker/Hayo Herrmann (Eds.)
Spatial Change and Interregional Flows in the Integrating Europe
2001. ISBN 3-7908-1344-3

Kirstin Hubrich
Cointegration Analysis in a German Monetary System
2001. ISBN 3-7908-1352-4

Nico Heerink et al. (Eds.)
Economic Policy and Sustainable Land Use
2001. ISBN 3-7908-1351-6

Friedel Bolle/Michael Carlberg (Eds.)
Advances in Behavioral Economics
2001. ISBN 3-7908-1358-3

Volker Grossmann
Inequality, Economic Growth, and Technological Change
2001. ISBN 3-7908-1364-8

Thomas Riechmann
Learning in Economics
2001. ISBN 3-7908-1384-2

Miriam Beblo
Bargaining over Time Allocation
2001. ISBN 3-7908-1391-5

Peter Meusburger/Heike Jöns (Eds.)
Transformations in Hungary
2001. ISBN 3-7908-1412-1

Claus Brand
Money Stock Control and Inflation Targeting in Germany
2001. ISBN 3-7908-1393-1

Erik Lüth
Private Intergenerational Transfers and Population Aging
2001. ISBN 3-7908-1402-4

Nicole Pohl
Mobility in Space and Time
2001. ISBN 3-7908-1380-X

Pablo Coto-Millán (Ed.)
Essays on Microeconomics and Industrial Organisation
2002. ISBN 3-7908-1390-7

Ludwig Schätzl/Javier Revilla Diez (Eds.)
Technological Change and Regional Development in Europe
2002. ISBN 3-7908-1460-1

GPSR Compliance

The European Union's (EU) General Product Safety Regulation (GPSR) is a set of rules that requires consumer products to be safe and our obligations to ensure this.

If you have any concerns about our products, you can contact us on ProductSafety@springernature.com

In case Publisher is established outside the EU, the EU authorized representative is:

Springer Nature Customer Service Center GmbH
Europaplatz 3
69115 Heidelberg, Germany

Batch number: 10732714

Printed by Printforce, the Netherlands